The Spleen

Ahmed H. Al-Salem

The Spleen

Anatomy, Physiology and diseases

Springer

Ahmed H. Al-Salem
Consultant Pediatric Surgeon and Pediatric Urologist
Al Sadiq Hospital
Saihat, Saudi Arabia

ISBN 978-981-99-6190-0 ISBN 978-981-99-6191-7 (eBook)
https://doi.org/10.1007/978-981-99-6191-7

This Springer imprint is published by the registered company Springer Nature Singapore Pte Ltd.
The registered company address is: 152 Beach Road, #21-01/04 Gateway East, Singapore 189721, Singapore

Paper in this product is recyclable

Preface

The spleen is a fascinating and mysterious organ with multiple functions. Our understanding of the spleen and its functions has improved over the years, and efforts were made to study the structure, functions, and anatomy of the spleen. The spleen is known to have a unique anatomy and physiology, and because of this, the spleen is considered an important organ for the body. The spleen acts as a filter for the blood but also has important immunological functions. Malpighi, the founder of microscopic anatomy, provided a comprehensive account of the histology and physiological functions of the spleen; over the past 50 years, significant advances have been made in our understanding of the spleen, its functions, and splenic surgery. The first splenectomy described in detail was carried out by Adrian Zaccarelli, in 1549. The work of Campos Christo in 1962 about the segmental anatomy of the spleen helped surgeons perform a partial splenectomy, thereby avoiding complications of post splenectomy infection. With the recent advances of minimal invasive surgery, laparoscopic splenectomy and partial splenectomy have been shown to be feasible and safe both in adults and children.

Over more than 20 years of experience in the field of pediatric surgery, I decided to write this book on the spleen. This is different from any of the already written textbooks. It is actually a well-illustrated, easy to read and quick reference book. It is written in a simple, point-by-point manner, making it useful for consultant surgeons, hematologists, pediatricians, fellows, specialists, residents, and nurses. The book covers most important areas of the spleen including its history, physiology, functions, and various diseases that affect the spleen. Splenectomy, partial splenectomy, and post splenectomy overwhelming sepsis are also covered, with emphasis on the most important points relevant to patient management including clinical, operative, radiological, and hand-drawn illustrations.

Saihat, Saudi Arabia　　　　Ahmed H. Al-Salem

Acknowledgments

I would like to express my special thanks of gratitude to my family who supported me all these years as well as all my patients and their families.

Secondly, I would also like to thank all my friends who helped me a lot in finishing this project.

Contents

Introduction and History

1

- The word spleen is derived from the Greek σπλήν (*splḗn*), and it represents the equivalent of the heart in English.
- It is stated in the past if a person is to be good-spleened (εὔσπλαγχνος, *eús-plankhnos*) then he is a good-hearted or compassionate person.
- For many years the spleen is known to be an organ of interest for physicians and they considered the spleen as a fascinating organ that is full of mysterious.
- The spleen is known to have many more functions than any other organ in the human body (Figs. 1.1 and 1.2).
- Many facts and functions of the spleen however remains unknown today and the spleen still remains a fascinating organ as it was in 400 BC, the time of Hippocrates.
- Galen (in the second century) described the spleen as "an organ full of misteries", "mysterii pleni organon" (Fig. 1.3).
- Plato on th other hand stated that the spleen was created in order "to maintain the liver bright and pure" (Fig. 1.4).

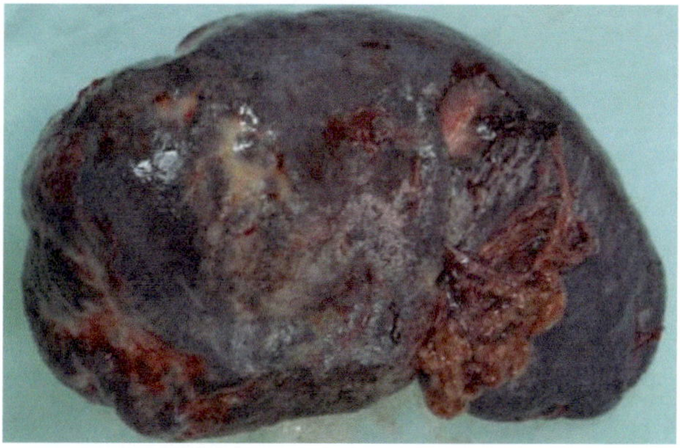

Fig. 1.1 Clinical photographs showing enlarged pathological spleens from a patient with hemoglobinopathy

A. H. Al-Salem, *The Spleen*, https://doi.org/10.1007/978-981-99-6191-7_1

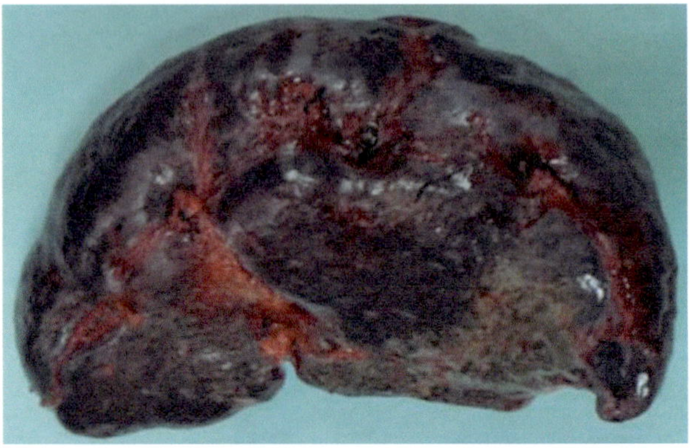

Fig. 1.2 Clinical photographs showing enlarged pathological spleens from a patient with hemoglobinopathy

Fig. 1.3 Aelius Galenus or Claudius Galenus (Greek: Κλαύδιος Γαληνός; September 129—c. AD 216), often Anglicized as Galen (/ˈgeɪlən/) or Galen of Pergamon. He was a Greek physician, surgeon and philosopher in the Roman Empire. He was also considered to be one of the most accomplished medical researchers of antiquity. Galen influenced the development of various scientific disciplines, including anatomy, physiology, pathology, pharmacology and neurology as well as philosophy and logic (Portrait painting of Galen by Ferdinand Georg Waldmllen)

Fig. 1.4 Plato (/ˈpleɪtoʊ/ PLAY-toe; Greek: Πλάτων *Plátōn*; was a Greek philosopher born in Athens. He established the the first institute of higher learning in Europe (Platonist school of thought and the Academy). He was considered a central figure in the history of Ancient Greek philosophy (History and Society by Constance C. Meinwald)

- Hippocrates described the spleen to be soft and fibrous.
- Aristotle (384–322 BC) described the anatomical position of the spleen in the human body (Fig. 1.5).
- Hippocrates stated that the spleen is essential to life and it produces the "black bile", one of four vital humors. He also described with remarkable accuracy the anatomy of the spleen.
- Celsus described the spleen as "a completely unknown organ". He also stated that the main function of the spleen is to secret melancholy humors of the blood through the splenogastric vessels.
- There were others who considered that the spleen function is "to balance the left part of the body with the right one".
- On the other hand, the spleen was also considered by many to bee "a useless Organ and should be removed" (Fig. 1.6).
- One of the main functions of the spleen that was considered is tht the spleen was responsible for producing "black bile". Black bile is one of the four humors of the body that are responsible to make person healthy. Any imbalance f these humors will result in sickness. If a spleen made too much black bile, it would make someone sad or depressed.
- The spleen is also responsible for cleansing the bile and it is thought that this function is responsible for the feeling of happiness and laughter.
- Hippocrates was the first to describe with accuracy the anatomy of the spleen.

Fig. 1.5 Hippocrates of Kos (/hɪˈpɒkrətiːz/; Greek: Ἱπποκράτης ὁ Κῷος, translit. *Hippokrátēs ho Kôios*; c.460—c.370 BC), also known as Hippocrates II, was a Greek physician. He was considered as one of the most outstanding physcians in the history of medicine. He is also known as the "Father of Medicine". He contributed to the growth and development of medicine including the prognosis and clinical observation of diseases, the systematic categorization of diseases, and the humoral theory. He established medicine as a profession and revolutionized it through the Hippocratic School of medicine (Science and Tech byWesley D. Smith)

Fig. 1.6 Aulus Cornelius Celsus (25 BC-50 AD) was a Roman encyclopedist. He claimed and stated that the spleen is an unknown organ and the main function of the spleen is to produce and secrete melancholy humors of the blood. This is secreted through the splenogastric vessels (Science and Tec by Encyclopedia Britanica)

- He was also the first to emphasize the importane of the spleen and stated that the spleen plays a vital role in life. Accordig to Hipporates the spleen produces the *"black bile"*, one of four vital humors of the body.
- The spleen was described as an important site to remove black choler from the body (Gray, 1854; Burton, 1989). Black choler is considered by Hippocrates (c. 460–380 bc;) as one of the four humors of the body. These humors regulate bodily functions.
- Black choler, or also known as black bile, when in acess is responsible to the state of unhappiness and inactivity. Excess of black bile results from failure of the spleen to absorb and remove it from the body.
- This possibly represents the earliest concept of hyposplenism.
  - While blood (hot, sweet and red)
  - Phlegm (cold and moist)
  - These two were considered beneficial humors to the human body.
  - Black choler (cold, dry and thick)
  - Yellow bile (hot, dry and bitter)
  - These were considered non desirable humors to the human body.
- The yellow bile was known to be excreted into the small intestine via the gallbladder.
- The route of excretion of black choler on the other hand was unknown. Black choler function was to belived to create a balance and counteract the effects of the two hot humors. Black choler was also belived to give mourishment to the bones of the body.
- Hippocrates claimed that the spleen function is to absorb the watery elements from the food in the stomach while the gallbladdwe takes the yellow bile from the liver.
- Celsus on the other hand stated that the main function of the spleen is to secret the melancholy humors of the blood and this is done via the splenogastric vessels.
- There were others who considered the spleen a useless organ, and should be removed.
- Paracelsus was one of them and he wrote that the spleen was a superfluous organ that should be removed when it becomes diseased (Fig. 1.7).
- Vesalius was born in Brussels, which was then part of the Habsburg Netherlands.
- Vesalius was another physician who considered the spleen unessential to life. To prove this, he removed the spleens of many animals without adverse effects (Fig. 1.8).
- Vesalius was a professor at the University of Padua (1537–1542) and later became Imperial physician at the court of Emperor Charles. He was also considered as the founder of modern human anatomy.
- Erasistratus (310–250 BC) was a Greek anatomist and physician. He was the first to observe and document that internal organs of the human body appeared to be organized symmetrically to either side of the midline.
- Aristotle and Erasistratus and because of this symmetrical arrangement of the human internal organs thought that the spleen represented a left-sided equivalent of the liver.

Fig. 1.7 Paracelsus (full name Philippus Aureolus Theophrastus Bombastus von Hohenheim), was a Swiss physician and considered the father of toxicology. He described the spleen as a superfluous organ that should be removed when it becomes diseased (History and Science by John G. Hargrave)

Fig. 1.8 Andreas Vesalius (Latinized from Andries van Wezel) (/vɪˈseɪliəs/; 31 December 1514–15 October 1564). He lived in the sixteenth-century and was considered an anatomist and a physician. He also published one of the most influential books on human anatomy, De Humani Corporis Fabrica Libri Septem (*On the fabric of the human body in seven books*) (Science and Tech written by Marcel Florkin)

Fig. 1.9 Erasistratus (/ˌɛrəˈsɪstrətəs/; Greek: Ἐρασίστρατος; c. 304—c. 250 BC) was a Greek anatomist and royal physician under Seleucus I Nicator of Syria. Erasistratus and Herophilus were the founder of the school of anatomy in Alexandria. And he also helped establish the methodic school of teachings of medicine in Alexandria. He also together with Herophilus establish neuroscience and studied nerves and their roles in motor control through the brain and skeletal muscles (Science and Tech by Encyclopedia Britanica)

- Erasistratus also stated the spleen is not a useful organ and had no function of its own (Fig. 1.9).
- Galen was the first to identify and describe communicating structures between the spleen and the stomach. These were called Galen's gastro-splenic communications.
  - He also described nourishment for the spleen via thick juices produced by the liver.
  - The remaining of these secretions which can not be handeled by the spleen were delivered to the stomach via the gastro-splenic communications and subsequently excreted.
  - He also described the spleen as "an organ full of misteries", "mysterii pleni organon".
- Vesalius in 1725 challenged Galen's gastro-splenic communications and described the true nature of the vasa brevia of the stomach.
- Vesalius showed that the vascular supply of the liver and the spleen are different and the spleen can not be considered as a supplementary liver.
- Nonetheless Vesalius supported Hippocratic concept with the spleen acting as a filter of black bile from blood (Figs. 1.10, 1.11 and 1.12).
- In the Ancient Babylonian, Egyptian, Greek and Roman times the spleen was considered an organ that weakened men and horse's athletic capacity.
  - One contributing factor for this is the fact that many of these patients had splenomegaly as a result of malaria infection and this will impair their athe-letic ability.
  - To improve the running ability of horses and men a variety of techniques were adopted to reduce the size of the spleen.

Fig. 1.10 *Aristotle* and Erasistratus thought that the spleen represented a left-sided equivalent of the liver in the human body (History and Society written by Anthony J.P. Kenny Anselm H. Amadio

Fig. 1.11 *Aristotle* and Erasistratus thought that the spleen represented a left-sided equivalent of the liver in the human body (History and Society written by Anthony J.P. Kenny Anselm H. Amadio

Fig. 1.12 Galen (Aelius Galenus or Claudius Galenus) was a Greek physician, surgeon and philosopher in the Roman Empire Science and Tech written by Vivian Nutton)

- One of them was cauterizing the left flank of the body in an attempt to shrink the size of the spleen.
- A red-hot iron was used to achieve this procedure.
- Splenectomy was another adopted procedure and performed to improve the atheletic ability of men and horses and allow them to run faster.
- Macht and Finesilver in 1922 tested this theory on mice in the laboratory at Johns Hopkins University. They concluded that asplenic mice were able to run faster than mice with an intact spleen.

• The first study on the vascular pedicle of the spleen is attributed to Julius Caesar Arantius, in 1571 (Fig. 1.13).
• One of the important discoveries of Julius Caesar Arantius is the hippocampus.
• Marcelo Malpighi (1628–1694)
  - He studied the morphology of the spleen and described the spleen capsule and its intraparenchymatous insertions and the spleen microanatomy.
  - Malpighi accurately described the splenic capsule and the trabeculae that accompany the distributions of vessels.
  - This was done via injections of china ink into the vessel lumen.
  - He also identified and described the splenic follicles which are named after him (Malpighian corpuscles). These are known as lymphoid follicles of the spleen which form the white pulp of the spleen.
  - He described these follicles as excretory glands filled with liquids and suggested that they their contents were releaed into trabeculae and then into veins to be carried elsewhere in the body by the bloodstream.
  - He described accurately the organization of the arterial and venous system of the spleen.
  - Malpighi also postulated that these follicles played a role with secretion of bile by the liver.

Fig. 1.13 Julius Caesar Aranzi (Giulio Cesare Aranzio, Arantius) (1530—April 7, 1589) was well koown for his contributions to human anatomy. He was a famous anatomist and surgeon of the sixteenth century (The free social encyclopedia)

Fig. 1.14 Marcello Malpighi (10 March 1628–30 November 1694) was an Italian biologist and physician. He was known as the "Founder of microscopical anatomy, histology & Father of physiology and embryology". He also identified the splenic follicles which are now called the "Malpighian bodies of the spleen" or Malpighian corpuscles (Science and Tech by Ettore Toffoletto and Alfredo Riva

– Malpighi performed a splenectomy on a dog with no adverse effects apart from increased appetite and weight gain.
– Henry Gray on the other hand considered that the function of Malpighian corpuscles was to regulate and balance the level of albumen in the blood. He also noted that they became atrophied in starved animals. Gray considered that the function of Malpighian corpuscles was to add albumen to the blood (Fig. 1.14, 1.15 and 1.16).

Fig. 1.15 Henry Gray FRS (1827–13 June 1861) was a British anatomist and surgeon. He published the famous anatomy book which is named after him (Gray's Anatomy) (From Wikimedia Commons)

- In 1777, Hewson classified the spleen as part of the lymphatic system of the body.
- Hewson in the 1780s suggested that the spleen played a role in haemopoiesis.
- Hewson considered the spleen to play an important role in the production of red blood cells which were initially colorless and subsequently becomes red in color.
- Some of the patients with splenomegaly were found to increased white cells in their blood suggested that the spleen was also responsible for white blood cell production.
- Julian Evans
  - In 1844, he gave a medical presentation to the Royal Society in London.
  - At the end of his presentation, he made the conclusion that Malpighian corpuscles were lymphatic glands and that the spleen had two sets of functions.

    The first function is that the spleen act as a multiloculated reservoir.

    The second function of the spleen is to separate fluid from blood, a function carried by the Malpighian corpuscles and this then secreted into efferent lymphatics (Fig. 1.17).
- The spleen is considered part of the lymphatic system of the body and resembles a large lymph node.
- The spleen infact is the largest lymphoid tissue of human body and it accounts for 25% of total body lymphocytes.

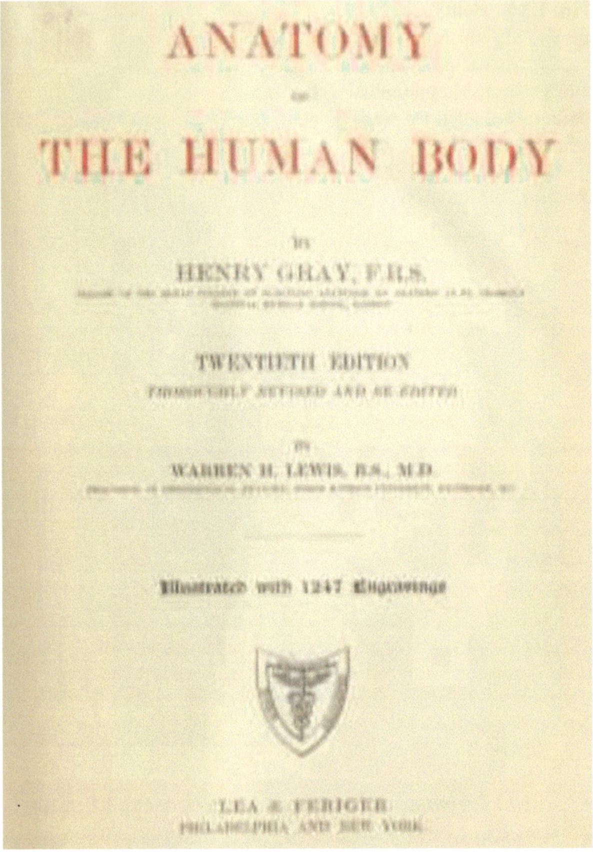

Fig. 1.16 Henry Gray FRS (1827–13 June 1861) was a British anatomist and surgeon. He published the famous anatomy book which is named after him (Gray's Anatomy) (From Wikimedia Commons)

- The spleen is a valuable organ and splenectomy should not be done unless the spleen becomes pathological (Figs. 1.18 and 1.19).
- In 1701, Ruysch studied the microanatomy of the spleen. He injected spleens with wax and concluded that the bulk of the spleen is made up of a vascular network (Fig. 1.20).
- In the 1820s, erythrophagocytosis function of the spleen was described. The spleen was considered an important organ responsible for removal of senescent red cells from the circulation.
- Virchow suggested that cell death was not a normal natural process and was always a sign of disease.
  - Virchow contributed a lot to clinical medicine and was the first to describe and name diseases such as leukemia, chordoma, ochronosis, embolism, and thrombosis.
  - Virchow was the first physician to establish a link between the origin of cancers from otherwise normal cells.
  - He also described Virchow's node which are supraclavicuar lymph nodes.

Fig. 1.17 William Hewson (14 November 1739–1 May 1774) was a British surgeon, anatomist and physiologist. He was known as the "father of haematology"(Science and Tech by Encyclopedia Britanica)

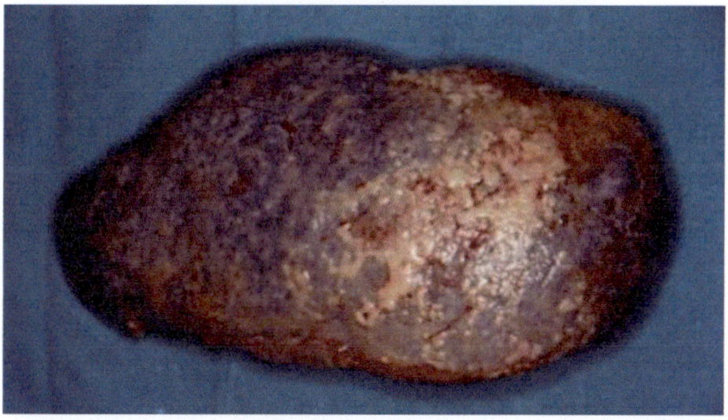

Fig. 1.18 Clinical photographs showing enlarged pathological spleens with multiple small infarcts

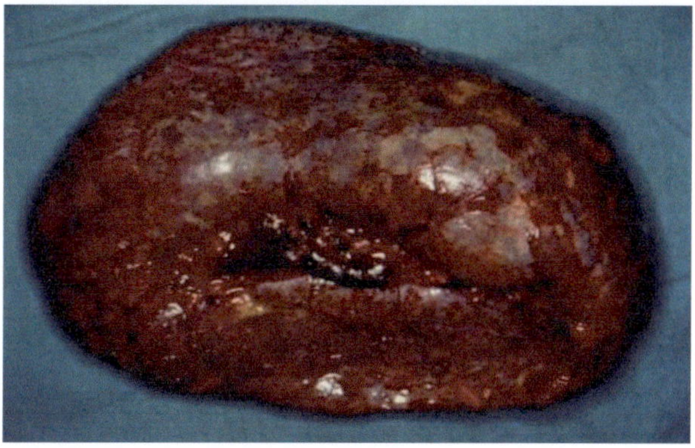

Fig. 1.19 Clinical photographs showing enlarged pathological spleens with multiple small infarcts

Fig. 1.20 Frederik Ruysch (Dutch: March 28, 1638—February 22, 1731) was a Dutch botanist and anatomist. Ruysch, in 1701, injected spleens with wax to study and evaluate their microanatomy (Wikipedia, the free encyclopedia)

  - The supraclavicular lymph nodes when found on the left side are called Virchow's nodes.
  - These are indicative of a mass that can be recognized and when this is the case, clinically tis is called Trosier sign.
- Theodor Billroth, a famous surgeon who developed the technique of gastrectomy. He became famous because of this technique which bear his name today.
  - He described the microanatomical structure of red pulp cords of the spleen and emphasized the idea that a major function of the splenic red pulp is haemopoiesis (Figs. 1.21 and 1.22).
- Ambroise Paré de Laval described the first case of a traumatic hernia of the spleen.
- Otis in 1876 described traumatic hernia of the spleen in more details. He reported 16 cases from the American War of Rebellion.
- Jaboulay in 1893 introduced the term "exosplenopexy". In this technique he extracts the spleen partially into the wound and sture it to the edges of the wound. This way he was iatroenically creating hernia of the spleen in the left upper quadrant of the body. This made a traumatized spleen accessible and easier to control bleeding.
- Subsequently, Pallavecchio used the term "Exosplenolisi". He used to use thermocautery to destroy the herniated part of the spleen rather waiting for spontaneous atrophy of the extruded part of the spleen (Fig. 1.23 and 1.24).
- Splenosis:

Fig. 1.21 Christian Albert Theodor Billroth (26 April 1829–6 February 1894) was a German and Austrian surgeon and a musician. He is known as the founding father of modern abdominal surgery (Science and Tech by Encyclopedia Britanica)

Fig. 1.22 Rudolf Ludwig Carl Virchow (13 October 1821–5 September 1902) was a German physician. He was also known to be an anthropologist, pathologist, prehistorian, biologist, writer, editor, and politician. He is known as "the father of modern pathology"and as the founder of social medicine. He was also called the "Pope of medicine" (Science and Tech by E. Ashworth Underwood)

Fig. 1.23 Ambroise Paré (c. 1510–20 December 1590) was a French surgeon who served kings Henry II, Francis II, Charles IX and Henry III. His main contribution was in surgical techniques, forensic pathology and battlefield medicine. He also contributed to the treatment of wounds and invented several surgical instruments (Science and Tech by Encyclopedia Britanica)

Fig. 1.24 Mathieu Jaboulay (5 July 1860–4 November 1913) was a French surgeon. His main contribution was in techniques of vascular anastomosis. Jaboulay also was the first to introduce the term proposed "exosplenopexy" (Wikipedia, The free encyclopedia)

- Following splenic trauma and injurey, part of the spleen will scatter throughout the abdominal cavity. Some of these cells and in the presence of adequate blood supply they will grow and form small spleens splenunculi.
- This process is called splenosis.
- Splenosis is not rare and about 1 in 5 people have accessory spleens.
- H. Albrecht (1896), in Germany was the first describe splenic implant in the peritoneum, after a trauma.

THE FIRST DESCRIPTION OF SPLENIC IMPLANT IN THE PERITONIUM AFTER HUMAN TRAUMA WAS REPORTED BY H ALBRECHT (1896) IN GERMANY (SPLENOSIS).

- In 1911, Kocher advocated excision of the spleen following trauma. This is to stop the bleeding. He also stated that no side effects will follow its excision.
- Following this everybody advocated splenectomy for every indication.
- The first documented splenectomy for trauma was performed in 1590.
- Following this many splenectomies were performed following injury to the spleen.
- The first splenectomy in America for trauma was performed in 1816.
- In 1962, Campos Christo established the segmental anatomy of the spleen.
- His work encouraged surgeons to perform partial splenectomy.
- In 1590, Dr. Viard performed the first partial splenectomy. He used a piece of string to stitch a segment of the spleen following a small abdominal wound.
- The first successful partial splenectomy for trauma was reported by Franciscus Rosetti in 1590 (Fig. 1.25).

- **THE FIRST DOCUMENTED SPLENECTOMY FOR TRAUMA WAS PERFORMED IN 1959.**
- **THE FIRST AMERICAN SPLENECTOMY WAS PERFORMED IN 1816.**

THE FIRST SUCCESSFUL PARTIAL SPLENECTOMY FOR TRAUMA WAS REPORTED BY FRANCISCUS ROSETTI IN 1590.

- In 1910, William Mayo described splenorrhaphy (Fig. 1.26).
- In the sixteenth century, Paracelsus considered the spleen a superfluous organ. He also advocated excision of the spleen when diseased.
- Vesalius performed splenectomy on many animals without adverse effects. He concluded that the spleen was not necessary to life.
- Adrian Zaccarelli in 1549 performed the first splenectomy.

Fig. 1.25 Emil Theodor Kocher (25 August 1841–27 July 1917) was a Swiss physician. In 1909, he received the Nobel Prize in Medicine for his research in the field of physiology, pathology and surgery of the thyroid gland. His mortality following thyroidectomies was less than 1%. He was the first to advocate aseptic techniques in surgery (Science and Tech by Encyclopedia Britanica)

Fig. 1.26 William Worrall Mayo (May 31, 1819—March 6, 1911) was a British-American medical doctor and chemist. He was the founder of the Mayo Clinic (Wikipedia, The free encyclopedia)

  - Leonardo Fioravanti who was assisting Adrian Zaccarelli during the procedure described the operation which was performed in Italy.
  - The patient was a 24-year-old female who suffered from malarial splenomegaly.
  - The spleen weighed 1340 grams.
  - This however was questioned and it has been suggested that an ovarian cystectomy was performed rather than splenectomy.
- The first splenectomy for a tumorous spleen was performed by Quittembaum in 1826. This was done for a 22 years old female who died in the immediate postoperative period.
- In 1881, Franzolini, in Italy, performed splenectomy on a women with aplastic anemia.
- Philippe Charles Ernest Gaucher:
  - In 1855, he described the post-mortem examination of a 32-year-old female with massive splenomegaly.
  - Her spleen weighed more than 4·5 kg.
  - He examined the spleen histologically and described the spleen as being almost normal in shape and color but hard in consistency. The splenic tissue was replaced by by huge epithelial cells, accompanied by interstitial haemorrhages and complete absence of Malpighian corpuscles.
- In 1866, Spencer Wells performed the first splenectomy for splenomegaly in England.
- The first splenectomy for leukemia was performed by Thomas Bryant in 1866.
- In 1911, Micheli in Italy performed splenectomy for severe haemolytic anaemia.
- Kaznelson in 1916 performed splenectomy for severe thrombocytopenia and reported good recovery.
- Lord Berkeley Moynihan (Between 1905 and 1926), published four editions of his book Abdominal operations. In his book, he discussed and explained the indications for splenectomy (Figs. 1.27, 1.28, 1.29, 1.30, and 1.31).
- The first splenectomy described in detail was carried out by Adrian Zaccarelli, in Italy in 1549.
- Delaitre and Maignien in 1991 performed the first total laparoscopic splenectomy.

- **THE FIRST SPLENECTOMY DESCRIBED IN DETAIL WAS CRRIED OUT BY ADRIAN ZACCARELLI IN ITALY IN 1954.**
- **THE FIRST SPLENECTOMY FOR A TUMORUS SPLEEN WAS PERFORMED BY QUITTEMBAUM IN 1826.**
- **SPENCER WELLS IS CREDITED WITH PERFORMING THE FIRST SPLENECTOMY FOR SPLENOMEGALY IN ENGLAND IN 1866 (?LEUKEMIA).**

Fig. 1.27 Karl Friedrich Quittenbaum (1793–1852) was a German physician who performed tThe first splenectomy for a tumorous spleen (Wikipedia, The free encyclopedia)

Fig. 1.28 Philippe Charles Ernest Gaucher (July 26, 1854—January 25, 1918) was a French dermatologist. Gaucher is well known for his detailed description of the disease that become known as Gaucher's disease (Wikipedia, The free encyclopedia)

Fig. 1.29 Sir Thomas Spencer Wells, first Baronet (3 February 1818–31 January 1897). He was surgeon to Queen Victoria. He was also a professor and president of the Royal College of Surgeons of England. He performed the first splenectomy for splenomegaly in England in 1866 (Wikipedia, The free encyclopedia)

Fig. 1.30 Berkeley George Andrew Moynihan (2 October 1865–7 September 1936) was also known as Sir Berkeley Moynihan. He published four editions of his book on Abdominal operations in which he discussed the different indications for splenectomy (Wikipedia, The free encyclopedia)

Fig. 1.31 Paul Kaznelson was born in 1892 in Warsaw. He was a hematologist who described the first case of pure red cell aplasia and the therapeutic role of splenectomy in idiopathic thrombocytopenic purpura. He performed the first splenectomy for idiopathic thrombocytopenic purpura which led to complete recovery (Wikipedia, The free encyclopedia

- **THE FIRST DOCUMENTED SPLENECTOMY FOR TRAUMA WAS PERFORMED IN 1959.**
- **THE FIRST AMERICAN SPLENECTOMY WAS PERFORMED IN 1816.**

IN 1855, PHILIPP CHARLES ERNEST GAUHER DESCRIBED THE POST-MORTEM EXAMINATION OF A 32-YEAR-OLD WOMEN WITH MASSIVE SPLENOMEGALY (GAUCHER'S DISESE)

- **THOMAS BRYANT PERFORMED THE FIRST SPLENECTOMY FOR LEUKEMIA IN 1866.**
- **IN 1911, MICHELI CURED A PATIENT IN ITALY OF SEVERE HEMOLYTIC ANEMIA FOLLOWING SPLENECTOMYTHE.**
- **IN 1916, KAZNELSON REPORTED RECAVARY FROM SEVERE THROBOCYTOPENICPURPURAFOLLOWINGSPLENECTOMY.**

IN 1991, DELAITRE AND MAIGNIEN PERFORMED THE FIRST LAAROSCOPIC SPLENECTOMY.

- During the first years of the twentieth century, a link was demonstrated between splenectomy and increased susceptibility to infection. This was demonstrated on animals.
- The first reported case of lethal post-splenectomy sepsis in a human was made in 1929.
  - This was a child who splenectomy for haemolytic anaemia.
  - The child's father had also died from septicaemia following splenectomy.
  - This was reported in 1952 by King and Shumacker. They reported five children who developed severe sepsis among one hundred children who underwent splenectomy.
  - They stressed the fact that splenectomy increases the susceptibility to bacterial sepsis and leads to elevated mortality.
  - This was strongly reemphasized by Singer in 1973.
  - It I clear now that splenectomy leads to overwhelming post-splenectomy infection (OPSI) which is usually caused by encapsulated bacteria.

IN 1952, KING AND SHUMACKER REPORTED FIVE CASES OF FULINANT SEPSIS AMONG 100 SPLENECTOMIZED CHILDREN (OPSI)

- The development of OPSI lead to the development of vaccines to protect patients who are going to undergo splenectomy. These vaccines were developed against infection by encapsulated organisms, particularly Streptococcus pneumoniae, Haemophilus influenzae type B and Neisseria meningitidis (types A and C).
- To overcome this, alternatives to total splenectomy were developed. This resulted in reduction in the number of total splenectomies performed and adaption of more conservative techniques such as splenorrhaphy and partial splenectomy.
- Total splenectomy should be avoided wherever possible. This is specially so for children..

Further Reading

1. Wilkins BS. Historical Review. The spleen. Br J Haematol. 2002;117(2):265–74.
2. Sherman R. Perspectives in Management of Trauma to the Spleen: 1979 Presidential Address, American Association for the Surgery of Trauma. J Trauma. 1980;20(1):1–12.
3. Wells TS. Excision of enlarged spleen, with a case in which the operation was performed. Med Times. Gazette. 1866:2–5.
4. Carstens JH. A short history of splenectomy. New York: A.R. Elliott Publishing Co.; 1905. p. 7.

5. Senn N. The surgical treatment of traumatic hemorrhage of the spleen. J Amer Med Assoc. 1903;41(21):1241–5.
6. Morris DH, Bullock FD. The importance of the spleen in resistance to infection. Ann Surg. 1919;70(5):513–21.
7. King H, Shumacker HB. Splenic studies. I. Susceptibility to infection after splenectomy performed in infancy. Ann Surg. 1952;136(2):239–42.
8. Haller JA Jr, Jones EL. Effect of Splenectomy on Immunity and Resistance to Major Infections in Early Childhood: Clinical and Experimental Study. Ann Surg. 1966;163(6):902–8.
9. Christo MC. Segmental resections of the spleen: Report on first eight cases operated on. Hospital. 1962 Sep;62:575–90.
10. Wells TS. Remarks on Splenectomy, with a Report of a successful Case. Med Chir Trans. 1888;71:255–63.
11. Mayo WJ. Principles underlying surgery of the spleen: With a report of ten splenectomies. J Amer Med Assoc. 1910;54(1):14–8.
12. Evans C. Ductless glands: Rupture of the spleen from external violence. Trans Pathol Soc London. 1866;17:299–301.

Embryology and Anatomy

2

- Embryologically, the gastrointestinal tract is derived from endoderm. The spleen on the other hand is derived from mesenchymal tissue. The spleen embryologically develops within and from the dorsal mesentery.
- The spleen receives its blood supply from the celiac trunk. The splenic artery which arises from the celiac trunk is the main vessel supplying the spleen. The spleen also receives blood supply from the short gastric blood vessels. These maily supply the upper pole of the spleen which is in close proximity to the stomach.
- Embryologically:
  - The spleen starts to develop at around the fifth week of intrauterine life.
  - It starts as a localized condensation of mesenchymal tissue on the left side of the body.
  - This condensation of mesenchymal tissue appears between the two layers of the dorsal mesogastrium above the tail of the pancreas. This explains the close proximity of the spleen to the tail of the pancreas.
- The spleen develops as a lobulated organ in the fetus. These lobulations normally disappear before birth but sometimes they can persist (Figs. 2.1 and 2.2). These lobulations have no clinical significance.
- The spleen starts to move to the left side as the stomach changes its position. As a result of this movement the spleen is positioned behind the stomach and lies in close contact with the left kidney. The final location of the spleen on the left side of the body appears to result from preferential development of the spleen precursor cells on the left side of the embryo.
- The gastrosplenic ligment develops from the part of the dorsal mesogastrium which lies between the spleen and the greater curvature of the stomach.
- The development of the spleen is dependant on the activity of the transcription factors BAPX1, HOX11, TCF21.
- This initiates condensation of mesenchymal cells and ultimately form thickenings in the coelomic epithelium of the dorsal mesogastrium near the dorsal pancreatic bud. This also explains the close proximity of the spleen to the tail of the

A. H. Al-Salem, *The Spleen*, https://doi.org/10.1007/978-981-99-6191-7_2

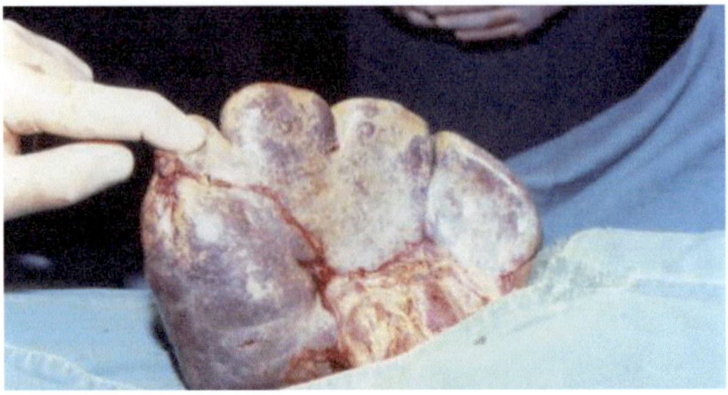

Fig. 2.1 A clinical intraoperative photograph showing splenic lobulations. These were seen in a child at the time of splenectomy for hematological cause

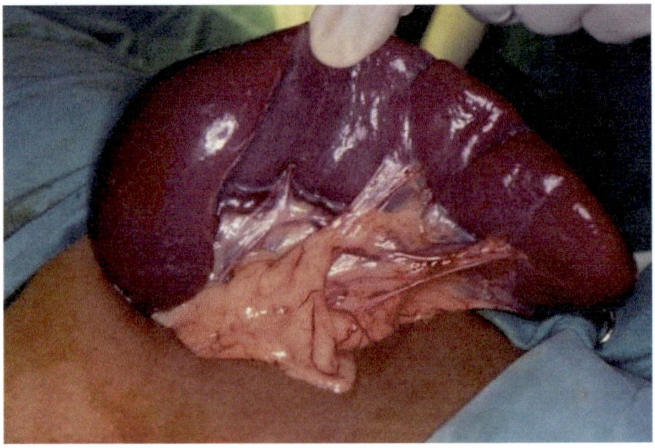

Fig. 2.2 A clinical intraoperative photograph of a child spleen undergoing splenectomy for a hematological condition showing mild splenic lobulations

pancreas. This will result in a bulge in the dorsal mesogastrium representing the spleen and this will appear around 5 weeks of intrauterine life.

- Embryologically, the developmental of the spleen is divided into two stages:
  - First stage:
    This stage lasts up to the 14th week of gestation.
    During this stage the spleen primordium develops a mesenchymal character and aquire primary vascular reticulum.
  - Second stage:
    During this second stage, the red and white pulp develops. These form the spleen's main architecture.
    The red and white pulp of the spleen develop after the development of the spleen primordium.
    The spleen development occurs in two stages:
    The preliminary stage:
    This is also called the "primary vascular reticulum stage
    This lasts up to the 14th gestational week.
    It is characterized by the development of a network of mesenchymal cells and argyrophilic fibers.

Among this, numerous erythrocytes, normoblasts and macrophages are seen.
This stage is also known by the development of hematopoiesis especially erythropoiesis.
The transformation stage:
During this stage of development, the splenic structure becomes developed and well established.
This stage starts from the 15th gestational week.
Splenic lobules: They begin to take shape during the 15th to 17th gestational week.
Splenic lobules consist of a central artery which is surrounded by a sheath of cells which resemble myofibroblasts.
The red pulp of the spleen develops at the periphery of these lobules.
The venous sinuses of the spleen: These develop from lacunae among the reticular network of the spleen and subsequently come in close contact with the venous system.
These sinuses are surrounded by an endothelial wall which remains discontinuous. This is supporting the theory of the open vascularization of the spleen.
Subsequently, the larger veins of the spleen develop as a result of the differentiation of the splenic trabeculae.
The whit pulp of the spleen:
The development of the white pulp of the spleen is related to lymphoid colonization within the spleen.
The white pulp of the spleen starts to develop around the 18th gestational week.
During the 19th and 20th gestational week, lymphocytes start to accumulate around the central arteries of the spleen.
These lymphocytes are morphologically similar to T-precursor cells and also have similar immunohistochemical characteristics.
The periarterial lymphoid sheath (PALS) starts to be well established and a few precursors of interdigitating cells (IDC) can be recognized at this stage. This gives rise to the the differentiation of the T-cell region.

- The primary follicles of the spleen:
  These appear around the 23rd gestational week.
  They become assembled at the periphery of the periarterial lymphoid sheath.
  During this stage, precursors of the follicular dendritic reticulum cell (FDRC) start to be recognized.
  The specific stationary cell of the B-cell region will be recognized at this stage also.
  These primary follicles represent the beginning formation of the B-cell region.

• The spleen has a characteristic fibrous capsule with elstic tissue and is covered by a mesothelial lining.

- The splenic capsule is continuous with the fibrous trabeculae. These give support to the spleen.
- The spleen has little if any contractile function. This is because there is no smooth muscles in the spleen.
- Histologically, 'the spleen is made of two parts:
  - The red pulp
  - The white pulp
  - These are separated from each other by the marginal zone
  - Each of these two parts has different and important functions.
- The spleen ha important functions and splenectomy should always be avoided. Every attempt should be made to delay necessary splenectomy and always try to preserve the spleen even traumatic rupture of the spleen by doing partial splenectomy or splenorrhaphy.
- The spleen size variable and is often described as being about the size of a small fist.
- The spleen is purple in color and it is surrounded and covered by a fibroelastic capsule. This allows the spleen to significantly expand and increase its size when necessary.
- The spleen is an intraperitoneal organ and all of its surfaces are covered with visceral peritoneum. Only the hilum of the spleen is not covered by peritoneum. This is the site through which the splenic artery and vein pass.
- The spleen is positioned under the rib cage, below the diaphragm, and above the left kidney.
- The spleen varies widely in size and weight.
  - The average weight of the spleen is 150 g (range 50 to 250 g).
  - The weight of the spleen is attributed mostly to the blood within the splenic tissue.
  - The spleen, in healthy adults, is approximately 11 centimeters (4.3 in) in length.
- The spleen is considered the largest lymphoid organ in the body. The spleen is basically made up of a mass of lymphoid tissue and vascular channels.
- The spleen is divided into many lobules.
- The spleen consists of two types of tissue:
  - The red pulp and white pulp (Fig. 2.3).
  - The white pulp is made up of lymphatic tissue and consists of lymphocytes. There two types of lymphocytes called B-lymphocytes and T-lymphocytes. These lymphocytes are located around the arteries.
  - The red pulp consists of venous sinuses and splenic cords.
- Splenic venous sinuses:
  - These are cavities within the splenid tissue filled with blood.
- Splenic cord:
  - These are connective tissues containing red blood cells and certain white blood cells including lymphocytes and macrophages.
- The spleen has several important functions including:
  - Phagocytosis

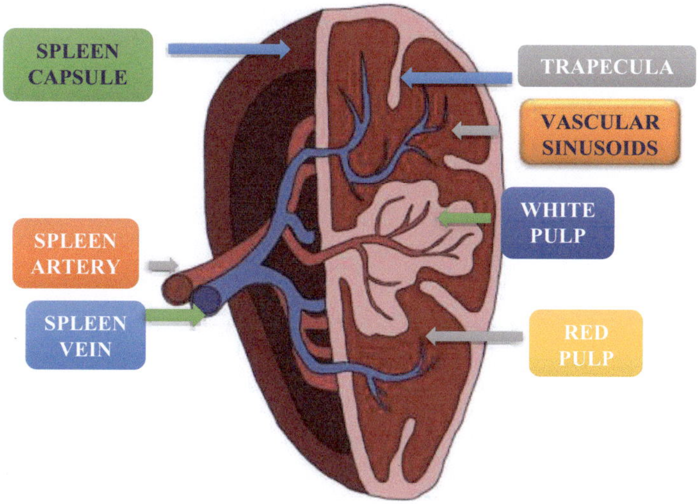

Fig. 2.3 Diagramatic epresentation of the antomy of the spleen

  - Immune responses
  - Lymphopoiesis
  - Blood cell storage
- The spleen is surrounded by peritoneum and is suspended by multiple ligaments. These ligments protect and keep the spleen in its position.
- There are three ligaments which originate from the surrounding structures and attach to the spleen.
- Two of these ligaments attach to the splenic hilum and are traversed by the splenic vessels.
  - The gastrosplenic ligament:

    The gastrosplenic ligament extends from the hilum of the spleen to the greater curvature of the stomach.

    It contains the short gastric vessels and left gastroepiploic arteries and veins.

    It also contains the associated lymphatics and sympathetic nerves.
  - The splenorenal ligament:

    The splenorenal ligament extends from the hilum of the spleen to the anterior surface of the left kidney.

    It transmits the splenic artery and vein

    It also contains the tail of the pancreas.
  - The phrenicocolic ligament:

    The phrenicocolic ligament is a horizontal fold of peritoneum that extends from the splenic flexure of the colon to the diaphragm along the midaxillary line.

    It forms the upper end of the left paracolic gutter.
- The arterial supply and venous drainage of the spleen (Fig. 2.4)
  - The arterial supply of the spleen comes from the tortuous splenic artery.

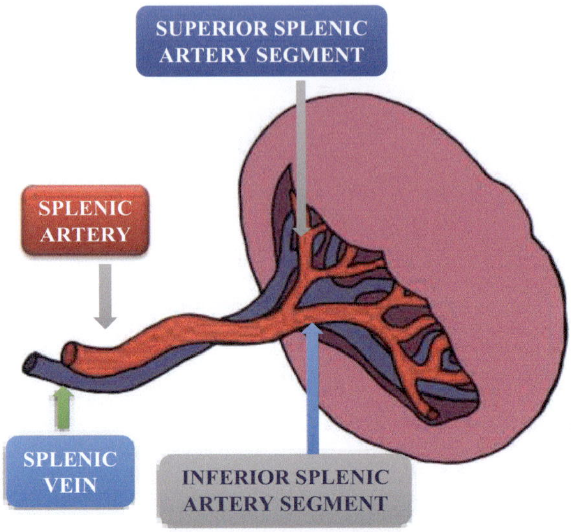

Fig. 2.4 Diagramatic representation of the arterial supply and venous drainage of the spleen. Note the division of the splenic artery into superior and inferior segments at the hilum

- The splenic artery is the largest branch of the celiac trunk which is a branch of the abdominal aorta.
- The celiac trunk also gives off branches to the pancreas, 5–7 short gastric branches, and the left gastroepiploic artery.
- The splenic artery reaches the spleen as it passes through the splenorenal ligament into the hilum of the spleen.
- The splenic artery divides into multiple branches at the hilum.
- These branches are the straight vessels and called penicillin, ellipsoids, and arterial capillaries in the spleen.
- It is also important to note that the spleen has superior and inferior vascular segments. These two segments are separated by an avascular plane. This is important when it comes to the issue of partial splenectomy.
- The venous drainage of the spleen is via the splenic vein. The splenic vein also drains blood from the inferior mesenteric vein.
- The splenic vein provides the principal venous drainage of the spleen.
- The splenic vein after formining in the hilum of the spleen runs behind the pancreas before joining the superior mesenteric vein behind the neck of the pancreas to form the portal vein.
- The short gastric veins, left gastroepiploic, pancreatic, and inferior mesenteric veins are tributaries of the splenic vein.

- Lymphatic drainage of the spleen
  - The spleen has a rich lymphatic drainagr and contains efferent lymphatic vessels, which transport lymph away from the spleen.
  - These efferent lymphatic vessels collect and direct lymph from the spleen toward lymph nodes.

  - The splenic lymph nodes lie at the hilum and receive lymph via perivascular and subcapsular lymphatic vessels. It is then drained to the superior pancreatic (pancreaticosplenic) lymph nodes found at the superior surface of the pancreas.
  - These subsequently drain the lymph to the celiac lymph nodes.
- The role of 1 × 3 × 5 × 7 × 9 × 11:
  - This is an easy way to remember the anatomy of the spleen.
  - 1, 3, and 5: The spleen is 1“ by 3” by 5″ in size.
  - 7: The spleen weighs approximately 7 oz.
  - 9, 11: The spleen lies between the ninth and 11th ribs on the left side.
- Accessory spleens
  - Accessory spleens (lienes accesorii, splenunculi) are considered as anatomic variants during the development of the spleen.
  - They are made up of notmal splenic tissue nodules that fail to fuse with the normally developing spleen.
  - They result from the incomplete fusion of original splenic primordias in the dorsal mesogastrium.
  - They may be single or multiple, but there are seldom more than six.
  - Accessory spleens are accidental finding but they can be found in 10–30% of patients at autopsy or during splenectomy (Figs. 2.5 and 2.6).
  - Accessory spleens are usually small in size (1–1.5 cm in diameter).
  - Accessory spleens may be found in one of the peritoneal folds, commonly in the hilum of the spleen or in the tail of the pancreas.
  - The accessory spleens can be found in various locations mainly:
    The gastrosplenic ligament
    The splenorenal ligament
    The gastrophrenic ligament

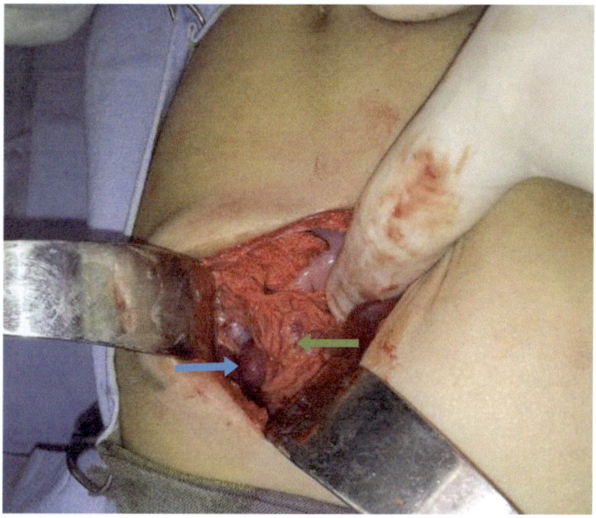

Fig. 2.5 Clinical intraoperative photographs showing accessory spleens. Note the variable sizes of accessory spleens

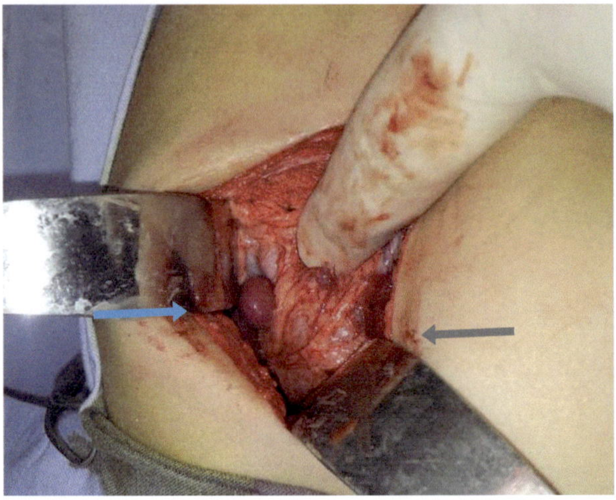

Fig. 2.6 Clinical intraoperative photographs showing accessory spleens. Note the variable sizes of accessory spleens

The gastrocolic ligament
They have also been reported to occur in the broad ligament of the uterus and in the spermatic cord.

- Asplenia:
  - This is a rare condition in which there is a congenital absence of the spleen.

Further Reading

1. Gasser RF. Atlas of human embryos. 1st ed. Hagerstown: Harper and Row; 1975.
2. Gasser RF, Cork RJ, Stillwell BJ, McWilliams DT. Rebirth of human embryology. Dev Dyn. 2014;243:621–8.
3. Endo A, Ueno S, Yamada S, Uwabe C, Takakuwa T. Morphogenesis of the spleen during the human embryonic period. Anat Rec. 2015;298:820–6.
4. Drake RL, Vogl AW, Mitchell AWM. Gray's anatomy for students. 3rd ed. Philadelphia, PA: Churchill Livingstone; 2015.
5. Moore KL, Dalley AF, Agur AMR. Clinically oriented anatomy. 7th ed. Philadelphia, PA: Lippincott Williams & Wilkins; 2014.
6. Sadler TW. Chapter 14: Digestive system. langman's medical embryology. 11th ed. Philadelphia, PA: Lippincott Williams & Wilkins; 2009. p. 215–6.
7. Gray H. Chapter 88: The spleen. In: Standring S, editor. Gray's anatomy: the anatomical basis of clinical practice. 39th ed. Edinburgh, UK, Churchill Livingstone Elsevier. p. 1239–44.
8. Snell RS. Chapter 5: The abdomen: part II. The abdominal cavity. Clinical anatomy by regions. 8th ed. Baltimore, MD: Lippincott, Williams & Wilkins; 2007. p. 259–60.
9. Lee McGregor A, Decker GAG, du Plessis DJ. Chapter 8: The spleen. In: Lee McGregor's synopsis of surgical anatomy. 12th ed. Oxford: Butterworth-Heinemann; 1986. p. 106–13.
10. Romanes GJ. Abdomen: spleen. Cunningham's manual of practical anatomy. In: Vol II: thorax and abdomen, vol. 2. 15th ed. New York, NY, Oxford Medical Publications, Oxford University Press; 1986.
11. Guyton AC, Hall JE. Chapter 15: Vascular distensibility and functions of arterial and venous systems. In: Guyton and hall textbook of medical physiology. 11th ed. Philadelphia, PA: Saunders; 2005. p. 179–80.

3 Pathophysiology and Functions of the Spleen

- The spleen consists of two parts, stroma and parenchyma.
  - The stroma of the spleen:

    This is composed mainly of a network of reticular connective tissue.

    This a network of reticular connective tissue provides support for red blood cells, white blood cells and cells of the immune system which include lymphocytes, macrophages, and dendritic cells.
  - The parenchyma of the spleen:

    This is made up of two components.

    These two components are functionally and morphologically distinct.

    They include the red pulp and white pulp of the spleen.

    The red pulp and white pulp are separated by a tissue layer called the marginal zone.
- The perifollicular zone:
  - This lies outside the marginal zone.
  - It is made up of capillaries and blood-filled spaces which have no endothelial lining.
- In general, the spleen is made up of 4 components:
  - Supporting tissue (Stroma)
  - The white pulp
  - The red pulp
  - The vascular system
- The supporting tissue of the spleen is fibroelastic and forms:
  - The capsule of the spleen
  - Coarse trabeculae
  - A fine reticulum
- There several septa in the spleen and these are called trabeculae.
- These trabeculae extend from the fibroelastic connective tissue of the capsule into the parenchyma of the spleen.
- Both the capsule of the spleen and trabeculae contain myoepithelial cells which have the ability to contract.

A. H. Al-Salem, *The Spleen*, https://doi.org/10.1007/978-981-99-6191-7_3

- This is important as the contraction of these myoepithelial cells help pump stored blood in the spleen into the circulatory system when the body is in need.
- This is seen during intense physical activity or when there is massive hemorrhage and the body is in need for more blood.
- The white pulp:
  - Consists of lymphatic nodules, which are arranged around an eccentric arteriole called the Malpighian corpuscle. They mostly consist of immune cells (T cells and B cells).
- The red pulp:
  - Contains venous sinuses which are cavities filled with blood, and splenic cords (connective tissues containing red blood cells and white blood cells).
  - It is formed by a collection of cells in the interstices of the reticulum, in between the sinusoids.
  - The cell population includes all types of lymphocytes, blood cells, and fixed and free macrophages.
  - The lymphocytes are freely transformed into plasma cells, which can produce large amounts of antibodies and immunoglobulins.
- Blood supply of the spleen:
  - The spleen is supplied mainly by the splenic artery which is a branch of the celiac axis.
  - The splenic artery enters the spleen through the hilum.
  - At the splenic hilum, the splenic artery divides into usually two branches which subsequently divide into smaller branches that enter the splenic parenchyma.
  - These smaller branches follow the course of trabeculae of the spleen.
  - The smaller arteries then branch throughout the parenchyma of the spleen and gradually they become smaller in size.
  - These smaller arterioles of the spleen leave the dense connective tissue of splenic trabeculae and enter the parenchyma of the spleen.
  - At this stage, these smaller arterioles become surrounded by periarteriolar lymphatic sheath (Fig. 3.1).
  - The artery in the center of this region is called the central artery.
  - The endothelial cells lining the central artery have finger-like extensions. These extensions have an important function as they spirally wrap around the lumen protecting the periarteriolar lymphatic sheath from a direct antigen invasion.
  - The central artery divides into branches that enter the marginal zone of the white pulp and then it continues into the red pulp where it divides further and eventually forms the capillaries.
  - These capillaries are surrounded by clusters of macrophages and are called sheathed capillaries.
  - Blood travels freely through the venous sinuses of the red pulp. The endothelial cells of the sinusoids have special histological features:

    They are elongated, spindle-shaped cells and don't have the normal characteristic cellular junctions to adjacent cells.

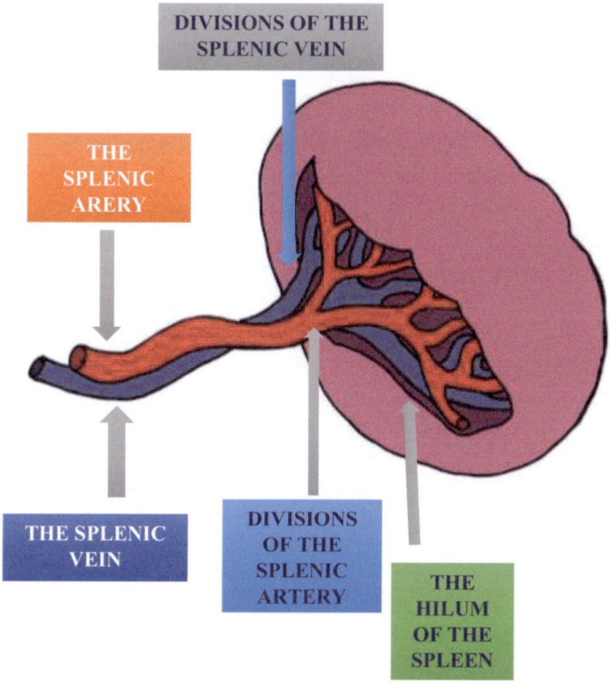

Fig. 3.1 Diagrammatic representation of the anatomy of the splenic artery, splenic vein and their division at the hilum of the spleen

The lack of these junctions creates wide gaps between adjacent cells and this arrangement serve as a mechanical filter between the blood and splenic cords.

These gaps are important and as microorganisms, cellular debris and aged and damaged red blood cells cross these gaps they can be phagocytosed and destroyed by macrophages.

This type of blood flow arrangement is unique to the spleen and it is called the open circulation.

- The blood in the venous sinuses of the spleen flows into trabecular veins and finally exits the spleen through the splenic vein.
- The central arterioles of the spleen subdivide and these are surrounded by the localized areas of B lymphocytes and these make up the lymphoid follicles of the spleen.
- There are two types of lymphoid follicles: depending on the features of the B lymphocytes that comprise them:
  - Primary follicles
  - Secondary nodules
- This division is based on the characteristics of the B lymphocytes that comprise them:

- A follicle that is made up mainly of small, immature lymphocytes is called a primary follicle.
- As the lymphocytes mature and increase in size, they form the secondary nodule.
- Most nodules found in the spleen are secondary nodules.
- Secondary nodules are different from primary follicles. They have a distinctive centrally positioned zone called the germinal center.

• The germinal centers:
- These are the sites where lymphocytes mature and acquire the ability to produce antibodies.
- The presence of these germinal center is a sign that lymphoid tissue is responding to an antigen.
- The germinal centers also contain follicular dendritic cells (FDC).
- The number of these follicular dendritic cells increases in response to antigen stimulation.
- The dendritic cells play a role in initiating and modulate the immune response of B lymphocytes.

• The marginal zone:
- This is found on the periphery of the lymphoid follicles.
- The also help initiate an appropriate immune response as they contain different immune cells.

• The Red pulp:
- The red pulp occupies the majority of the stromal tissue of the spleen.
- It consists of:
  The cords of Billroth
  The splenic sinusoids
- The cords of Billroth (splenic cords) are cellular aggregations which are supported by the reticular connective tissue (Fig. 3.2).
- The splenic sinusoids are located between the cords of Billroth which contain macrophages.
  The splenic sinusoids are made up of thin-walled venous sinusoids.
  They are lined by special endothelial cells with a discontinuous wall.
  This arrangement allows red blood cells to pass between the sinuses and splenic cords.
- The splenic sinusoid act as reservoir and are filled with blood and this what makes the red pulp appear as red in color.
- The blood flows through these sinusoids slowly and at this stage blood will be exposed to macrophages from the cords of Billroth. This stage is important as at this stage these macrophages will be exposed to potential foreign antigens in the blood.
- One of the important functions of the red pulp is to filter the blood form various toxins and destroy them before they enter systemic circulation and cause damage to other organs of the body.
- The central artery of periarterial lymphoid sheath passes from the white pulp and subdivides to subsequently enters the red pulp as a capillary.

Fig. 3.2 Christian Albert Theodor Billroth (26 April 1829–6 February 1894) was a German and Austrian surgeon. He described the splenic cords or red pulp cords and now are called The Cords of Billroth

- These capillaries empty into the splenic cords.
- The macrophages of the splenic cords are important as they phagocyte old and damaged erythrocytes.
- From the splenic cords, blood diffuses into the splenic sinuses and subsequently return to the venous circulation.
- The red pulp of the spleen has the following important functions:

  It filters red blood cells and ingests old red blood cells, damaged red blood cells such as sickle cells or spherocytes, and antibody coated red blood cells.

  It also removes Heinz bodies and other red blood cells inclusion bodies such as Howell-Jolly bodies.

- The White pulp:
  - The white pulp of the spleen is made of three different compartments:

    Periarterial lymphoid sheath (PALS)

    Lymphoid follicles

    The marginal zone
- The PALS:
  - This is made up of a central artery which is a branch of the splenic artery. This central artery is surrounded by a sheath of lymphoid tissue.
- The lymphoid tissue is organized into two layers:
  - The inner layer and outer layer.

    The inner layer is made up mainly of T lymphocytes and this why it is also called the T-zone.

    The outer layer on the other hand is made up of more diverse cellular morphology, containing both T and B lymphocytes.

- These sheaths of aggregates of lymphoid cells are located around arteries (periarteriolar lymphatic sheath). They act as defense line as they are the first to react and defend the body if microorganisms reach the spleen through the bloodstream.
- These sheaths are made up of T cells and B cells lymphocytes which trap antigens from the blood stream for processing.
- The white pulp has the following function:

  The white pulp is responsible for the active immune response through both humoral and cell-mediated pathways.

• lymphocyte recirculation:
  - A large number of lymphocytes in the spleen are considered as migratory cells.
  - These lymphocytes are carried to the spleen via the blood.
  - They travel into a compartment to settle down for some time and then leave the spleen again into the blood stream.
  - Their main role is to spot and detect the various antigens and spread the immune responses in the body.

• In general, the splenic white pulp is made up of three components:
  - The Periarterial lymphoid sheath (PALS)

    This is made up of a central artery which is surrounded by a sheath of lymphoid tissue.

    The lymphoid tissue is made up of an inner and an outer layer.

    The inner layer is called the T-cell zone because it is made up mainly of T- lymphocytes

    The outer layer contains both T-lymphocytes and B-lymphocytes

    Macrophages are found in both the inner and outer layers.

    There are two further compartments primarily inhabited by B- lymphocytes. These are the follicles and the marginal zone (MZ).

    Most of the T cells in the PALS are CD4+, while CD8+ T cells form a smaller population.

    In addition to T lymphocytes, MHC class II+ dendritic cells which are derived from the bone marrow cells are found evenly distributed in the periarteriolar lymphatic sheath. These cells have long cytoplasmic extensions.

    These cells are called the interdigitating dendritic cells (IDCs).

    Their main function is to present the peptide antigens for the induction of primary T-cell immune response.

    Macrophages are also found in every primary follicle of the spleen.

    In secondary follicles, the macrophages develop into large macrophages. These are called 'tingible body macrophages. These macrophages have their cytoplasm filled with the nuclear remnants of phagocytosed apoptotic B cells.

• Splenic follicles or lymphoid follicle:
  - These are hemispherical aggregates of B lymphocytes
  - They are attached to the PALS at larger intervals.
  - These are divided into primary and secondary follicles

- Primary follicles:
  - These are made up of small B lymphocytes expressing immunoglobulin M (IgM) and IgD.
  - These B cells pass through a network of follicular dendritic cells (FDCs).
  - The follicular dendritic cells are B lymphocyte stimulating cells
  - Once the migratory B lymphocytes meet a specific antigen which is recognized by their cell surface immunoglobulin, they form immune complexes.
  - These immune complexes are retained on the surface of FDCs and remain within the follicle.
  - FDCs have an extraordinary capacity to retain antigen–antibody complexes on their cell surface for prolonged periods of time.
  - The specific B lymphocytes then proliferate and after some time they form what is called 'germinal center' within the follicle.
  - These B lymphocytes are different than the recirculating B lymphocytes. They much larger in size and have pale-staining nuclei and abundant cytoplasm.
  - The expanding germinal center displaces the small, dark recirculating B lymphocytes to the periphery of the follicle. As a result of this expansion, these small dark B-lymphocytes form a dark ring-like structure called the mantle zone or corona.
- Secondary follicles:
  - Once the follicles are formed within the germinal centers, these follicles are called secondary follicles.
  - Secondary follicles develop only when specific B-cell immune reactions occur in the spleen.
  - CD4+ T cells are also seen evenly distributed within primary follicles while in secondary follicles CD4+ T cells are localized more at the circumference of the germinal center close to the mantle zone.
  - Germinal center B cells are surface IgD2, while mantle zone B cells coexpress IgM and IgD.

• Germinal centers:
  - A full-blown germinal center is made up of two zones:
    - The dark zone
    - The light zone
  - The dark zone is oriented towards the PALS
  - The dark zone contains centroblasts or proliferating B cells with sparse cytoplasm and large nuclei.
  - The light zone is oriented towards the red pulp
  - The light zone:
    - This is full of centrocytes.
    - These subsequently become plasma cells or memory B cells.
    - They are characterized by their smaller nuclei, more cytoplasm, and are less densely arranged than centroblasts.
  - The PALS and the follicles are surrounded by the Macrophages Marginal zone (MZ). These separates both compartments from the splenic red pulp.

- Macrophages Marginal zone (MZ):

  The MZ is occupied mainly by relatively large memory B cells.

  These memory B cells are characterized by a paler nucleus and more cytoplasm than the T cells of the PALS or the B cells of primary follicles.

  The MZ is separated from the PALS and the follicles by a very irregular capillary blood vessel called the marginal sinus.

  The endothelium of the marginal sinus towards the MZ is rather leaky. This allows access of particulate materials, antigens, some red blood cells and recirculating lymphocytes to the MZ.

  The MZ memory B cells have a distinctive cell surface immunoglobulin that are capable to express IgM, but no or only minor quantity of IgD.

  There are two types of macrophages which are associated with the MZ:

  The MZ macrophages: These are scattered macrophages in the MZ.

  The marginal metallophilic macrophages: These macrophages form a row of cells accompanying the marginal sinus.

  The MZ represents the entry compartment for both antigens and migratory lymphocytes to the white pulp of the spleen. This as well as the recirculating T and B lymphocytes as they cross the MZ on their way to the PALS or to the follicles respectively.

- The splenic red pulp is made up of:
  - The splenic cords
  - The splenic sinuses (sinusoids)
  - The splenic cords:

    The splenic cords are formed by strands of loose connective tissue occupied by lymphocytes, plasma cells, macrophages, granulocytes, red cells and thrombocytes.

    They represent the 'open' part of the splenic

    The terminal arterioles of the red pulp arise from the central arterioles which are branches of the splenic artery and directly open into the splenic cords.

    In the splenic cords the blood moves in irregular spaces. These irregular spaces have no endothelial cells and finally enter the splenic sinuses of the spleen or the red pulp vein.
  - The splenic sinuses:

    The splenic sinuses are specialized vessels.

    They are lined by specialized endothelial cells. These cells have intercellular slits which allow entry of red and white cells from the cords.

    These sinuses are connected to other terminal branches of central arterioles.

    They are considered as the 'closed' part of the splenic circulation.

    The splenic sinuses drain into the trabecular veins of the spleen.
- Other functions of the spleen:
  - The spleen is an important organ for the production of:

    Opsonins

    Properdin

    Tuftsin

- As a blood filter:

  One of the most important functions of the spleen is that it acts as a blood filter.

  It removes old and damaged red blood cells from circulation.

  The spleen's primary function is to filter the blood. As blood flows into the spleen, the spleen will detect any red blood cells in the circulation that are old or damaged. These damaged and unhealthy red blood cells are broken down by macrophages.

  This function is attributed to the unique structure of the blood vessels and macrophages which are present in the red pulp of the spleen.

  The macrophages in the red pulp of the spleen phagocyte red blood cells from the circulation and destroy them.

  The iron from the hemoglobin in these red blood cells is recycled, stored and reused in the bone marrow.

  Hematopoiesis:

  Normally, the bone marrow is the primary site of hematopoiesis. The spleen has also important hematopoietic functions up until the fifth month of gestation.

  During fetal development the spleen act as an important site of hematopoiesis.

  After birth this function is taken totally by the bone marrow.

  Subsequently blood is produced in the spleen only during some pathological conditions such as severe bacterial infections or in certain diseases such as chronic myeloid leukemia and myelosclerosis.

  lymphopoiesis as a function of the spleen continues throughout life. These lymphocytes take part in immune responses of the body.

  Storage of blood:

  The spleen acts as storage for blood.

  The spleen stores red blood cells and platelets.

  Approximately 8% of the circulating RBCs are present within the spleen.

  It has been estimated that up to 236.5 mL of red blood cells can be stored in the spleen and released in cases of hypovolemia.

  This amount of stored blood in the blood vessels of the spleen can be released into the circulation during hemorrhagic blood shock as a result of acute and severe blood loss.

  It has been estimated that up to a quarter of lymphocytes can be stored in the spleen at any one time.

  In healthy individuals, approximately one-third of total platelets are stored in the spleen. This increases in case of splenomegaly
- Immunological function:

  The spleen is the largest lymphoid organ in the body.

  The spleen is one of the major sites of initiating and modulating the immune response. The spleen stimulates the maturation and activation of lymphocytes.

The spleen is important immunologically as it can detect and present specific intruding pathogens in the blood and initiate an immune response to help defend the body against these pathogens.

This is especially against encapsulated bacteria such as pneumococci. This is why people who had splenectomy are more susceptible to be infected with capsulated bacterial infections like pneumonia that is caused by pneumococci.

Phagocytosis requires natural antibodies. These immunoglobulins facilitate phagocytosis either directly or by complement deposition on the capsule. These immunoglobulins are produced by IgM memory B cells in the marginal zone of the spleen.

The spleen is also important in the development and production of mature immune cells. These B and T-lymphocytes. T-lymphocytes are capable of identifying and destroying pathogens.

The T-lymphocytes are derived from the thymus gland and travel to the spleen via blood stream.

The T-lymphocytes are responsible for cell-mediated immunity. These T-cells have receptors on their membrane. These receptors are capable of recognizing various types of antigens.

The B-lymphocytes bone develop in the bone marrow stem cells.

The B-lymphocytes develop antibodies that are specific to a specific antigen. These antibodies bind to the antigens which are subsequently destroyed by other immune cells. This is the function of macrophages which are responsible for destruction of antigens, dead cells, and debris by engulfing and digesting them.

The initial step of a primary antibody response against a protein antigen in the spleen is the 'priming' or stimulation of native CD4+ T lymphocytes by antigen presenting MHC class II+ dendritic cells in the PALS.

Besides dendritic cells, B lymphocytes are can also present antigen to CD4+ T lymphocytes.

FUNCTIONS OF THE SPLEEN

- **ONE OF THE MOST IMPORTANT FUNCTIONS OF THE SPLEEN IS TO ACT AS BLOOD FILTER.**
- **THE MACROPHAGES PHAGOCYTE AND DESTROY ERYTHROCYTES AND THEN RECYCLE THEIR IRON FROM HEMOGLOBIN TO BE STORED AND REUSED IN THE BONE MARROW.**
- **PRODUCTION OF OPSONINS, PROPERDIN AND TUFTSIN**
- **DURING FETAL DEVELOPMENT, THE SPLEEN IS THE PRIMARY SITE OF HEMATOPOIESIS. AFTER BIRTH**

ERYTHROPOIETIC FUNCTION OF THE SPLEEN CEASES, EXCEPT IN CERTAIN CONDITIONS.
- **SPLEEN LYMPHOPOIESIS FUNCTION CONTINUES THROUGHOUT LIFE. THE SPLEEN RETAINS THE ABILITY TO PRODUCE LYMPHOCYTES.**
- **STORAGE OF RED BLOOD CELLS, LYMPHOCTES, PLATELETS AND OTHER FORMED ELEMENTS.**
- **THE SPLEEN IS ONE OF THE MAJOR SITES OF INITIATING AND MODULATING THE IMMUNE RESPONSE.**
- **PHAGOCYTOSIS**

- Phagocytosis
  - One of the spleen's most important functions is phagocytosis.
  - The spleen is considered part of the reticuloendothelial system.
  - The splenic different types of phagocytes including:
  - Reticular cells
- Free macrophages of the red pulp of the spleen
- Modified reticular cells of the ellipsoids
  - The functions of splenic phagocytes include filtering the blood and removal of:
- Debris
- Old and damaged red blood cells (RBCs)
- Abnormal blood cells
- Microorganisms
- Phagocytosis of circulating antigens initiates the humoral and cellular immune responses.
- Measuring splenic function:
  - It is important to measure splenic function and there are different ways to measure the function of the spleen.
  - Blood markers, such as Howell–Jolly bodies (HJB) or pitted cells (PIT) have been used to assess the filtering function of the spleen.
  - These are used as a marker of splenic dysfunction (Figs. 3.3, 3.4 and 3.5).
  - Howell–Jolly bodies:
- These are nuclear remnants found in circulating mature red cells. These are normally removed by the spleen.

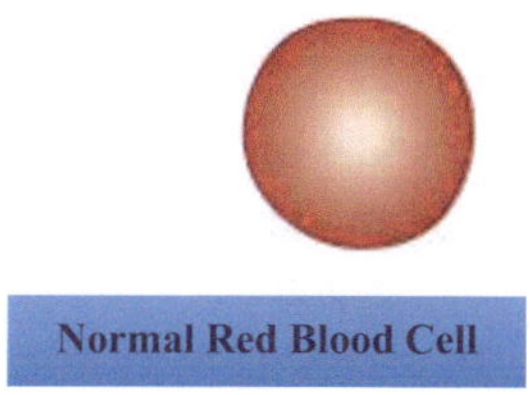

Fig. 3.3 Normal red blood cell and red blood cell with Howell-Jolly body

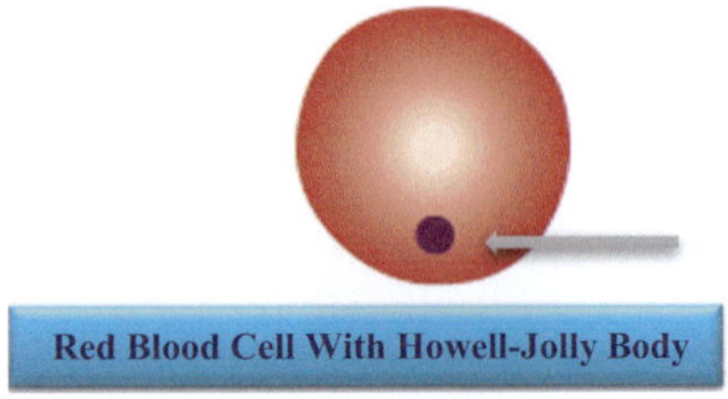

Fig. 3.4 Normal red blood cell and red blood cell with Howell-Jolly body

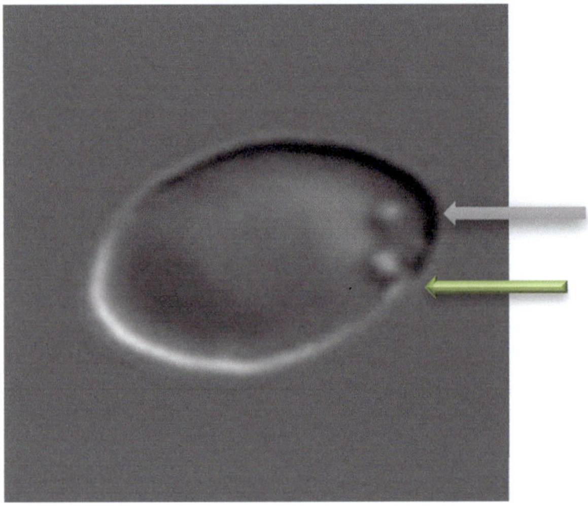

Fig. 3.5 Pitted Red Blood Cell

- The number of circulating HJB can be counted on blood smears or by flow cytometry.
- A HJB count of ≥665/10^6 RBC is considered abnormally high and diagnostic of asplenia in patients with Sickle Cell Anemia.
  - Counts of specific B cell subsets or measures of immune response following vaccination have been used to assess the spleen immunological function.
  - PIT or pocked RBC:
- These are seen with interference light microscopy (Normarsky optics)
- They have been used to measure splenic hypofunction.
- The percentage of PIT red blood cells has been found to be <1·2% in patients with Sickle Cell Anemia with normal splenic function vs. ≥ 4·5% in asplenic patients.
  - Measurement of circulating IgM memory B cells after splenectomy has also been used as a measure to evaluate hypersplenism.
  - Measuring spleen function is important when splenectomy is considered as an option in young children in order to assess the value of the spleen or the residual function and also following hematopoietic stem cell transplantation (HSCT) in order to safely stop antibiotic prophylaxis.

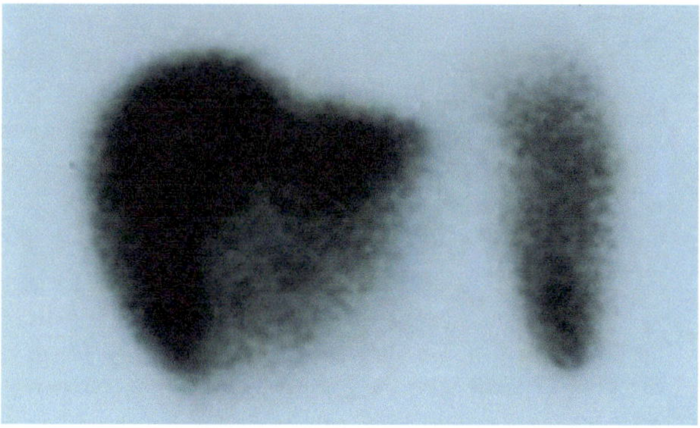

Fig. 3.6 Isotope scintigraphy showing normal splenic function

- Radio isotopic scanning of the spleen can also be used to evaluate the function of the spleen. This will give both morphological and functional evaluation of the spleen (Fig. 3.6).

  Spleen scintigraphy with 99mTc-labelled colloids allows measurement of the phagocytic function of the spleen and liver.

  This gives a spleen/liver uptake ratio.

  It is important to take in consideration that only 10% of injected Sulphur colloids accumulate in the spleen, as they are mainly phagocytosed in the liver.

  A better alternative is spleen scanning with 99mTc-labelled heated autologous RBC.

  These labelled red blood cells are trapped in the spleen due to their rigidity and this gives quantitative measurement of the spleen uptake and anatomy.

  This method of spleen evaluation is however invasive, costly and technically difficult.

MEASURING SPLENIC FUNCTION

- **HOWELL-JOLLY BODIES (HJB)**
- **PITTED CELLS OR POCKED RBC (PIT)**
- **COUNTS OF SPECIFIC BCELL SUBSETS OR MEASURING THE IMMUNE RESPONSE FOLLOWING VACCINATION**
- **MEASURMENT OF CIRCULATING IgM MEMORY B CELLS**
- **SPLEEN SCINTIGRAPHY WITH ^{99m}Tc-LABELLED COLLOIDS (LIVER/SPLEEN SCAN)**
- **SPLEEN SCINTIGRAPHY WITH ^{99m}Tc-LABELLED HEATED AUTOLOGUS RBC**

- Splenectomy:
  - Splenectomy is well known to be associated with the followings:

    An increase in the number of white blood cells (Leukocytosis).

    An increase in the number of platelets (Thrombocytosis).

    The post-splenectomy platelet count may rise to abnormally high levels (Thrombocytosis).

    Thrombocytosis is known to be associated with an increased risk of potentially fatal clot formation or portal vein thrombosis.

    This must be kept in mind in patients following splenectomy and the level of thrombocytosis should be monitored following splenectomy.

    An increase in the number of Howell-Jolly bodies and less commonly Heinz bodies in the blood smears.

    Heinz bodies are usually found in patients with G6PD (Glucose-6-Phosphate Dehydrogenase) deficiency and those with chronic liver disease.

    A decreased response to some vaccines. It is preferable to give the vaccines prior to splenectomy unless splenectomy ids done for an emergency reason.

    An increased susceptibility to infections by bacteria and protozoa. Splenectomized patients are at an increased risk of post splenectomy sepsis from polysaccharide encapsulated bacteria such as pneumococci.
- Hypersplenism:
  - This result from massive enlargement of the spleen with sequestration of red blood cells, leucocytes and thrombocytes.
  - Causes of Hypersplenism:

    Macrophage activation and proliferation may contribute to hypersplenism.

    Blockade of cell passage through the open circulation pathways of the spleen by abnormal red cells, as seen in patients with spherocytosis, malaria or sickle cell disease.

    Hypersplenism may also develop as a consequence of increased pressure in the portal vein. This is called portal hypertension.

    Hypersplenism may also develop when blood passage through the red pulp of the spleen is impeded. This is seen in patients with leukemia, lymphoma and inherited metabolic diseases.

    Hypersplenism may be complicated by a severe reduction of all types of cellular elements of the blood (Red blood cells, White blood cells and platelets). This is called pancytopenia and will require partial splenic embolization or splenectomy to treat it.

    Total splenectomy should be avoided whenever possible as this known to be associated with severe consequences.

    Partial splenectomy or partial arterial embolization of the spleen has been advocated as alternatives to total splenectomy.

    Total splenectomy may lead to an increase in the number of leucocyte and thrombocyte (Thrombocytosis) in the blood. This however may not be seen in all the patients following splenectomy.

    Thrombocytosis is associated with an increased risk of thromboembolic complications.

CHANGES FOLLOWING SPLENECTOMY

- **AN INCREAS IN THE UMBER OF WBC**
- **AN INCREASE IN THE UMBER OF PLATELTS**
- **AN INCREAS IN THE Hb**
- **AN INCREASE IN THE UMBER OF HOWELL-JOLLY BODIES**
- **AN INCREASE IN THE NUMBER OF HEINZ BODIES**
- **AN INCREASED SUSCEPTIBILITY TO INFECTION BY BACTERIA**

- Splenectomy will lead to a defect in IgM antibody production against TI-2 antigens such as capsular polysaccharides of bacteria.
- There are also reports that CD27 + IgM + IgD+/2 memory B cells are severely reduced following splenectomy.
- Splenectomy is known to be associated with an increased risk of sudden overwhelming sepsis caused by Streptococcus pneumoniae, Neisseria meningitidis and Haemophilus influenzae. This more commonly seen in children following splenectomy.
- There are several explanations for the overwhelming postsplenectomy infection (OPSI) syndrome.
  - The splenic marginal zones are the most important sites in the body having B-lymphocytes that are specific for bacterial capsular polysaccharides. Following splenectomy, the antibody responses against encapsulated bacteria will be reduced.
  - The spleen is known to have direct contact of blood borne antigens with the memory B cells in the perifollicular zone.
  - Splenectomy also loses one of the largest area for phagocytosis of opsonized blood borne bacteria (bacteria covered with immune complexes and/or complement).

RECOMMONDATIONS TO REDUCE THE RISK OF OPSI

- **DELAY SPLENECTOMY AS MUCH AS POSSIBLE EXCEPT IN EMERGENCY CASES**
- **POLYVALENT VACCINATION AGAINST PNEUMOCOCCI AT LEAST 2 WEEKS BEFORE SPLENECTOMY OR PRIOR TO DISCHARGE FROM THE HOSPITAL FOLLOWING EMERGENCY SPLENECTOMY**
- **POLYVALENT VACCINATION AGAINST PNEUMOCOCCI BOSSTER DOSES EVERY 5–6 YEARS**
- **CONJUGATE POLYVALENT VACCINES AGAINST PNEUMOCOCCI ARE MORE RECOMMENDED**

- **VACCINATION AGAINST H. INFLUENZA**
- **VACCINATION AGAINST MENINGOCOCCI**
- **LONG-TERM PROPHYLAXIS WITH ANTIBIOTICS FOR CHILDREN MAINLY FOR 5 YEAS OR LONGER**
- **INSTRUCT THE PARENTS TO REPORT TO THE HOSPTAL IN CASE OF FEBRILE INFECTIONS**

- Recommendations to reduce the risk of OPSI:
  - Total splenectomy should be avoided as much as possible or delayed as much as possible.
  - Alternatives to total splenectomy such as partial splenectomy or partial splenic embolization should be adopted whenever possible.
  - Polyvalent vaccination against pneumococci should be given either before splenectomy or afterwards in cases of emergency splenectomy.
  - Polyvalent vaccination against pneumococci booster dose should be given after 5–6 years following splenectomy.
  - Conjugate Polyvalent vaccines against pneumococci are more recommendable, because they provoke T-cell-dependent antibody production.
  - Vaccination against H. influenza should be given.
  - Vaccination against meningococci should be given.
  - Long-term prophylaxis with antibiotics is recommended for children following splenectomy.
- There is a natural defect in antibody production against encapsulated bacteria under the age of 2 years. This is due to incomplete development of splenic MZs. It is recommended that children should receive long-term antibiotic treatment following splenectomy with penicillin until the age of 5 years or longer, depending on the cause necessitating splenectomy.
- OPSI may, however, occur at all ages in spite of vaccinations and antibiotic prophylaxis.
- It is recommended that splenectomized children and their relatives should be instructed to report to the hospital in case of febrile infection.

Further Reading

1. Mebius RE, Kraal G. Structure and function of the spleen. Nat Rev Immunol. 2005;5(8):606–16.
2. Patterson KD, Drysdale TA, Krieg PA. Embryonic origins of spleen asymmetry. Development. 2000;127(1):167–75.
3. Varga I, Galfiova P, Adamkov M, Danisovic L, Polak S, Kubikova E, Galbavy S. Congenital anomalies of the spleen from an embryological point of view. Med Sci Monit. 2009;15(12):RA269–76.
4. Barnhart MI, Lusher JM. Structural physiology of the human spleen. Am J Pediatr Hematol Oncol. 1979 Winter;1(4):311–30.

5. Bohnsack JF, Brown EJ. The role of the spleen in resistance to infection. Annu Rev Med. 1986;37:49–59.
6. Bronte V, Pittet MJ. The spleen in local and systemic regulation of immunity. Immunity. 2013;39(5):806–18.
7. Mebius RE, Nolte MA, Kraal G. Development and function of the splenic marginal zone. Crit Rev Immunol. 2004;24(6):449–64.
8. Espersen K, Frandsen H, Lorentzen T, Kanstrup I-L, Christensen NJ. The human spleen as an erythrocyte reservoir in diving-related interventions. J Appl Physiol. 2002;92(5):2071–9.
9. William B, Corazza G. Hyposplenism: a comprehensive review. Part I: basic concepts and causes. Hematology. 2007;12(1):1–3.
10. Barmparas G, Lamb A, Lee D, Nguyen B, Eng J, Bloom M, et al. Postoperative infection risk after splenectomy: a prospective cohort study. Int J Surg. 2015;17:10–4.
11. Swirski F, Nahrendorf M, Etzrodt M, Wildgruber M, Cortez-Retamozo V, Panizzi P, et al. Identification of splenic reservoir monocytes and their deployment to inflammatory sites. Science. 2009;325(5940):612–6.
12. Cesta MF. Normal structure, function, and histology of the spleen. Toxicol Pathol. 2006;34(5):455–65. Accessed August 2019
13. van Krieken JH, te Velde J. Normal histology of the human spleen. Am J Surg Pathol. 1988;12:777.
14. Wilkins BS. The spleen. Br J Haematol. 2002;117:265.
15. Di Sabatino A, Carsetti R, Corazza GR. Post-splenectomy and hyposplenic states. Lancet. 2011;378:86.
16. Corazza GR, Ginaldi L, Zoli G, et al. Howell-Jolly body counting as a measure of splenic function. A reassessment. Clin Lab Haematol. 1990;12:269.
17. Corazza GR, Tarozzi C, Vaira D, et al. Return of splenic function after splenectomy: how much tissue is needed? Br Med J (Clin Res Ed). 1984;289:861.
18. McKenzie CV, Colonne CK, Yeo JH, Fraser ST. Splenomegaly: pathophysiological bases and therapeutic options. Int J Biochem Cell Biol. 2018 Jan;94:40–3.
19. Dijkstra CD, Dopp EA, Joling P, Kraal G. The heterogeneity of mononuclear phagocytes in lymphoid organs: distinct macrophage subpopulations in the rat recognized by monoclonal antibodies. Immunology. 1985;54:589–99.
20. Kraal G. Cells in the marginal zone of the spleen. Int Reviews Cytol. 1992;132:31–74.
21. Kruetzmann S, Rosado MM, Weber H, et al. Human immunoglobulin M memory B cells controlling Streptococcus pneumoniae infections are generated in the spleen. J Exp Med. 2003;197:939–45.
22. Shatz DV. Vaccination practices among north American trauma surgeons in splenectomy for trauma. J Trauma. 2002;53:950–6.
23. Shaw JHF, Print CG. Postsplenectomy sepsis. Br J Surg. 1989;76(10):1074–81.
24. Steiniger B, Barth P. Microanatomy and function of the spleen. Adv Anat Embryol Cell Biol. 2000;151:1–101.
25. Steiniger B, Barth P, Hellinger A. The perifollicular and marginal zones of the human splenic white pulp. Do fibroblasts guide lymphocyte immigration? Am J Pathol. 2001;159:501–12.
26. Steiniger B, Ruttinger L, Barth PJ. The three-dimensional structure of human splenic white pulp compartments. J Histochem Cytochem. 2003;51:655–63.
27. Steinman RM, Pack M, Inaba K. Dendritic cells in the T cell areas of lymphoid organs. Immunol Rev. 1997;156:25–37.
28. Van Krieken JHJM, te Velde J. Immunohistology of the human spleen: an inventory of the localization of lymphocyte subpopulations. Histopathology. 1986;10:285–94.
29. Veerman AJP, van Ewijk W. White pulp compartments in the spleen of rats and mice. A light and electron microscopic study of lymphoid and non-lymphoid cell types in T- and B-areas. Cell Tissue Res. 1975;156:417–41.
30. Weller S, Braun MC, Tan BK, et al. Blood IgM 'memory' B cells are circlating splenic marginal zone B cells in humans. Blood. 2004;104:3647–54.

4 Splenomegaly

4.1 Introduction

- The spleen is an important organ to the body and has several functions.
- The most important functions of the spleen include:
  - The spleen act as a filter in the body. It Filters and remove abnormal and senescent red blood cells, particulates and microorganisms.
  - The spleen has important immunological functions including synthesis of immunoglobulin M (IgM), properdin and tuftsin.
  - Erythropoiesis: This function of the spleen is important specially early in fetal life. This may also be effective as a response to bone marrow failure.
  - The spleen is an important reservoir of blood cells, including red blood cells and platelets. This reserved blood is valuable and can be released under certain stressful conditions such as hypovolemic shock.
  - Iron metabolism: The spleen is a source of iron that is released from the breakdown of senescent red blood cells and abnormal red blood cells. This iron is released into the blood stream and then stored in the bone marrow.
- The spleen lies within the left upper quadrant of the peritoneal cavity and it is protected by the 9–12 ribs (Fig. 4.1).
- The average weight of a normal spleen is about 150 g and measures approximately 11 cm in craniocaudal length.
- The normal spleen is usually not palpable, but mild to moderate splenomegaly can be detected radiologically.
- In the past, splenomegaly was considered a clinical diagnosis, but in recent years, imaging studies specially ultrasonography have been used to assess or confirm splenomegaly. This is especially so in mild splenomegaly.
- Radiological studies such as sonography and abdominal CT scan are useful in detecting subclinical splenomegaly.
- Splenomegaly:
  - This is defined as enlargement of the spleen, measured by size or weight.

A. H. Al-Salem, *The Spleen*, https://doi.org/10.1007/978-981-99-6191-7_4

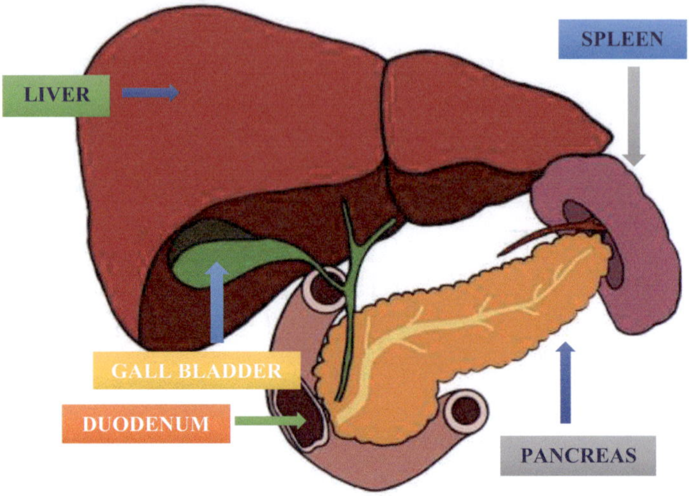

Fig. 4.1 Diagrammatic representation of the anatomy of the spleen

  - Splenomegaly is defined as a spleen greater than 12 cm (approximately 5 inches) in length or over 400 grams (approximately 14 ounces) in weight.
  - A spleen weight of 400–500 g indicates splenomegaly.
  - Massive splenomegaly is defined as a spleen weighing more than 1000 g.
  - Splenomegaly is also defined clinically when the spleen edge is palpable more than 2 cm below the left costal margin.
- A palpable spleen tip is considered a normal finding in up to 30% of neonates. This frequency drops to around 10% in healthy school-age children and it is seen in less than 3% in young adults.
- Moderate splenomegaly is defined if the largest dimension of the spleen is 11–20 cm.
- Severe splenomegaly is considered if the largest dimension of the spleen is greater than 20 cm.
- Splenomegaly is known to be associated with complications and can lead to cytopenia (Low red blood cells, low white blood cells and low platelets). This is due to sequestration and destruction of cells in the spleen particularly platelets.
- Massive splenomegaly: This denotes a greater enlargement of the spleen, with a spleen size larger than 20 cm (approximately 8 inches) in length or over 1 kg (approximately 2.2 pounds) in weight (Fig. 4.2).
- The causes of splenomegaly is diverse and may result from:
  - Increased splenic activity
  - Decreased venous drainage of the spleen
  - Infiltration of the spleen
- The causes of splenomegaly include infectious and non-infectious conditions.
- These causes include:
  - Infections:

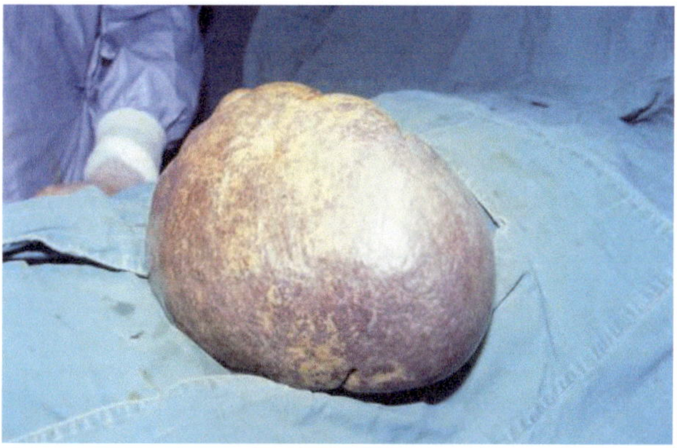

Fig. 4.2 A clinical intraoperative photograph showing massive splenomegaly

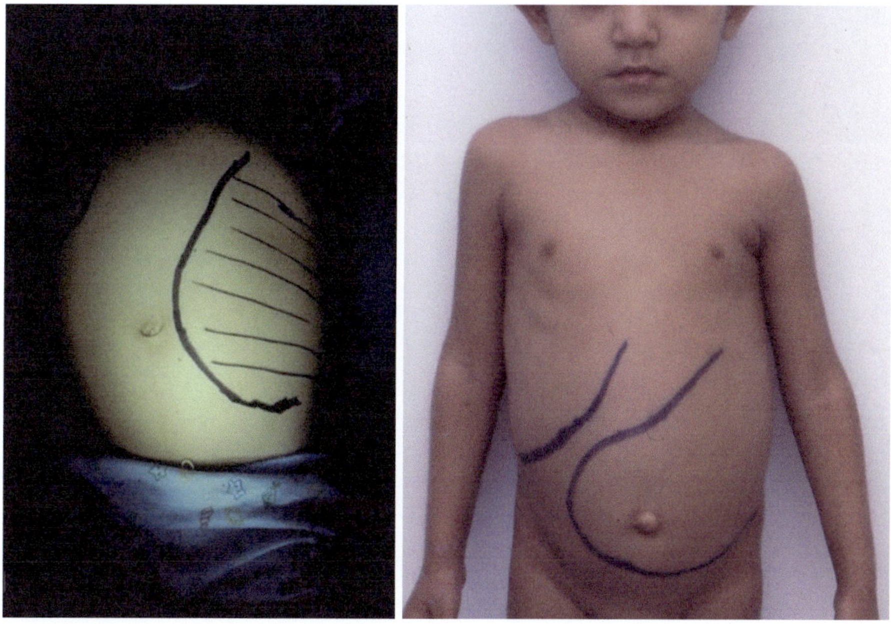

Figs. 4.3 and 4.4 Clinical photographs showing two children with splenomegaly secondary to hematological disorders. Note the marked enlargement of the spleen

Viral infections, such as infectious mononucleosis.
Bacterial infections, such as syphilis or subacute bacterial endocarditis.
Parasitic infections, such as malaria and hydatid cyst.

- Liver cirrhosis and other diseases affecting the liver.
- Hemolytic anemias: These are characterized by premature destruction of red blood cells such as sickle cell anemia, thalassemia and hereditary spherocytosis (Figs. 4.3 and 4.4).
- Leukemia, and lymphomas, such as Hodgkin's disease.

- Metabolic disorders, such as Gaucher's disease and Niemann-Pick disease.
- Increase pressure on the veins in the spleen or liver (liver cirrhosis) or thrombosis in these veins (portal vein thrombosis).

- Splenomegaly rarely causes symptoms apart from heaviness and protrusion of the abdomen. The symptoms of splenomegaly are due to its hematologic effects on the different cell lines, which can manifest as:
  - Anemia
  - Leucopenia
  - Thrombocytopenia
  - A combination of all of these (Pancytopenia)

CAUSES OF SPLENOMEGALY

- **INFECTIONS**
  - **BACTERIAL (SUBACUTE BACTERIA ENDOCARDITIS SYPHILIS)**
  - **VIRAL (INFECTIOS MONONEUCLEIOSIS)**
  - **PARASITIC (HYDATID CYST, MALARIA)**
- **LIVER CIRRHOSIS AND OTHER DISEASES AFFECTING THE LIVER**
- **HEMOLYTIC ANEMIAS**
- **LYMPHOMAS AND LEUKEMIAS**
- **METABOLIC DISEASES (GAUCHER'S DISEASE AND NIEMAN-PICK DISESE)**
- **INCREASED PRESSURE IN THE VEINS OF THE LIVER OR SPLEEN**
  - **LIVER CIRRHOSIS**
  - **PORTAL VEIN THROMBOSIS**

- Splenomegaly can be caused by a wide range of disorders.
- Splenomegaly is most commonly caused by infections.
- Infectious causes of splenomegaly include:
  - Viral infections, such as infectious mononucleosis.
  - Parasitic infections, such as malaria and leishmania.
  - Bacterial infections, such as bacterial subacute endocarditis.
- Normally the body tries to fight these infections and as a result of this the spleen works hard to produce more antibodies. This leads to an increase in the number and size of splenic cells causing splenic enlargement (Splenomegaly).
- Splenomegaly can also develop as a complication of certain blood malignancies such as leukemias and lymphomas. In these conditions, the malignant cells infiltrate the spleen causing its enlargement.
- Splenomegaly can develop secondary to portal hypertension> Portal hypertension results from liver cirrhosis or liver fibrosis. As a result of this, the blood flow

through the liver is decreased and portal hypertension develops. This will cause blood engorgement in the splenic vein and backpressure on the spleen leading to splenomegaly.

- Other less common conditions that can cause splenomegaly include:
  - Hemolytic anemias such as sickle cell anemia, hereditary spherocytosis and beta thalassemia. In these conditions the abnormal and defective red blood cells are rapidly destroyed in the spleen causing the spleen to work harder than usual leading to splenomegaly. These abnormal red blood cells especially sickled red blood cells can stick together and block the splenic capillaries, preventing blood flow out of the spleen and causing splenic enlargement.
  - Chronic inflammatory diseases such as systemic lupus erythematosus (SLE) and rheumatoid arthritis. In these conditions, there increased demand for white blood cells. This will put extra burden on the spleen to over work, which in turn, may result in splenomegaly.
  - Metabolic diseases, such as Gaucher's disease and Niemann–Pick disease.

    In Gaucher disease, fatty substances tend to accumulate in various organs and tissues, including the bone marrow, liver, and spleen. The accumulation of these in the liver and spleen will lead to their enlargement (Hepatomegaly and Splenomegaly).

    Niemann-Pick disease is secondary to an enzyme deficiency. This enzyme deficiency will lead to accumulation of sphingomyelin in various organs, including the spleen leading to splenomegaly.
- Splenomegaly is a serious condition that is known to be associated with complications.
  - An enlarged spleen is susceptible to trauma and can rupture easily.
  - In some conditions an enlarged spleen can also rupture spontaneously.
  - Splenic rupture can lead to a life-threatening hypovolemic shock.
  - Splenomegaly can lead to a hyperactivity of the spleen and this can cause destruction of circulating blood cells, such as white blood cells, red blood cells, and platelets.

    Decreased number of white blood cells (Leukopenia) can result in an increased susceptibility to infections.

    Excessive destruction of red blood cells in the spleen may result in anemia.

    Decreased number of platelets (Thrombocytopenia) may increase the risk of bleeding. This is serious if it leads to intracranial hemorrhage.

4.2 Pathophysiology and Etiology of Splenomegaly

- There are several pathological causes of Splenomegaly. These include the followings:
  - Immune response work hypertrophy
  - Red blood cell (RBC) work hypertrophy
  - Congestive splenomegaly
  - Infiltrative splenomegaly

  - Neoplastic splenomegaly
  - Miscellaneous causes
- Immune response work hypertrophy
  - Splenomegaly can develop a result of increased splenic immunological activity in an attempt to fight various infections.
  - This acute enlargement of the spleen results from an increase in the defense activities of the spleen to overcome various infections or inflammatory diseases. These include:
    Subacute bacterial endocarditis (SBE)
    Systemic Lupus Erythematosus (SLE)
    Felty syndrome
    Infectious mononucleosis
  - The spleen and in an attempt to fight and clear invading antigens from the blood, works harder. This will result in increased numbers of reticuloendothelial cells and lymphoid hyperplasia which ultimately will lead to splenomegaly.
- Red blood cell work hypertrophy
  - One of the main functions of the spleen is to remove and get rid of abnormal blood cells from the circulation. This includes cells with intrinsic defects such as sickle cell anemia and thalassemia or cells coated with antibodies.
  - This will put extra work on the spleen and lead to hypertrophy and splenomegaly.
  - Splenomegaly is seen in patients with:
    Thalassemia major
    Hereditary spherocytosis
    Pyruvate kinase deficiency
    Sickle cell anemia
    Patients with hemoglobin SC disease
    In patients with Thalassemia major extramedullary hematopoiesis will also contribute to splenomegaly.
- Congestive splenomegaly
  - This results from blood engorgement in the splenic vein and spleen.
  - This will develop as a result from decreased blood flow through the liver and portal hypertension.
  - This is caused by several conditions including the followings:
    Liver cirrhosis with portal hypertension
    Splenic vein thrombosis
    Congestive heart failure (CHF) with increased venous pressure
    Patients receiving oxaliplatin-based chemotherapy which can lead to hepatic sinusoidal obstructive syndrome.
    Banti disease
- Infiltrative splenomegaly
  - Infiltrative splenomegaly results from engorgement of splenic macrophages with indigestible materials.
  - Splenomegaly is seen in the following conditions:

Sarcoidosis
Gaucher's disease
Amyloidosis

- Neoplastic splenomegaly
  - Hematologic neoplasms are the main cause of cancer-related causes of splenomegaly.
  - These include:
    Chronic lymphocytic leukemia
    Hairy cell leukemia
    Lymphomas
    Chronic myeloid leukemia
    Primary myeloid fibrosis
    Essential thrombocythemia
    Polycythemia vera
    Rarely, sarcoma can occur in the spleen
    Rarely, primary solid tumors can metastasize to the spleen
- Miscellaneous causes of splenomegaly
  - Additional causes of splenomegaly include:
    Trauma
    Splenic cysts
    Splenic hemangioma
    Splenic abscess
    Acute splenic sequestration crisis in patients with sickle cell anemia
    Tropical splenomegaly syndrome occurs most often in persons indigenous to the malarial belt of tropical Africa.

ETIOLOGY OF SPLENOMEGALY

- **IMMUNE RESPONSE WORK HYPERTROPHY**
- **RED BLOOD CELLS WORK HYPERTROPHY**
- **CONGESTIVE SPLENOMEGALY**
- **INFILTRATIVE SPLENOMEGALY**
- **NEOPLASTIC SPLENOMEGALY**
- **MISCELLANEOUS CAUSES**

4.3 Diagnosis

- The diagnosis of splenomegaly is usually made clinically when the spleen is palpable during a physical examination.
- In patients with massive splenomegaly the spleen might be palpated across the midline of the abdomen. Sometimes the splenomegaly may extend to the right lower quadrant of the abdomen and in the pelvis.

- The diagnosis of associated hypersplenism in made by blood investigation which will reveal anemia, leukopenia, thrombocytopenia or all combined (Pancytopenia).
- Abdominal ultrasound and CT scans can be used to confirm the diagnosis of splenomegaly. These investigations will also give accurate measurement of the size of the spleen.
- Liver-Spleen Colloid Scans
  - Labeled red blood cells are used for spleen scanning.
  - The infused red blood cells can be labeled with chromium-51 (^{51}Cr), mercury-197 (^{197}Hg), rubidium-81 (^{81}Rb), or technetium-99 m (^{99m}Tc).
  - These cells are also treated with heat, antibody, chemicals, or metal ions to alter their shape. The spleen will recognize them as foreign, uptake them and sequester them.
  - A splenic length greater than 14 cm is considered enlarged on liver-spleen scan
  - A spleen scan is a good technique to measure the size of the spleen.
  - A spleen scan is not only useful to determine the size of the spleen but also helpful for:
    Detecting space-occupying lesions in the spleen
    Evaluation of splenic function
    Evaluation for asplenia
    Determining the presence of an accessory spleen

4.4 Massive Splenomegaly

- Massive splenomegaly is arbitrary defined as a spleen that weighs >1000 grams (Figs. 4.5, 4.6 and 4.7).
- Massive splenomegaly is also defined as a spleen:
  - That is palpable more than 8 cm below the left costal margin
  - That is >20 cm in diameter
  - weight > 1000 g
- Diseases that can cause massive splenomegaly include:
  - Different types of malignancies such as chronic myelogenous leukemia, myelofibrosis and splenic marginal zone lymphoma.
  - Certain infections such as malaria
- The causes of massive splenomegaly include:
  - Chronic myeloid leukemia
  - Chronic lymphocytic leukemia
  - Myelofibrosis
  - Polycythemia Vera
  - Hairy cell leukemia
  - Hodgkin lymphoma
  - Non-Hodgkin lymphoma
  - Gaucher's disease
  - Hairy cell leukemia
  - Marginal zone B cell lymphoma

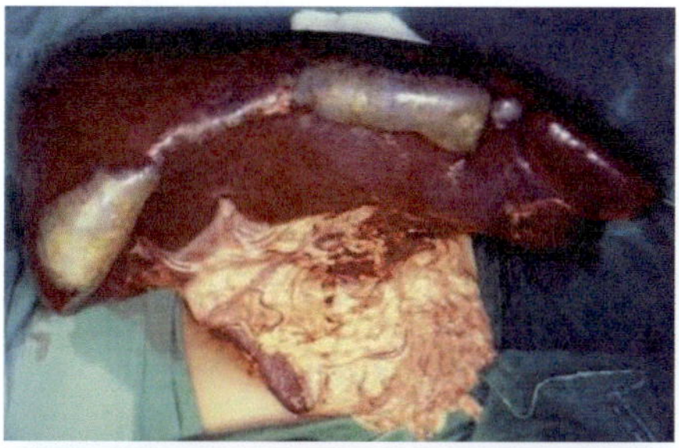
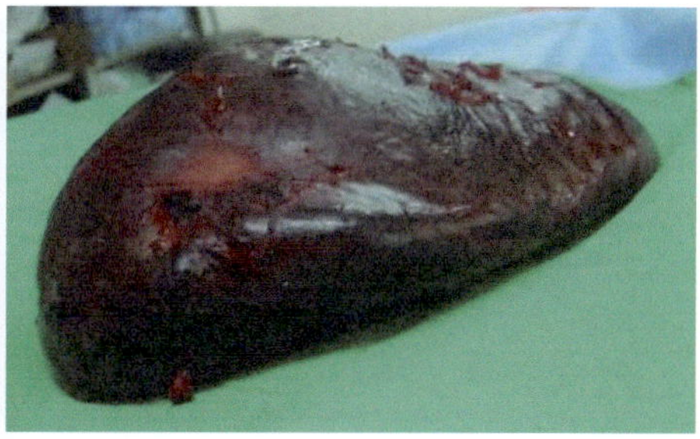
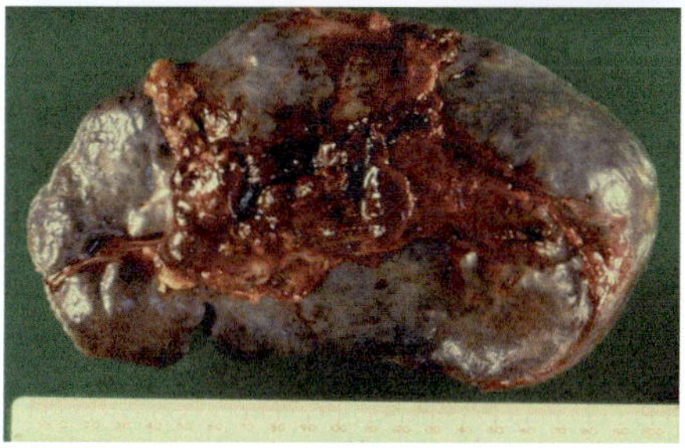

Figs. 4.5–4.7 Clinical photographs showing massive splenomegaly in two children. Note also the splenic infarcts in the first photograph

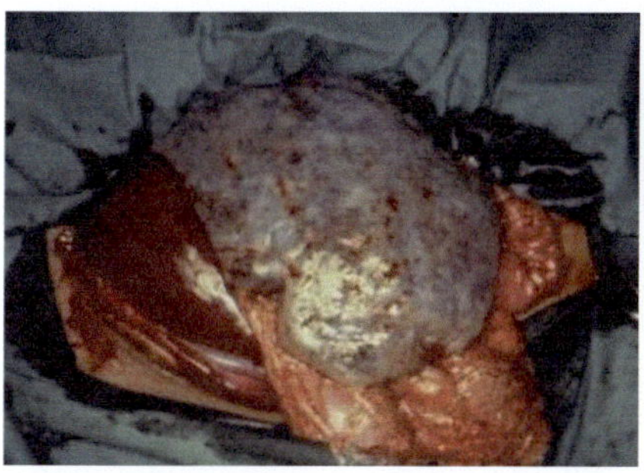
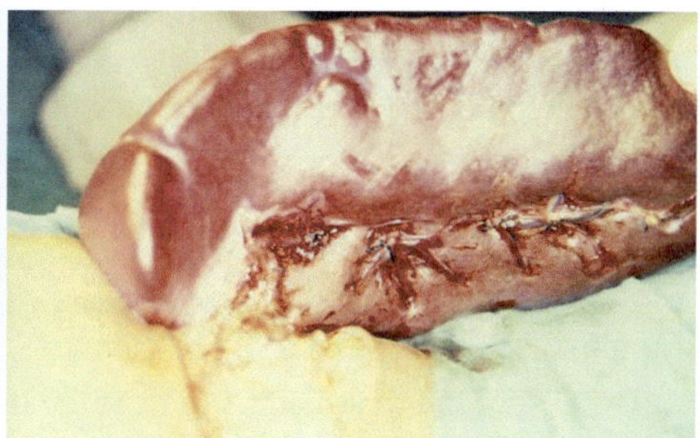

Figs. 4.8 and 4.9 Intraoperative photographs showing massive splenomegaly in a child with sickle cell disease and another child with thalassemia

- Myelofibrosis
- Plasmacytoma
- Prolymphocytic leukemia
- Infections such as malaria
- Rarely thalassemia and sickle cell disease (Figs. 4.8 and 4.9)

The criteria for the diagnosis of hypersplenism include the following:

- Splenomegaly
- Anemia, leukopenia, thrombocytopenia, or a combination of all of them (Pancytopenia)
  - Anemia results from blood sequestration and hemodilution.

  - Leukopenia results from increased destruction of white blood cells or sequestration of leukocytes.
  - Thrombocytopenia results from increased splenic platelet pooling in the spleen.
- Cellular bone marrow
- Improvement of the blood parameters after splenectomy. Splenectomy is therapeutic in individuals with severe pancytopenia due to splenomegaly. Thrombocytosis however may develop post-splenectomy and require treatment when the platelet count exceeds 1 million/μL. Multiple treatment modalities have been used to reduce the number of platelets or inhibit their thrombotic effects.
- These modalities include:
  - Hydroxyurea
  - Aspirin
  - Plateletpheresis

THE CRITERIA FOR THE DIAGNOSIS OF HYPERSPLENISM

- **SPLENOMEGALY**
- **ANEMIA, THROMBOCYTOPENIA, LEUKOPENIA OR A COMBINATION OF ALL OF THEM (PANCYTOPENIA)**
- **IMPROVEMENT OF BLOOD PARAMETERS AFTER SPLENECTOMY**
- **CELLULAR BONE MARROW**

4.5 Treatment of Splenomegaly

- The aim of the treatment is to treat the underlying cause of splenomegaly.
- All patients with splenomegaly are susceptible to trauma and abdominal injuries should be avoided to prevent rupture of spleen.
- Splenectomy may be a necessary treatment option in certain conditions.
- Splenectomy however is known to be associated with increased risk of infection and sepsis from encapsulated bacteria mainly.
  - Fulminant, life-threatening infection represents a major long-term risk after splenectomy.
  - Overwhelming post splenectomy infection (OPSI), also known as post splenectomy sepsis syndrome, begins as a nonspecific, flulike symptoms that is followed by a rapid deterioration to full-blown bacteremia septic shock. This accompanied by hypotension, anuria, and disseminated intravascular coagulation.
  - The most serious organisms causing OPSI are encapsulated bacteria, such as pneumococci (e.g., Streptococcus pneumoniae).

  - Pneumococcal infections account for 50–90% of cases reported and may be associated with a mortality rate of up to 60%.
  - H influenza type B, meningococci, and group A streptococci account for an additional 25% of OPSI.
  - It has been estimated that the incidence of OPSI is approximately 1 case occurring per 500 person-years of observation.
  - These patients must be vaccinated against Pneumococcus, Meningococcus, and Haemophilus influenzae. These vaccines should be given two weeks prior to elective splenectomy.
- Splenectomy
  - Currently, the majority of splenectomies are performed laparoscopically. This known to be associated with less morbidity and mortality.
  - Laparoscopic splenectomy is safe and is associated with decreased morbidity and reduced hospital stays.
  - Currently, laparoscopic splenectomy can be performed safely even on patients with massive splenomegaly.
  - A reactive thrombocytosis may develop following splenectomy. This is relatively common and maybe associated with an increased risk of venous thrombosis and embolism. This reactive thrombocytosis typically peaks in 2–3 weeks following splenectomy but can persist for months to years.
- Splenectomized children younger than 5 years, especially infants splenectomized for trauma, may have an infection risk greater than 10%.
- Splenectomy performed for a hematologic disorder, such as thalassemia, hereditary spherocytosis, or lymphoma, appears to carry a higher risk for OPSI than splenectomy performed as a result of trauma.
- This is attributed to the development of splenic implants or accessory spleens following trauma.
- Post-splenectomy patients are at risk for overwhelming post splenectomy infection (OPSI) due to infection with encapsulated organisms such as:
  - Haemophilus influenzae
  - Neisseria meningitidis
  - Streptococcus pneumoniae
- Prevention of OPSI include:
  - Patients and their families should be educated about the risk of OPSI and they should notify their physician and report to the nearest hospital in the event of an acute febrile illness.
  - In some centers, asplenic patients are encouraged to wear a Medi-Alert bracelet and carry a card explaining that they had splenectomy.
  - Vaccination is also important in the prevention of OPSI.

    The pneumococcal vaccine should be administered at least 2 weeks before an elective splenectomy or as soon as possible after recovery and before discharge from the hospital when splenectomy is performed as an emergency.

    A booster dose is also recommended after 5 years.

Antibiotic prophylaxis is also recommended for children following splenectomy. These are given but the duration of prophylaxis antibiotics is still controversial. Prophylaxis antibiotics should be given at least for the first for the first 2 years after splenectomy. Others recommend continuing prophylaxis antibiotics in children for at least 5 years.
Vaccination against Haemophilus influenza and Neisseria meningitidis should be given

Further Reading

1. Hosea SW, Burch CG, Brown EJ, et al. Impaired immune response of splenectomised patients to polyvalent pneumococcal vaccine. Lancet. 1981;1:804.
2. Giebink GS, Le CT, Schiffman G. Decline of serum antibody in splenectomized children after vaccination with pneumococcal capsular polysaccharides. J Pediatr. 1984;105:576.
3. Heath HW, Pearson HA. Thrombocytosis in pediatric outpatients. J Pediatr. 1989;114:805.
4. McIntyre OR, Ebaugh FG Jr. Palpable spleens in college freshmen. Ann Intern Med. 1967;66:301.
5. Suttorp M, Classen CF. Splenomegaly in children and adolescents. Front Pediatr. 2021;9:704635.
6. O'Reilly RA. Splenomegaly in 2505 patients in a large university medical center from 1913 to 1995. 1913 to 1962: 2,056 patients. West J Med. 1998;169:78.
7. Tamayo SG, Rickman LS, Mathews WC, et al. Examiner dependence on physical diagnostic tests for the detection of splenomegaly: a prospective study with multiple observers. J Gen Intern Med. 1993;8:69.
8. Vancauwenberghe T, Snoeckx A, Vanbeckevoort D, et al. Imaging of the spleen: what the clinician needs to know. Singap Med J. 2015;56:133.
9. Rosenberg HK, Markowitz RI, Kolberg H, et al. Normal splenic size in infants and children: sonographic measurements. AJR Am J Roentgenol. 1991;157:119.
10. Pelizzo G, Guazzotti M, Klersy C, et al. Spleen size evaluation in children: Time to define splenomegaly for pediatric surgeons and pediatricians. PLoS One. 2018;13:e0202741.
11. Gielchinsky Y, Elstein D, Hadas-Halpern I, et al. Is there a correlation between degree of splenomegaly, symptoms and hypersplenism? A study of 218 patients with Gaucher disease. Br J Haematol. 1999;106:812.
12. Chapman J, Goyal A, Azevedo AM. Splenomegaly. Treasure Island, FL: StatPearls Publishing; 2022.
13. Eichner ER. Splenic function: normal, too much and too little. Am J Med. 1979;66(2):311–20.
14. Poulin EC, Mamazza J, Schlachta CM. Splenic artery embolization before laparoscopic splenectomy. An update. Surg Endosc. 1998;12(6):870–5.
15. Lewis SM, Williams A, Eisenbarth SC. Structure and function of the immune system in the spleen. Sci Immunol. 2019;4(33):eaau6085.
16. Wintrobe MM. Disorders of the spleen. In: Greer JP, Arber DA, Glader B, List AF, Means RT, Paraskevas F, Rodgers GM, Foerster J, editors. Wintrobe's clinical hematology. 13th ed. Lippincott Williams & Wilkins; 2014. p. 1369–83.
17. Eichner ER, Whitfield CL. Splenomegaly. An algorithmic approach to diagnosis. JAMA. 1981;246(24):2858–61.
18. Arakawa Y, Shimada M, Utsunomiya T, Imura S, Morine Y, Ikemoto T, et al. Bevacizumab improves splenomegaly and decreases production of hyaluronic acid after L-OHP based chemotherapy. Anticancer Res. 2014;34(4):1953–8.
19. Koduri PR, Kovarik P. Acute splenic sequestration crisis in an adult with sickle beta-thalassemia. Ann Hematol. 2006;85(9):633–5.

20. Reilly RA. Splenomegaly in 2,505 patients at a large university medical center from 1913 to 1995. 1963 to 1995: 449 patients. West J Med. 1998;169(2):88–97.
21. Butler JR, Eckert GJ, Zyromski NJ, Leonardi MJ, Lillemoe KD, Howard TJ. Natural history of pancreatitis-induced splenic vein thrombosis: a systematic review and meta-analysis of its incidence and rate of gastrointestinal bleeding. HPB (Oxford). 2011;13(12):839–4.
22. Zhu JH, Wang YD, Ye ZY, Zhao T, Zhu YW, Xie ZJ, et al. Laparoscopic versus open splenectomy for hypersplenism secondary to liver cirrhosis. Surg Laparosc Endosc Percutan Tech. 2009;19(3):258–62.
23. Anegawa G, Kawanaka H, Uehara H, Akahoshi T, Konishi K, Yoshida D, et al. Effect of laparoscopic splenectomy on portal hypertensive gastropathy in cirrhotic patients with portal hypertension. J Gastroenterol Hepatol. 2009;24(9):1554–8.
24. Motta I, Filocamo M, Poggiali E, Stroppiano M, Dragani A, et al. A multicentre observational study for early diagnosis of Gaucher disease in patients with splenomegaly and/or thrombocytopenia. Eur J Haematol. 2016;96:352.
25. Olson AP, Trappey B, Wagner M, Newman M, Nixon LJ, Schnobrich D. Point-of-care ultrasonography improves the diagnosis of splenomegaly in hospitalized patients. Crit Ultrasound J. 2015;7(1):13.
26. Goldstone J. Splenectomy for massive splenomegaly. Am J Surg. 1978;135(3):385–8.
27. Laws HL, Burlingame MW, Carpenter JT, Prchal JT, Conrad ME. Splenectomy for hematologic disease. Surg Gynecol Obstet. 1979;149(4):509–12.
28. Musser G, Lazar G, Hocking W, Busuttil RW. Splenectomy for hematologic disease. The UCLA experience with 306 patients. Ann Surg. 1984;200(1):40–5.
29. Wilhelm MC, Jones RE, McGehee R, et al. Splenectomy in hematologic disorders. The ever-changing indications. Ann Surg. 1988;207(5):581–9.
30. Flowers JL, Lefor AT, Steers J, et al. Laparoscopic splenectomy in patients with hematologic diseases. Ann Surg. 1996;224(1):19–28.
31. Wang KX, Hu SY, Zhang GY, Chen B, Zhang HF. Hand-assisted laparoscopic splenectomy for splenomegaly: a comparative study with conventional laparoscopic splenectomy. Chin Med J. 2007;120(1):41–5.
32. Mosquera-Klinger G, de la Serna Higuera C, Bazaga S, García-Alonso FJ, Sánchez Ocaña R, Antolín Melero B, et al. Endoscopic ultrasound-guided fine-needle aspiration for splenomegaly and focal splenic lesion: is it safe, effective and necessary? Rev Esp Enferm Dig. 2020;112(5):355–9.
33. Bezerra AS, D'Ippolito G, Faintuch S, Szejnfeld J, Ahmed M. Determination of splenomegaly by CT: is there a place for a single measurement? AJR Am J Roentgenol. 2005;184(5):1510–3.
34. Subhasis RC, Rajiv C, Kumar SA, Kumar AV, Kumar PA. Surgical treatment of massive splenomegaly and severe hypersplenism secondary to extrahepatic portal venous obstruction in children. Surg Today. 2007;37(1):19–23.
35. Owusu-Ofori S, Remmington T. Splenectomy versus conservative management for acute sequestration crises in people with sickle cell disease. Cochrane Database Syst Rev. 2015;9:CD003425.
36. Lavrenkov K, Krepel-Volsky S, Levi I, Ariad S. Low dose palliative radiotherapy for splenomegaly in hematologic disorders. Leuk Lymphoma. 2012;53(3):430–4.
37. Bruns F, Bremer M, Dettmer A, Janssen S. Low-dose splenic irradiation in symptomatic congestive splenomegaly: report of five cases with literature data. Radiat Oncol. 2014;27(9):86.
38. Zaorsky NG, Williams GR, Barta SK, Esnaola NF, Kropf PL, Hayes SB, et al. Splenic irradiation for splenomegaly: a systematic review. Cancer Treat Rev. 2017;53:47–52.
39. Verstovsek S, Mesa RA, Gotlib J, Levy RS, et al. A double-blind, placebo-controlled trial of ruxolitinib for myelofibrosis. N Engl J Med. 2012;366(9):799–807.
40. Mistry PK, Lukina E, Ben Turkia H, Amato D, Baris H, et al. Effect of oral eliglustat on splenomegaly in patients with Gaucher disease type 1: the ENGAGE randomized clinical trial. JAMA. 2015;313(7):695–706.

41. Kawanaka H, Akahoshi T, Kinjo N, et al. Technical standardization of laparoscopic splenectomy harmonized with hand-assisted laparoscopic surgery for patients with liver cirrhosis and hypersplenism. J Hepato-Biliary-Pancreat Surg. 2009;16(6):749–57.
42. Xu WL, Li SL, Wang Y, Shi BJ, Li M, Li YC, et al. Laparoscopic splenectomy: color Doppler flow imaging for preoperative evaluation. Chin Med J. 2009;122(10):1203–8.
43. Vannucchi AM, Barbui T. Thrombocytosis and thrombosis. Hematology Am Soc Hematol Educ Program. 2007:363–70.
44. Khan PN, Nair RJ, Olivares J, Tingle LE, Li Z. Postsplenectomy reactive thrombocytosis. Proc (Bayl Univ Med Cent). 2009;22(1):9–12.
45. Lee DH, Barmparas G, Fierro N, Sun BJ, Ashrafian S, Li T, et al. Splenectomy is associated with a higher risk for venous thromboembolism: a prospective cohort study. Int J Surg. 2015;24(Pt A):27–32.
46. Shaw JH, Print CG. Postsplenectomy sepsis. Br J Surg. 1989;76(10):1074–81.
47. Kim DK, Riley LE, Harriman KH, Hunter P, Bridges CB, Advisory Committee on Immunization Practices. Recommended immunization schedule for adults aged 19 years or older, United States, 2017. Ann Intern Med. 2017;166(3):209–19.
48. Davies JM, Barnes R, Milligan D, British Committee for Standards in Haematology. Working Party of the Haematology/Oncology Task Force. Update of guidelines for the prevention and treatment of infection in patients with an absent or dysfunctional spleen. Clin Med (Lond). 2002;2(5):440–3.
49. Davies JM, Lewis MP, Wimperis J, Rafi I, Ladhani S, Bolton-Maggs PH, et al. Review of guidelines for the prevention and treatment of infection in patients with an absent or dysfunctional spleen: prepared on behalf of the British Committee for Standards in Haematology by a working party of the Haemato-Oncology task force. Br J Haematol. 2011;155(3):308–17.
50. Ginzel AW, Kransdorf MJ, Peterson JJ, Garner HW, Murphey MD. Mass-like extramedullary hematopoiesis: imaging features. Skeletal Radiol. 2012;41:911.
51. Ezeofor SN, Obikili EN, Anyanwu GE, Onuh AC, Mgbor SO. Sonographic assessment of the normal limits of the spleen in healthy school children in South-East Nigeria. Niger J Clin Pract. 2014;17(4):484–8.
52. Kahramaner Z, Erdemir A, Arik B, Bilgili G, Tekin M, Genc Y. Reference ranges of liver and spleen dimensions in term infants: sonographic measurements. J Med Ultrason. 2015;42(1):77–81.
53. Snape J, Baker AR, Rees Y. Pseudo-splenomegaly as a result of subphrenic abscess. Postgrad Med J. 1986;62(723):29–30.
54. Brown NF, Marks DJ, Smith PJ, Bloom SL. Splenomegaly. Br J Hosp Med (Lond). 2011;72(11):M166–9.
55. Schlesinger AE, Hildebolt CF, Siegel MJ, Pilgrim TK. Splenic volume in children: simplified estimation at CT. Radiology. 1994;193(2):578–80.
56. Anand K, Munker R. Is a liver-spleen scan helpful in the evaluation of patients with thrombocytopenia? Ann Hematol. 2015;94(11):1921–2.
57. Larsson F, Ahlin A, Marshall Heyman M, Abrahamsson J. [Acute splenic sequestration in children with sickle cell disease--an overview]. Lakartidningen 2016;113:D4AI.
58. Ghmaird A, Alnoaiji MM, Al-Blewi S, Zaki S, El-Lewi A, Ahmad N. Splenectomy in patients with sickle cell disease in Tabuk. Open Access Maced J Med Sci. 2016;4(1):107–11.
59. Owusu-Ofori S, Remmington T. Splenectomy versus conservative management for acute sequestration crises in people with sickle cell disease. Cochrane Database Syst Rev. 2017;11:CD003425.
60. Costi R, Castro Ruiz C, Romboli A, Wind P, Violi V, Zarzavadjian Le Bian A. Partial splenectomy: who, when and how. A systematic review of the 2130 published cases. J Pediatr Surg. 2019;54(8):1527–38.
61. Gutierrez Diaz AI, Svarch E, Arencibia Nunez A, et al. [Partial splenectomy in sickle cell disease]. An Pediatr (Barc). 2015;82(4):228–34.

62. Lopez JM Jr, McGonagill PW, Gross JL, et al. Subcapsular hematoma in blunt splenic injury: a significant predictor of failure of nonoperative management. J Trauma Acute Care Surg. 2015;79(6):957–9; discussion 959–60.
63. Pugi J, Carcao M, Drury LJ, Langer JC. Results after laparoscopic partial splenectomy for children with hereditary spherocytosis: are outcomes influenced by genetic mutation? J Pediatr Surg. 2018;53(5):973–5.
64. Ates U, Tastekin NY, Gollu G, Ergun E, Yagmurlu A. Laparoscopic splenectomy in pediatric age: long-term follow-up. Arch Argent Pediatr. 2017;115(6):e385–90.
65. Shan GD, Chen WG, Hu FL, et al. A spontaneous hematoma arising within an intrapancreatic accessory spleen: a case report and literature review. Medicine (Baltimore). 2017;96(41):e8092.
66. Weledji EP. Benefits and risks of splenectomy. Int J Surg. 2014;12(2):113–9.
67. Rorholt M, Ghanima W, Farkas DK, Norgaard M. Risk of cardiovascular events and pulmonary hypertension following splenectomy - a Danish population-based cohort study from 1996-2012. Haematologica. 2017;102(8):1333–41.
68. Luu S, Spelman D, Woolley IJ. Post-splenectomy sepsis: preventative strategies, challenges, and solutions. Infect Drug Resist. 2019;12:2839–51.
69. Okabayashi T, Hanazaki K. Overwhelming postsplenectomy infection syndrome in adults - a clinically preventable disease. World J Gastroenterol. 2008;14(2):176–9.
70. Iolascon A, Andolfo I, Barcellini W, et al. Recommendations regarding splenectomy in hereditary hemolytic anemias. Haematologica. 2017;102(8):1304–13.
71. Rosenberg HK, Markowitz RI, Kolberg H, Park C, Hubbard A, Bellah RD. Normal splenic size in infants and children: sonographic measurements. AJR Am J Roentgenol. 1991;157(1):119–21.
72. Sills RH. Splenic function: physiology and splenic hypofunction. Crit Rev Oncol Hematol. 1987;7(1):1–36.
73. Pizzi M, Fuligni F, Santoro L, et al. Spleen histology in children with sickle cell disease and hereditary spherocytosis: hints on the disease pathophysiology. Hum Pathol. 2017;60:95–103.
74. Mebius RE, Kraal G. Structure and function of the spleen. Nat Rev. Immunol. 2005;5(8):606–16.
75. Ebaugh FG, McIntyre OR. Palpable spleens: ten-year follow-up. Ann Intern Med. 1979;90(1):130–1.
76. Ancliff P, Hann I. Splenomegaly. In: Sills RH, editor. Practical algorithms in pediatric hematology and oncology. Basel: Karger; 2003. p. 50–1.
77. Genton B, al-Yaman F, Beck HP, et al. The epidemiology of malaria in the Wosera area, East Sepik Province, Papua New Guinea, in preparation for vaccine trials. I. Malariometric indices and immunity. Ann Trop Med Parasitol. 1995;89(4):359–76.
78. Pitney WR. The tropical splenomegaly syndrome. Trans R Soc Trop Med Hyg. 1968;62(5):717–28.
79. Farley DR, Zietlow SP, Bannon MP, Farnell MB. Spontaneous rupture of the spleen due to infectious mononucleosis. Mayo Clin Proc. 1992;67(9):846–53.
80. American Academy of Pediatrics. In: Kimberlin DW, editor. Red book: 2015 report of the committee on infectious diseases. 30th ed. Elk Grove, IL: American Academy of Pediatrics; 2015.
81. Rice SG, American Academy of Pediatrics Council on Sports Medicine and Fitness. Medical conditions affecting sports participation. Pediatrics. 2008;121(4):841–8.
82. Goddard SL, Chesney AE, Reis MD, et al. Pathological splenic rupture: a rare complication of chronic myelomonocytic leukemia. Am J Hematol. 2007;82(5):405–8.
83. Amaki J, Sekiguchi T, Hiraiwa S, et al. Three cases of spontaneous splenic rupture in malignant lymphoma. Int J Hematol. 2018;108(6):647–51.
84. Dumic I, Patel J, Hart M, Niendorf ER, Martin S, Ramanan P. Splenic rupture as the first manifestation of babesia microti infection: report of a case and review of literature. Am J Case Rep. 2018;23(19):335–41.

85. Wilson DB, Nathan DG, Orkin SH, Ginsburg D, Look AT. Acquired platelet defects. Nathan and Oski's hematology of infancy and childhood, vol. 2. 6th ed. Philadelphia, PA: WB Saunders; 2003. p. 1599.
86. Peck-Radosavljevic M. Hypersplenism. Eur J Gastroenterol Hepatol. 2001;13(4):317–23.
87. Jandl JH, Aster RH. Increased splenic pooling and the pathogenesis of hypersplenism. Am J Med Sci. 1967;253(4):383–98.
88. Sarin SK, Kapoor D. Non-cirrhotic portal fibrosis: current concepts and management. J Gastroenterol Hepatol. 2002;17(5):526–34.
89. Beutler E, Gelbart T. Estimating the prevalence of pyruvate kinase deficiency from the gene frequency in the general white population. Blood. 2000;95(11):3585–8.
90. Tunnessen WW Jr. Splenomegaly. In: Roberts K, Tunnessen W, editors. Signs and symptoms in pediatrics. 3rd ed. Philadelphia, PA: Lippincott Williams & Wilkins; 1999. p. 475–83.
91. Baris HN, Cohen IJ, Mistry PK. Gaucher disease: the metabolic defect, pathophysiology, phenotypes and natural history. Pediatr Endocrinol Rev. 2014;12(Suppl 1):72–81.
92. Nixon RK Jr. The detection of splenomegaly by percussion. N Engl J Med. 1954;250(4):166–7.
93. Castell DO. The spleen percussion sign. A useful diagnostic technique. Ann Intern Med. 1967;67(6):1265–7.
94. Grover SA, Barkun AN, Sackett DL. The rational clinical examination. Does this patient have splenomegaly? JAMA. 1993;270(18):2218–21.
95. Fenando A, Tatineni S, Raziq FI, Alratroot A. Subcapsular haematoma of the spleen complicating acute pancreatitis. BMJ Case Rep. 2019;12(9):e231716.
96. Stylianos S. To save a child's spleen: 50 years from Toronto to ATOMAC. J Pediatr Surg. 2019;54(1):9–15.
97. Ingle SB, Hinge Ingle CR, Patrike S. Epithelial cysts of the spleen: a minireview. World J Gastroenterol. 2014;20(38):13899–903.
98. Pochedly C, Sills RH, Schwartz AD, editors. Disorders of the spleen: pathophysiology and management. New York, NY: Marcel Dekker; 1989.
99. Clarke RT, Van den Bruel A, Bankhead C, Mitchell CD, Phillips B, Thompson MJ. Clinical presentation of childhood leukaemia: a systematic review and meta-analysis. Arch Dis Child. 2016;101(10):894–901.
100. Di Giorgio A, De Angelis P, Cheli M, et al. Etiology, presenting features and outcome of children with non-cirrhotic portal vein thrombosis: a multicentre national study. Dig Liver Dis. 2019;51(8):1179.
101. Yang AC, Bier L, Overbey JR, et al. Early manifestations of type 1 Gaucher disease in presymptomatic children diagnosed after parental carrier screening. Genet Med. 2017;19(6):652–8.
102. McGovern MM, Wasserstein MP, Bembi B, et al. Prospective study of the natural history of chronic acid sphingomyelinase deficiency in children and adults: eleven years of observation. Orphanet J Rare Dis. 2021;16(1):212.
103. Kinney TR, Ware RE, Schultz WH, Filston HC. Long-term management of splenic sequestration in children with sickle cell disease. J Pediatr. 1990;117(2 Pt 1):194–9.
104. Castagnola E, Fioredda F. Prevention of life-threatening infections due to encapsulated bacteria in children with hyposplenia or asplenia: a brief review of current recommendations for practical purposes. Eur J Haematol. 2003;71(5):319–26.
105. Price VE, Dutta S, Blanchette VS, Butchart S, Kirby M, Langer JC, et al. The prevention and treatment of bacterial infections in children with asplenia or hyposplenia: practice considerations at the Hospital for Sick Children, Toronto. Pediatr Blood Cancer. 2006;46(5):597–603.
106. Altered immunocompetence - general best practice guidelines for immunization: best practices guidance of the Advisory Committee on Immunization Practices (ACIP). Vaccine recommendations and guidelines of the ACIP. https://www.cdc.gov/vaccines/hcp/acip-recs/general-recs/immunocompetence.html. Accessed 20 Aug 2019.
107. Lane PA. The spleen in children. Curr Opin Pediatr. 1995;7(1):36–41.

108. Rice HE, Oldham KT, Hillery CA, Skinner MA, O'Hara SM, Ware RE. Clinical and hematologic benefits of partial splenectomy for congenital hemolytic anemias in children. Ann Surg. 2003;237(2):281–8.
109. Rosman CWK, Broens PMA, Trzpis M, Tamminga RYJ. A long-term follow-up study of subtotal splenectomy in children with hereditary spherocytosis. Pediatr Blood Cancer. 2017;64.
110. Li S, Li M, Xu W, Sun C, Liu L. Single-incision laparoscopic splenectomy using the suture suspension technique for splenomegaly in children with hereditary spherocytosis. J Laparoendosc Adv Surg Tech A. 2015;25(9):770–4.
111. Ahad S, Gonczy C, Advani V, Markwell S, Hassan I. True benefit or selection bias: an analysis of laparoscopic versus open splenectomy from the ACS-NSQIP. Surg Endosc. 2013;27(6):1865.
112. Hassan ME, Al AK. Massive splenomegaly in children: laparoscopic versus open splenectomy. JSLS. 2014;18(3):e2014.00245.
113. Utria AF, Goffredo P, Keck K, Shelton JS, Shilyansky J, Hassan I. Laparoscopic splenectomy: has it become the standard surgical approach in pediatric patients? J Surg Res. 2019;240:109–14.
114. Van Der Veken E, Laureys M, Rodesch G, Steyaert H. Perioperative spleen embolization as a useful tool in laparoscopic splenectomy for simple and massive splenomegaly in children: a prospective study. Surg Endosc. 2016;30(11):4962–7.
115. Chapman J, Azevedo AM. Splenomegaly. Treasure Island, FL: StatPearls Publishing; 2019.
116. Arkles LB, Gill GD, Molan MP. A palpable spleen is not necessarily enlarged or pathological. Med J Aust. 1986;145(1):15–7.
117. Eichner ER. Sports medicine pearls and pitfalls—defending the spleen: return to play after infectious mononucleosis. Curr Sports Med Rep. 2007;6(2):68–9.
118. Sul HJ, Kang D. Congenital neuroblastoma with multiple metastases: a case report. J Korean Med Sci. 2003;18(4):618–20.

5 Splenic Rupture

5.1 Introduction

- The spleen is the largest lymphoid organ in the body. It is located in the left upper quadrant of the abdominal cavity and it is protected by the rib cage.
- Anatomically, the spleen is protected by the rib cage but in spite of this the spleen is of the commonly injured organs by blunt abdominal trauma.
- The normal weight of the spleen ranges from 75 to 150 g.
- The spleen a highly vascular organ and it is supplied by the splenic artery which is a branch of the celiac axis.
- One of the important functions of the spleen is filtration of blood. It is estimated that the spleen filters about 10–15% of total blood volume every minute.
- The spleen has a capsule and it is completely covered with peritoneum except at the hilum where the splenic artery and vein passes. The capsule of the spleen provides added protection against blunt injury.
- The spleen is also fixated to the posterior aspect of the left upper quadrant by the gastrosplenic and splenorenal ligaments. These ligaments also provide additional support to the spleen by fixing it to the left upper quadrant. Sometimes these ligaments are lax or abnormally long which makes the spleen more mobile making it more susceptible to traumatic injury or predisposes the spleen to torsion y twisting around its vascular pedicle.
- Blood supply:
  - The major arterial supply to the spleen is via the splenic artery.
  - This is a branch of the celiac artery and runs superior and posterior to the pancreas.
  - The splenic artery commonly bifurcates at the hilum giving an upper and a lower polar branch. These subsequently divide within the splenic parenchyma.
  - The splenic vein drains the spleen and together with the superior mesenteric vein empties into the portal vein.
  - The spleen is also supplied by the short gastric vessels. These are branches of the left gastroepiploic artery.

A. H. Al-Salem, *The Spleen*, https://doi.org/10.1007/978-981-99-6191-7_5

  - The short gastric vessels supply the upper pole of the spleen and they are very short. This is the reason they are called the short gastric vessels. This is important and must be taken in consideration at the time of splenectomy especially during emergency splenectomy.
- The tail of the pancreas is close to the spleen and often adjacent to the splenic hilum. Great care must be taken to avoid injury of the pancreas as the tail of the pancreas can be easily damaged during splenectomy.
- The spleen has important immunologic functions and every attempt should be made to preserve the spleen and splenectomy should be avoided as much as possible.
- The radiological advancement especially abdominal CT scanning has made possible to outline the extent of splenic injury following abdominal trauma. This made it possible to conserve and safe the spleen in certain types of splenic injuries. Conservative management of splenic injuries is now more feasible and safer for patients with splenic injury.
- The spleen is considered the most commonly solid organ to be affected following blunt abdominal trauma. This is both in children and adults. Others consider the liver as the most commonly injured solid intraabdominal organ from blunt abdominal trauma followed by the spleen.
- Injury to the spleen is most often the result of blunt abdominal trauma.
- Splenic injuries are commonly seen secondary to:
  - Car accidents
  - Domestic violence
  - Different types of sports
  - Bicycle handlebars injuries
- Splenic injuries can also be iatrogenic:
  - The spleen can be injured in emergency operations
  - This is especially so when there are preexisting adhesions. These adhesions make mobilization of intra-abdominal structures difficult.
  - The spleen can also be injured during colonoscopy.
- Rarely, spontaneous rupture of the spleen occurs. This is seen in certain types of infections and also some hematological conditions.
- A ruptured spleen is a medical emergency and may result in massive intra-abdominal hemorrhage and shock.
- Rupture of the spleen can be classified into acute or delayed
  - Acute splenic rupture: In this, the patient presents immediately following abdominal trauma because of severe pain and shock.
  - Delayed splenic rupture: In this, the patient presents with sudden onset of pain and shock following an interval of symptom-free. The duration of this interval is variable and can range from few days to weeks following abdominal trauma.
- The treatment of splenic injury depends on the severity.
  - Conservative therapy: This should be the first option in a stable patient. The patient should be admitted for close monitoring and observation in a high dependency unit.
  - Most patients with traumatic splenic injury will require surgical intervention.

- Every attempt should be made to preserve the spleen at the time of surgery.
- Partial splenectomy or splenorrhaphy should be an option.
- Splenectomy is indicated in severe extensive injury that is life threatening.

- In the past, physical examination and diagnostic peritoneal lavage (DPL) were the only diagnostic methods available to diagnose splenic injury.
- As a result of this, minor splenic injuries were frequently missed and, in some cases, splenectomy could be avoided.
- With the widespread availability of abdominal CT-scan, the diagnosis of splenic injuries became much easier and the this also changed the management approach to splenic trauma.

5.2 Etiology

- There are four types of splenic injuries:
  - Blunt traumatic splenic injury
  - Penetrating splenic injury
  - Iatrogenic splenic injury
  - Spontaneous rupture of the spleen
- Splenic injury is commonly seen in patients following blunt abdominal trauma.
- Penetrating abdominal injuries may also affect the spleen.
- Splenic injury occurs with explosive type injuries, as seen in warfare and civilian bombing.
- Iatrogenic splenic injuries are much less common and these as seen during abdominal surgery and following colonoscopy.
- Rarely splenic rupture occurs spontaneously
- Preexisting illness or disease leading to splenomegaly can markedly increase the risks and severity of splenic injury.
- This is seen in patients with infectious mononucleosis, malaria, and hematologic conditions. These will lead to acute or chronic enlargement of the spleen. As the spleen enlarges, the capsule will be thinned out making it more fragile and minor trauma in these patients with splenomegaly can lead a major splenic injury.
- Splenic rupture can be:
  - Traumatic
  - Rarely spontaneous
- Traumatic splenic rupture can be caused by:
  - Blunt abdominal trauma (Fig. 5.1)

    This is most frequently caused by motor vehicle accidents.

    Other types of injury include contact sports, physical altercations, and falls from great heights.

    Left-sided thoracic trauma with fractures of the lower ribs can be associated with splenic injury.
  - Penetrating abdominal trauma as seen in stab wounds, and gunshot wounds.
  - Explosion-related blunt and/or penetrating trauma as seen in warfare and civilian bombing.

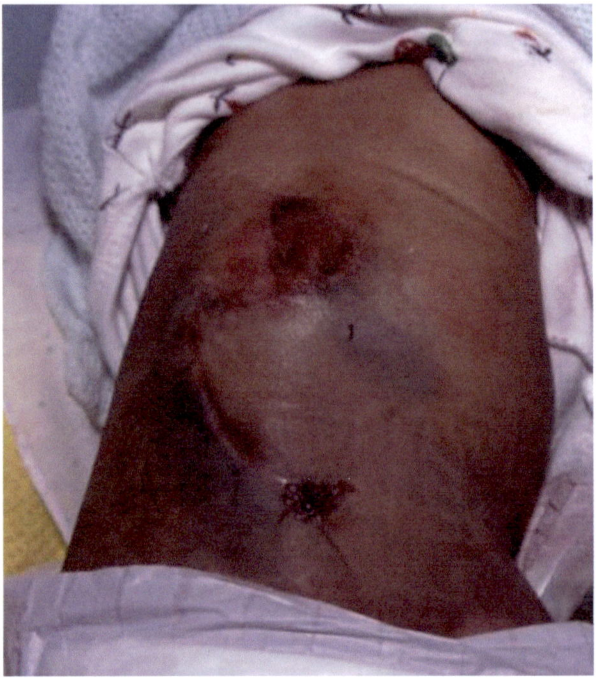

Fig. 5.1 A clinical photograph showing severe abdominal trauma with intraabdominal bleeding

- Spontaneous splenic rupture occurs in the following conditions:
  - Infections and inflammations such as:
    Infectious mononucleosis
    Malaria
    Typhoid fever
    HIV
    Acute splenitis
    Pancreatitis
    Subacute bacterial endocarditis
  - Leukemia and lymphoma
  - Peliosis lienis: Peliosis is a rare condition characterized by presence of multiple blood-filled cysts within the parenchyma of the spleen. It can also affect other organs. Peliosis affecting the spleen is usually seen in association with peliosis of liver.
  - Pregnancy
  - Amyloidosis and Glycogen storage disease
  - Congestion secondary to Portal hypertension
- Iatrogenic splenic rupture is seen:
  - During abdominal surgery especially those involving the stomach, pancreas or left hemidiaphragm.
  - During colonoscopy

5.3 Classification

- Acute splenic rupture:
  - This the commonest type
  - In this type there is acute injury of the splenic capsule and possibly the splenic parenchymal tissue.
  - This will result in acute intra-abdominal bleeding which may lead to hypovolemic shock if not treated.
- Delayed splenic rupture:
  - This result from injury of the splenic parenchymal tissue but an initially intact splenic capsule. The splenic capsule will contain the bleeding.
  - This will lead to central or subcapsular hematoma.
  - This will be associated with an asymptomatic interval lasting day to weeks. During this time the hematoma gradually build up and distends inside the capsule. Subsequent the splenic capsule ruptures leading to intra-abdominal bleeding.
  - In delayed splenic rupture, symptoms may not present until days to weeks after trauma. This must be kept in mind when a patient present to the hospital days or weeks following abdominal trauma with abdominal pain and shock.

5.4 Presentation

- The clinical presentation of splenic injury is variable. This depends on the degree of injury and the presence or absence of other associated intraabdominal injuries.
- The majority of patients with minor splenic injury present with left upper quadrant abdominal pain and tenderness.
- This may be associated with left shoulder pain and tenderness. This is due to subdiaphragmatic nerve root irritation with referred pain.
- Kehr's sign:
  - This referred left shoulder pain is called the Kehr's sign.
  - It results when blood from an injured spleen irritates the diaphragm and creates referred pain.
- Ballance's sign:
  - This is dullness on ***percussion*** in the left upper quadrant of the abdomen.
- Patients with free intraperitoneal blood present with:
  - Diffuse abdominal pain
  - Peritoneal irritation with tenderness and rebound tenderness.
- Once the intra-peritoneal bleeding exceeds 5–10% of the total blood volume, patients present with clinical signs of early shock. These include:
  - Tachycardia, tachypnea, restlessness, and anxiety
  - Pallor
  - Decreased capillary refill
  - Decreased pulse pressure
  - Abdominal distension

- Hypotension is a late sign that my require surgical intervention.
- There are no contraindications to surgical intervention in a hemodynamically unstable patient with a splenic injury and no time should be waisted in these patients trying to document the injury.
- On the other hand, hypotension or unstable vital signs are a contraindication to CT scanning. This will waste valuable time and there are reports of deaths in the radiology department while trying to document a splenic injury.
- Hypotensive and unstable patients can be assessed:
  - Clinically
  - By abdominal ultrasound
  - By DPL

5.5 Investigations

- A complete blood cell count
- Blood grouping and cross matching
- 4–6 units of blood should be arranged for surgery
- Liver function tests
- Renal function test
- Chest x-ray
- Abdominal X-ray
- Imaging studies
  - Abdominal ultrasound
    - This is a raid, noninvasive and relatively cheap investigation
    - Abdominal ultrasound is important to detect the presence of or absence of fluid in the peritoneal cavity.
    - The presence of fluid in the right upper quadrant, the left upper quadrant, and the pelvis suggests the possibility of liver injury, splenic injury or mesenteric injury.
  - Abdominal CT scanning (Figs. 5.2, 5.3, 5.4, and 5.5)
    - Abdominal CT scan is important to evaluate trauma to the abdomen including the spleen and surrounding organs.
    - Abdominal CT scan with intravenous contrast is more useful as it improves the diagnosis and help document the severity of injury.
    - It is also important to document active bleeding from the splenic parenchyma.
  - Angiography
    - Angiography is rarely used to evaluate patients with suspected splenic injury.
    - Currently, angiography is being used more frequently as a therapeutic modality of splenic injuries treatment.
    - Angiography is usually performed once abdominal CT scan images showed an arterial contrast blush or active extravasation from the spleen parenchyma.

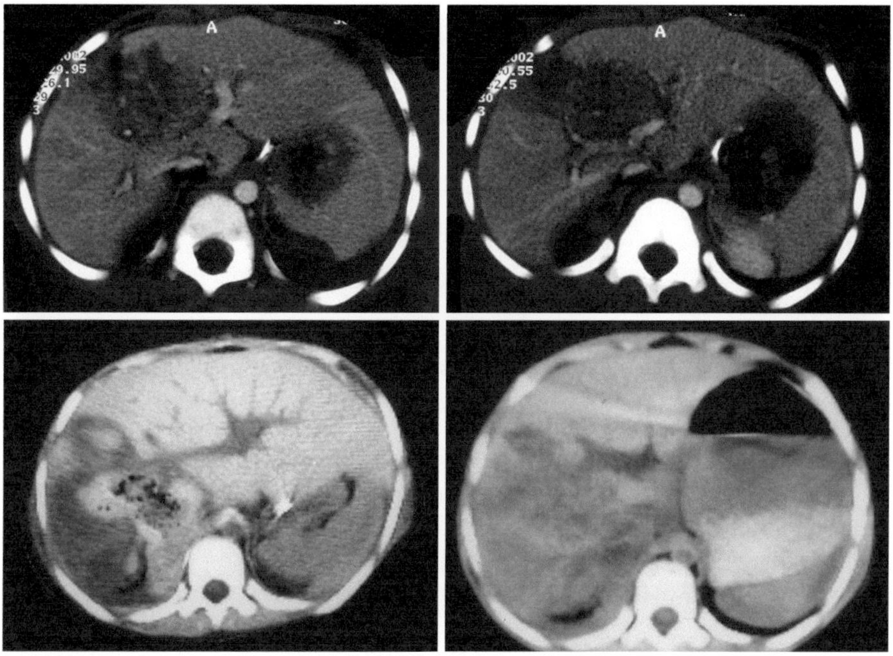

Figs. 5.2–5.5 Abdominal CT-scan with and without contrast in a traumatied child with intraperitoneal bleeding showing severe live injury. The possibility of an associated splenic injury must be kept in mind

  Once this is confirmed, angiography and angioembolization of active bleeding sites is performed.

  This technique is important as it helps preserve the spleen

  - MRI

    Rarely MRI is used to diagnose and assess the extent of splenic injury.

    MRI is used in patients with renal failure or those with well-known contrast allergy. In these patients abdominal CT scan with contrast is not an option.

  - Radioisotope studies

    These are rarely used to diagnose or assess splenic injury

    High-resolution CT scan is more superior.

- Diagnostic peritoneal lavage (DPL)
  - DPL is a rapid and inexpensive test that is used to rapidly determine if free intraperitoneal blood is present.
  - DPL is especially useful in hypotensive patients with suspected splenic injury who cannot undergo abdominal CT scan.
  - Currently, abdominal ultrasound and CT scan have replaced DPL in many institutions.
  - A positive DLP is an indication for exploration. Currently and with the aim of splenic preservation makes DLP not a valid option.

5.6 Staging and Grades

- Splenic injury is graded using the standards published by the Organ Injury Scaling Committee of the AAST (American Association for the Surgery of Trauma).
- These range from grade I (minor) to grade V (major) and this grading correlates to the need for laparotomy.
- This grading system is used in conjunction with:
  - Non-operative assessment including CT scan and angiography
  - Operative intervention by laparotomy
  - Postmortem studies
- This grading system of splenic injury is important. It helps physicians decide whether laparotomy or non-operative conservative management of splenic injury is an option for treatment.
- Splenic injury is classified according to the severity of splenic injury. This takes in consideration the level of splenic laceration, injury to the veins and arteries, and clotting.
- The American Association for the Surgery of Trauma grading system for spleen injury is as follows (Fig. 5.6):
  - Grade 1:

    In this grade there is a tear in the capsule of the spleen that is less than 1 centimeter deep into the spleen parenchyma, or a hematoma under the capsule (Subcapsular hematoma).

    The subcapsular hematoma covers less than 10 percent of the surface area of the spleen.
  - Grade 2:

    In this grade there is a 1-to-3-centimeter cm tear in the spleen but the tear does not involve the arterial branches of the spleen.

    Or a hematoma (Subcapsular hematoma) may develop under the capsule and covers between 10 and 50 percent of the surface area of the spleen.

    Or it involves a hematoma less than 5 cm in diameter in the parenchyma of the spleen.
  - Grade 3:

    In this stage there is a tear more than 3 cm deep in the spleen.

    Or a tear that involve the splenic artery.

    Or a hematoma that covers over half of the surface area of the spleen.

    A grade 3 injury involve a hematoma that is more than 5 cm in diameter or an expanding hematoma in the parenchyma of the spleen.
  - Grade 4:

    This grade involves a tear in the spleen that lacerates the segmental or hilar blood vessels of the spleen and causes the loss of more than 25 percent of the organ's blood supply.
  - Grade 5:

    This is an extremely severe injury that lacerates splenic blood vessels and causes a total loss of blood supply to the organ.

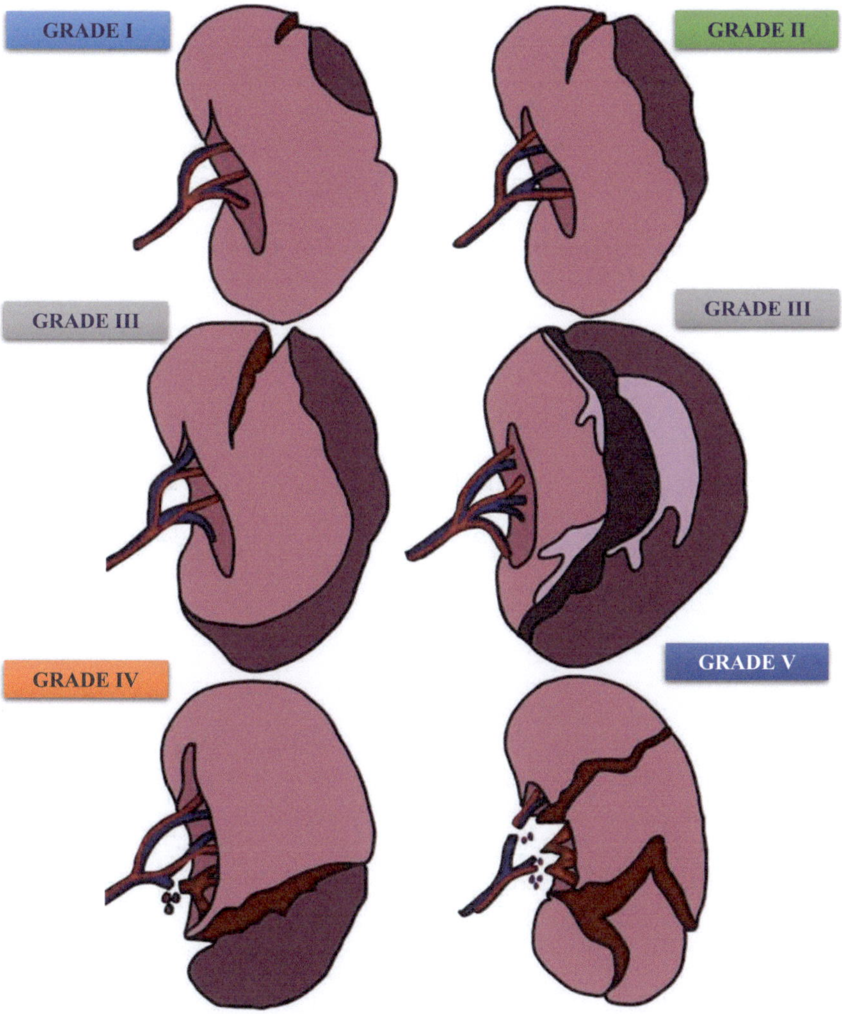

Fig. 5.6 Diagrammatic representation of splenic injury grades. The American Association for the Surgery of Trauma splenic injury grades

Or a hematoma that has completely shattered the spleen.

- Another classification of splenic trauma is the World Society of Emergency Surgery (WSES) classification.
- The WSES classification divides spleen injuries into four classes (Table 5.1):
  - Minor (WSES class I)
  - Moderate (WSES classes II and III)
  - Severe (WSES class IV)
- The management of splenic trauma has changed considerably in favor of non-operative management (NOM).

Table 5.1 Classification of splenic injuries according to WSES

	WSES class	Mechanism of injury	AAST	Hemodynamic status
Minor	WSES I	Blunt/Pentrating	I–II	Stable
Moderate	WSES II	Blunt/Pentrating	III	Stable
	WSES III	Blunt/Pentrating	IV–V	Stable
Severe	WSES IV	Blunt/Pentrating	I–V	Unstable

- NOM ranges from observation and monitoring alone to angiography/angioembolization (AG/AE) with the aim to preserve the spleen and its function, especially in children.

GRADES OF SPLENIC INJURIES

- **GRADE I:**
  - **LACERATION < 1 CM AND SUBCAPSULAR HEMATOMA < 10% OF SURFACE AREA**
- **GRADE II:**
  - **LACERATION 1–3 CM AND SUBCAPSULAR HEMATOMA 10–50% OF SURFACE AREA**
- **GRADE III:**
  - **LACERATION > 3 CM AND SUBCAPSULAR HEMATOMA > 50% OF SURFACE AREA**
  - **RUPTURED SUBCAPSULAR OR PARENCHYMAL HEMATOMA**
- **GRADE IV:**
  - **SEGMENTAL OR HILAR VASCULAR INJURY**
  - **DEVASCULARIZATION OF > 25% OF THE SPLEEN**
- **GRADE V:**
  - **SHATTERED SPLEEN**

5.7 Management

- The management of patients with splenic injury has changed over the years and it depends on the hemodynamic stability of the patient.
  - A hemodynamically unstable patient suspected of splenic injury and intra-abdominal bleeding should undergo exploratory laparotomy and either splenorraphy or splenectomy should be done. This will depend on the extent of splenic injury.
  - Patients who respond to initially to resuscitative measures only to deteriorate again are called transient responders. They are likely to have solid abdominal organ injury with ongoing hemorrhage and should undergo exploratory laparotomy and splenorraphy or splenectomy depending on the extent of splenic injury.

  - Patients who respond to resuscitative measures and continue to do so can be managed conservatively with or without splenic angioembolization. Angioembolization is not readily available but it has been reported as a safe method of splenic salvage when immediately available in the treating hospital.
- DPL is valuable diagnostic technique if other methods of investigations are not readily available or the patient status does not permit further investigations.
- Abdominal ultrasound is valuable in documenting the presence or absence of blood in the peritoneal cavity, which highly suggests the possibility of splenic injury. Abdominal ultrasound is readily available, cheap and non-invasive investigation and should be the first modality to investigate intraperitoneal bleeding.
- A stable patient (commonly defined as a patient with systolic blood pressure greater than 90 mm Hg and a heart rate less than 120 beats per minute) should have abdominal CT scan.
  - This is important not only to document intraperitoneal bleeding but also assess the extent of splenic injury.
  - It is also important to detect other associated intraabdominal organ injuries.
  - This will help the treating physician to decide whether to go for operative or non-operative management.
  - The presence of a splenic injury documented on CT scan does not exclude other associated abdominal injuries such as a mesenteric tear which may not be detected by CT scan.
  - It is important that all patients with abdominal trauma should have a thorough examination of the abdomen during laparotomy.
- MRI is a valuable investigation for patients with renal problems and an elevated creatinine level. Abdominal CT scan with contrast is contraindicated for these patients and should not be done. This is also the case for patients with known allergies to contrast media.
- The followings are determining factors for operative intervention in stable patients with a splenic injury:
  - The grade of injury as outlined by the American Association for the Surgery of Trauma (AAST).
  - The presence of intraperitoneal blood.
  - The presence of a blush on abdominal CT scan.
  - The calculated risk of rebleeding.
  - The presence and severity of concomitant abdominal injuries.
  - The options regarding the amount of blood transfusions required.
- Signs of persistent bleeding in a hemodynamically unstable patient are clear indications for exploratory laparotomy.
- The current management of splenic injury is nonoperative or conservative management and every attempt should be made to preserve the spleen. This is specially for patients with the following criteria:
  - Hemodynamically stable patients
  - Patients with stable hemoglobin levels over 12–48 h
  - Patients who require minimal blood transfusion (2 Units or less)
  - Patients with CT scan injury grade of 1 or 2 without a blush

  - Patients younger than 55 years.
    The risk of complications or failure of conservative nonoperative management appears to be worse in patients older than 55 years.
    Women older than 55 years are significantly more likely to fail nonoperative management with an increased mortality.
  - Patients with significant injury to other organs should be treated surgically.
  - Multisystem injury or concomitant liver, pancreas, or bowel injury increases the likelihood of splenectomy.
  - Patients on anticoagulants, such as warfarin and antiplatelet drugs are clinically considered to be at an increased risk for delayed bleed and should be evaluated and monitored closely.
- Splenic angioembolization
  - This mode of therapy is increasingly being used in both stable patients and transient responders.
  - A close cooperation between the treating surgeon and the interventional radiologist is important in this regard.
  - This is procedure is not readily available in every hospital and requires the presence of an interventional radiologist. A catheter is passed via the femoral artery and embolization of the splenic artery or its branches can be done using gel foam or metal coils.
  - This technique may salvage splenic injuries that previously required laparotomy and splenectomy.
- Surgical Therapy
  - Prior to laparotomy, the patient must be prepared including:
    Two wide-bore (16F or larger) IV lines for vascular access. This is important for fluid resuscitation and blood transfusion.
    4–6 units of blood should be available for surgery
    A nasogastric or orogastric tubes for stomach decompression should be inserted prior to surgery.
    A Foley catheter to monitor urine output should be placed prior to surgery
  - Every attempt should be made to preserve the spleen.
  - Patients with grade V splenic injuries will require an exploratory laparotomy.
  - In patients who are stable intraoperatively and where feasible, splenorrhaphy rather than splenectomy is the preferred surgical procedure.
  - Recurrent bleeding in those who had splenorrhaphy or new bleeding from missed or inadequately ligated blood vessels must be kept in mind in the first 24–48 h postoperatively.
  - Partial splenectomy is also a valuable alternative to splenectomy.
  - In patients with capsular injury, the electrocautery or argon beam coagulator device are useful and they may provide adequate hemostasis and avoid total splenectomy.
- Postoperatively, splenectomized patients are at increased risk of post splenectomy sepsis.

- The lifelong incidence of post splenectomy sepsis has been estimated to be 0.03–0.8% and the mortality rate of those developing post splenectomy sepsis may be as high as 60–70%.
- Postoperatively and to overcome the risk of post splenectomy sepsis, these patients should be immunized against Pneumococcal organisms, Haemophilus and Meningococcal species.
- These immunizations can be given from 24 h after injury to two weeks or just prior to their discharge from the hospital.
- These vaccinations include:
  - Pneumococcal vaccine
  - Haemophilus influenza vaccine
  - Meningococcal vaccine
- A pneumococcal vaccine booster dose should be given after 4–5 years.

- Accessory splenic tissue and reimplantation of splenic tissue have never been reliably proven to minimize the risk of post splenectomy sepsis.
- These patients should be educated about the risk of post splenectomy sepsis and the signs and symptoms of pneumococcal infection should be stressed.
- Complications of splenic trauma:
  - Nonoperative management of splenic trauma is associated with complications including:
    - Delayed bleeding
    - Posttraumatic splenic pseudocysts
    - Splenic necrosis and abscess formation
  - Complications of splenorrhaphy include:
    - Rebleeding
    - Thrombosis and necrosis of the residual spleen
  - Complications of splenectomy include:
    - Bleeding from the short gastric vessels
    - Bleeding from the splenic vessels
    - Increased risk of post splenectomy sepsis
  - Complications of angioembolization of the spleen include:
    - Noninfectious-related febrile reactions
    - Sympathetic pleural effusions
    - Splenic abscesses
    - Pancreatitis
    - Rarely, femoral arteriovenous fistulas and ilio-femoral pseudo aneurysms
  - Thrombocytosis:
    - This is a well-known complication following splenectomy.
    - The counts may reach above 1 million/mm^3 in some patients.
    - This may be complicated by thrombotic vascular occlusion such as deep vein thrombosis, pulmonary embolism, or stroke.
    - To avoid these potential complications, these patients should be treated with a daily baby aspirin.

Further Reading

1. Harbrecht BG, Franklin GA, Miller FB, Richardson JD. Is splenectomy after trauma an endangered species? Am Surg. 2008;74(5):410–2.
2. Fishback SJ, Pickhardt PJ, Bhalla S, Menias CO, Congdon RG, Macari M. Delayed presentation of splenic rupture following colonoscopy: clinical and CT findings. Emerg Radiol. 2011;18(6):539–44.
3. Ha JF, Minchin D. Splenic injury in colonoscopy: a review. Int J Surg. 2009;7:424.
4. Rozycki GS, Knudson MM, Shackford SR. Surgeon-performed bedside organ assessment with sonography after trauma (BOAST): a pilot study from the WTA Multicenter Group. J Trauma. 2005;59(6):1356–64.
5. Hedrick TL, Sawyer RG, Young JS. MRI for the diagnosis of blunt abdominal trauma: a case report. Emerg Radiol. 2005;11(5):309–11.
6. Lin WC, Chen YF, Lin CH, Tzeng YH, Chiang HJ, Ho YJ. Emergent transcatheter arterial embolization in hemodynamically unstable patients with blunt splenic injury. Acad Radiol. 2008;15(2):201–8.
7. Willmann JK, Roos JE, Platz A. Multidetector CT: detection of active hemorrhage in patients with blunt abdominal trauma. AJR Am J Roentgenol. 2002;179(2):437–44.
8. Nwomeh BC, Nadler EP, Meza MP. Contrast extravasation predicts the need for operative intervention in children with blunt splenic trauma. J Trauma. 2004;56(3):537–41.
9. Marmery H, Shanmuganathan K, Mirvis SE, Richard H 3rd, Sliker C, Miller LA, et al. Correlation of multidetector CT findings with splenic arteriography and surgery: prospective study in 392 patients. J Am Coll Surg. 2008;206(4):685–93.
10. Vick LR, Islam S. Recombinant factor VIIa as an adjunct in nonoperative management of solid organ injuries in children. J Pediatr Surg. 2008;43(1):195–8; discussion 198–9.
11. Scarborough JE, Ingraham AM, Liepert AE, Jung HS, O'Rourke AP, Agarwal SK. Nonoperative management is as effective as immediate splenectomy for adult patients with high-grade blunt splenic injury. J Am Coll Surg. 2016;223(2):249–58.
12. Wu Y, Wan L, Li P, Zhang Y, Li M, Gong J, et al. Application of radiofrequency ablation for splenic preservation. J Surg Res. 2015;193(2):781.
13. Killeen KL, Shanmuganathan K, Boyd-Kranis R. CT findings after embolization for blunt splenic trauma. J Vasc Interv Radiol. 2001;12(2):209–14.
14. Ekeh AP, McCarthy MC, Woods RJ. Complications arising from splenic embolization after blunt splenic trauma. Am J Surg. 2005;189(3):335–9.
15. Wu HM, Kortbeek JB. Management of splenic pseudocysts following trauma: a retrospective case series. Am J Surg. 2006;191(5):631–4.
16. Hamers RL, Van Den Berg FG, Groeneveld AB. Acute necrotizing pancreatitis following inadvertent extensive splenic artery embolisation for trauma. Br J Radiol. 2009;82(973):e11–4.
17. Bjerke S, Pohlman T, Saywell RM. Evolution, not revolution: splenic salvage for blunt trauma in a statewide voluntary trauma system—a 10-year experience. Am J Surg. 2006;191(3):413–7.
18. Carlin AM, Tyburski JG, Wilson RF. Factors affecting the outcome of patients with splenic trauma. Am Surg. 2002;68(3):232–9.
19. Amonkar SJ, Kumar EN. Spontaneous rupture of the spleen: three case reports and causative processes for the radiologist to consider. Br J Radiol. 2009;82(978):e111–3.
20. Bain IM, Kirby RM. 10-year experience of splenic injury: an increasing place for conservative management after blunt trauma. Injury. 1998;29(3):177–82.
21. Barone JE, Burns G, Svehlak SA. Management of blunt splenic trauma in patients older than 55 years. Southern Connecticut Regional Trauma Quality Assurance Committee. J Trauma. 1999;46(1):87–90.
22. Caplan ES, Boltansky H, Snyder MJ. Response of traumatized splenectomized patients to immediate vaccination with polyvalent pneumococcal vaccine. J Trauma. 1983;23(9):801–5.
23. Cathey KL, Brady WJ Jr, Butler K. Blunt splenic trauma: characteristics of patients requiring urgent laparotomy. Am Surg. 1998;64(5):450–4.

24. Clancy TV, Ramshaw DG, Maxwell JG. Management outcomes in splenic injury: a statewide trauma center review. Ann Surg. 1997;226(1):17–24.
25. Coburn MC, Pfeifer J, DeLuca FG. Nonoperative management of splenic and hepatic trauma in the multiply injured pediatric and adolescent patient. Arch Surg. 1995;130(3):332–8.
26. Cocanour CS, Moore FA, Ware DN. Delayed complications of nonoperative management of blunt adult splenic trauma. Arch Surg. 1998;133(6):619–24; discussion 624-5.
27. Ekeh AP, Izu B, Ryan M, McCarthy MC. The impact of splenic artery embolization on the management of splenic trauma: an 8-year review. Am J Surg. 2009;197(3):337–41.
28. Garber BG, Yelle JD, Fairfull-Smith R. Management of splenic injuries in a Canadian trauma center. Can J Surg. 1996;39(6):474–80.
29. Gaunt WT, McCarthy MC, Lambert CS. Traditional criteria for observation of splenic trauma should be challenged. Am Surg. 1999;65(7):689–91; discussion 691–2.
30. Guth AA, Pachter HL, Jacobowitz GR. Rupture of the pathologic spleen: is there a role for nonoperative therapy? J Trauma. 1996;41(2):214–8.
31. Harbrecht BG, Peitzman AB, Rivera L, et al. Contribution of age and gender to outcome of blunt splenic injury in adults: multicenter study of the eastern association for the surgery of trauma. J Trauma. 2001;51(5):887–95.
32. Ivatury RR, Simon RJ, Guignard J. The spleen at risk after penetrating trauma. J Trauma. 1993;35(3):409–14.
33. Kilic N, Gurpinar A, Kiristioglu I. Ruptured spleen due to blunt trauma in children: analysis of blood transfusion requirements. Eur J Emerg Med. 1999;6(2):135–9.
34. Krause KR, Howells GA, Bair HA. Nonoperative management of blunt splenic injury in adults 55 years and older: a twenty-year experience. Am Surg. 2000;66(7):636–40.
35. Krupnick AS, Teitelbaum DH, Geiger JD. Use of abdominal ultrasonography to assess pediatric splenic trauma. Potential pitfalls in the diagnosis. Ann Surg. 1997;225(4):408–14.
36. Lawson DE, Jacobson JA, Spizarny DL. Splenic trauma: value of follow-up CT. Radiology. 1995;194(1):97–100.
37. Lucas CE. Splenic trauma. Choice of management. Ann Surg. 1991;213(2):98–112.
38. Morrell DG, Chang FC, Helmer SD. Changing trends in the management of splenic injury. Am J Surg. 1995;170(6):686–9.
39. Novelline RA, Rhea JT, Bell T. Helical CT of abdominal trauma. Radiol Clin N Am. 1999;37(3):591–612, vi–vii.
40. Pachter HL, Guth AA, Hofstetter SR. Changing patterns in the management of splenic trauma: the impact of nonoperative management. Ann Surg. 1998;227(5):708–17.
41. Pisters PW, Pachter HL. Autologous splenic transplantation for splenic trauma. Ann Surg. 1994;219(3):225–35.
42. Poulin EC, Thibault C, DesCoteaux JG. Partial laparoscopic splenectomy for trauma: technique and case report. Surg Laparosc Endosc. 1995;5(4):306–10.
43. Ransom KJ, Kavic MS. Laparoscopic splenectomy for blunt trauma: a safe operation following embolization. Surg Endosc. 2009;23(2):352–5.
44. Rose AT, Newman MI, Debelak J. The incidence of splenectomy is decreasing: lessons learned from trauma experience. Am Surg. 2000;66(5):481–6.
45. Rutherford EJ, Livengood J, Higginbotham M. Efficacy and safety of pneumococcal revaccination after splenectomy for trauma. J Trauma. 1995;39(3):448–52.
46. Schurr MJ, Fabian TC, Gavant M. Management of blunt splenic trauma: computed tomographic contrast blush predicts failure of nonoperative management. J Trauma. 1995;39(3):507–12.
47. Shatz DV. Vaccination practices among North American trauma surgeons in splenectomy for trauma. J Trauma. 2002;53(5):950–6.
48. Wasvary H, Howells G, Villalba M. Nonoperative management of adult blunt splenic trauma: a 15-year experience. Am Surg. 1997;63(8):694–9.
49. Williams RA, Black JJ, Sinow RM. Computed tomography-assisted management of splenic trauma. Am J Surg. 1997;174(3):276–9.

50. Renzulli P, Hostettler A, Schoepfer AM, Gloor B, Candinas D. Systematic review of atraumatic splenic rupture. Br J Surg. 2009;96(10):1114–21.
51. Dave A, Dhand H, Mujalde A. Spontaneous rupture of spleen during pregnancy. J Obstet Gynecol India. 2012;62(6):692–3.
52. Aydin Y, Ulas AB, Eroglu A. Subphrenic abscess. Springer International Publishing; 2020. p. 745–50.

Splenosis

6

6.1 Introduction

- There are two types of ectopic splenic tissue:
  - Accessory spleen
  - Splenosis
- Splenosis:
  - This is a clinical condition where parts of splenic tissue undergo auto transplantation.
  - It is defined as the heterotopic transplantation of splenic tissue following:
    - Abdominal trauma
    - Splenectomy
  - It is an acquired benign condition.
- Splenosis is the result of implantation and seeding of splenic tissue anywhere in the abdominal cavity or pelvis.
- Accessory spleen:
  - This is a congenital developmental abnormality where there are accessory spleens beside the proper spleen.
  - A true accessory spleen is a common congenital anomaly occurring in 10–20% of autopsy series.
  - Accessory spleens develop within the dorsal mesentery of the embryo, and are therefore found at sites adjacent to the posterior wall of the stomach or the pancreas.
  - Accessory spleens are most commonly found near the splenic hilum and about 16% are found in or adjacent to the tail of the pancreas.
  - Accessory spleens are variable in size and often measure several centimeters in diameter. Accessory spleens as small as a few millimeters have also been described.
- Ectopic splenic tissue was first described in 1896 by Albrecht in Germany.
- Splenosis was first described in 1937 by Shaw and Shafi.

A. H. Al-Salem, *The Spleen*, https://doi.org/10.1007/978-981-99-6191-7_6

- The term splenosis was introduced in 1939 by Buchbinder and Lipkoff. They described an intraabdominal splenic deposit in a young woman following splenectomy for splenic rupture.
- The average interval between trauma and development of abdominal or pelvic splenosis is around 10 years (range 5 months- 32 years).
- Most cases of splenosis are secondary to splenectomy due to trauma (93%).

- **ECTOPIC SPLENIC TISSUE WAS FIRST DESCRIBED IN 1896 BY ALBRECHT IN GERMANY**

- **SPLENOSIS WAS FIRST DESCRIBED IN 1937 BY SHAW AND SHAFI**

- **THE TERM SPLENOSIS WAS INTRODUCED IN 1939 BY BUCHBINDER AND LIPKOFF**

- Splenosis is seen most commonly:
  - Following abdominal trauma with rupture of the spleen. Auto transplantation of splenic tissue following traumatic injury of the spleen is well known. This is the commonest cause of splenosis.
  - Following splenectomy
- Splenosis is slightly more common in males than females. This is probably due to the fact that traumatic injury to the spleen is more common in males than females.
- The splenic tissue fragments and in order to be auto transplanted and survive must be implanted on well vascularized surfaces such as the omental surface.
- Splenosis are seen commonly:
  - In the abdominal cavity or pelvis
  - Rarely, in the thorax. These are seen if the left diaphragm is ruptured at the time of trauma with a concomitant rupture of the spleen.
- The splenic tissue can regenerate through various mechanisms.
- The ectopic splenic foci are typically small, sessile and multiple.
- In the abdominal cavity ectopic splenic tissue usually grow on peritoneal surfaces. There are however reports of intrahepatic splenosis. These can cause serious diagnostic problems.
- The ectopic splenic foci may grow over time to reach a large size.
- Rarely, splenosis can be intrathoracic (Thoracic splenosis).
  - Thoracic splenosis is a very rare condition.
  - They usually present as multiple nodules over the pleural surface of the thoracic cavity.

  - They are commonly seen in the left hemithorax.
  - These are seen usually following abdominal trauma causing splenic injury that is associated with rupture of the let hemidiaphragm.
- The incidence of splenic auto transplantation and regeneration correlates with the severity of splenic injury.
- Currently, splenectomy for trauma is indicated for those with the greatest splenic injury which favors auto transplantation.
- Rarely, some surgeons intentionally auto transplant pieces of splenic tissue in the greater omentum at the time of splenectomy for trauma. The intension of this is for these patients to subsequently develop splenosis which may take over some of the splenic functions.
- This is one way to protect against post splenectomy sepsis if the auto transplanted tissue becomes functional.
- The auto transplanted splenic tissues gains blood supply from newly formed blood vessels.
- It has been estimated that splenosis may occur in up to 65% of traumatic ruptures of the spleen.
- Splenosis:
  - Splenosis is commonly seen in any part of the abdominal cavity including the pelvis.
  - Other rare locations of splenosis include:
    - The thoracic cavity
    - In subcutaneous tissues: Splenosis in subcutaneous tissues may result from abdominal surgery or traumatic gunshot wounds.
    - In the liver
    - In the cranial cavity
  - Splenosis in the cranial cavity or the liver develop as result of dissemination of tiny pieces of spleen tissue that travel through the bloodstream and settle in the liver or brain.
  - Splenosis must be distinguished from accessory spleen which is a congenital developmental abnormality.
  - Although splenosis is a benign condition, it must be differentiated from malignant tumors which may look similar radiologically. This must be kept in mind when evaluating a patient with an abdominal or liver tumor following splenectomy especially those following traumatic injury to the spleen.
- Grossly, splenosis appear as reddish-blue nodules with variable shape and size.
- Their number can be as few as one splenosis and as many as 300.
- They can appear as separate nodules or several nodules attached to each other by a thin stem.
- Their size is variable ranging from a few millimeters to as large as 12 cm but commonly they are less than 3 cm.
- Histologically, they resemble normal splenic tissue made up of red and white pulp.

6.2 Clinical Features

- Splenosis is a fairly common condition following splenic injury.
- It is reported to occur in 16–67% of patients following splenic trauma or history of abdominal surgery.
- Most patients with splenosis are asymptomatic.
- They are commonly discovered accidently or during the evaluation of other intraabdominal conditions.
- Splenosis in the peritoneal cavity can affect the following areas:
  - Small bowel serosa
  - Parietal peritoneum
  - Serosa of the large intestine
  - Mesentery
  - Diaphragm
- Rarely hematogenous dissemination of splenic tissue may occur. This mechanism explains the finding of hepatic and intracranial splenosis.
- Patients with splenosis are usually asymptomatic but may occasionally present with:
  - Nonspecific abdominal pain: This is most likely due to infarction of splenosis nodules as they enlarge and outgrow their blood supply.
  - Abdominal mass
  - Intestinal obstruction due to adhesive bands of the splenosis
  - Gastrointestinal hemorrhage

    Splenosis should be considered in the differential diagnosis of gastrointestinal hemorrhage in patients who had prior splenic trauma or splenectomy.
  - Hydronephrosis due to extrinsic compression of the ureters by the splenosis nodules.
  - Occasionally splenosis can be confused with malignant tumors, leading to invasive and extensive investigations which are unnecessary.
  - Splenosis can grow in size and undergo infarction due to limited blood supply and present with abdominal pain.
  - Thoracic splenosis can present with Pleurisy and hemoptysis or chest pain.
  - Recurrence of Felty's syndrome has been reported as a complication of splenosis. In some patients, splenic implants can grow and resume splenic function in 1–3 months following auto transplantation.
- It is important to keep splenosis in mind when evaluating patients with a history of splenic trauma or splenectomy and newly appearing peritoneal lesions.
- Splenectomized patients due to hematological diseases or hypersplenism can develop recurrence of the underlying disease due to the activity of ectopic splenic tissue.
- Splenosis can be confused with other conditions including metastasis, endometriosis and lymphoma.

- The average interval reported between trauma and development of abdominal or pelvic splenosis is 10 years, with a range of 5 months to 32 years.
- The average interval reported between trauma and development of thoracic splenosis is 21 years, with a range of 3–45 years.
- It is important to distinguish splenosis from accessory spleen, both of which are conditions of ectopic splenic tissue.
  - Splenosis is an acquired condition while accessory spleen is a congenital developmental condition.
  - Accessory spleen has normal splenic tissue histology while splenosis has distorted histological architecture consisting of poorly formed white pulp with normal appearing red pulp, and lacking trabecular structures.
  - Accessory spleen is supplied by the splenic artery, in contrast to splenosis, where the ectopic tissue gets its blood supply from the surrounding vessels.
  - The splenic tissue in splenosis has less elastic tissue, no hilum, and a poorly formed capsule.
  - Accessory spleens are found near the splenopancreatic or gastrosplenic ligament, whereas splenosis may be found anywhere in the peritoneal cavity, pelvis or even in extraperitoneal locations.
- The development of splenosis starts at the time of splenic rupture or splenectomy, when the splenic pulp disperses into the peritoneal cavity.
- The number of splenosis nodules that develop in the peritoneal cavity correlates with the severity of the splenic injury.
- Another mechanism of splenic tissue transplantation is via splenic vein emboli or hematogenous spread of splenic pulp. This is the most likely explanation for the development of intrahepatic and intracranial splenosis.
- Thoracic splenosis usually occurs when splenic rupture is accompanied by simultaneous diaphragmatic rupture. This is seen most commonly when there is simultaneous rupture of the left hemidiaphragm.
- Subcutaneous splenosis is a very rare condition and results most likely from mechanical implantation of splenic tissue at the time of laparotomy. All cases of subcutaneous splenosis have been reported in or at the site of surgical or traumatic scar.

SITES OF SPLENOSIS

- **INTRAABDOMINAL**
- **PELVIC**
- **INTRAHEPATIC**
- **INTRACRANIAL**
- **THORACIC**
- **SUBCUTANEOUS**

6.3 Radiological Features

- Ectopic splenic tissue is sometimes found incidentally during US, CT, or MRI examinations of the abdomen for other unrelated complaints.
- When the imaging features of a mass in a patient with splenic trauma or splenectomy are similar to those of normal splenic tissue, the diagnosis of splenosis should be considered.
- The commonest site for splenosis is within the greater omentum. The greater omentum has a good blood supply and this ideal for regeneration and neovascularization of splenic implants.
- Additional potential sites for the development of splenosis include:
  - The serosal surface of the small intestines
  - The Parietal peritoneum
  - Mesentery
  - Diaphragm
  - Intrahepatic
  - Intrathoracic
- Splenosis when present can be confused with other much more common conditions which include:
  - Renal tumors
  - Adrenal tumors
  - Abdominal tumor
  - Metastases from other tumors
  - Abdominal lymphoma
  - Endometriosis
  - Ectopic testis
- The radiological investigations for splenosis are not specific.
- They cannot distinguish splenosis from other conditions such as metastatic deposits.
- A valuable investigation for splenosis is the use of Technetium-99m heat-damaged erythrocytes (RBC) or Indium 111-labeled platelets scintigraphy.
- These are noninvasive and reliable investigations for the diagnosis of splenosis.
- Splenosis has the ability to absorb radio-labeled, damaged red blood cells.
- RBC scintigraphy is known to have increased sensitivity in the diagnosis of splenosis.
- Splenosis on ultrasound will appear as a well-demarcated, hypoechoic to isoechoic mass with non-specific arterial and venous color Doppler signals.
- CT scan:
  - On non-contrast CT scan, splenosis will appear as hypodense mass. On contrast CT, splenosis will appear as a hyperdense mass in the arterial phase, isodense in the portal phase, and hypodense in the equilibrium phase. There is also a hypodense rim around the splenosis nodule.
  - They appear as rounded or sessile nodules.
  - They have density and enhancing characteristics similar to the spleen
- MRI:

- Pre-contrast MRI shows splenosis as a homogenously hypointense on T1-weighted images, and hyperintense on T2-weighted images.
- There is also a hypointense rim around the mass on T1-weighted images.
- Post-contrast MRI shows splenosis as a hyperintense mass as compared to the liver.
- The presence of a rim surrounding the lesion has been described as a characteristic finding of splenosis. This rim has low signal intensity on T1- and T2-weighted images, representing a thin layer of fat or fibrous capsule around the lesion.
- They have signal characteristics similar to normal spleen
- On T1 they are hypointense
- On T2 they are hyperintense
- On T1 C+ (Gd) they show heterogeneous enhancement
- MRI is a specific investigation for the diagnosis of splenosis. This is done with intravenous administration of superparamagnetic iron oxide (SPIO), which is used to outline hepatic and splenic diseases.

• Nuclear medicine:
  - Scintigraphy is the method of choice for the diagnosis of splenosis.
    Scintigraphic agents, such as Tc-99m sulfur colloid, Tc-99m heat-damaged erythrocytes, and In-111 labeled platelets are taken by the reticuloendothelial system of the liver, spleen and bone marrow.
    They are very sensitive to diagnose ectopic splenic tissue, especially small or multiple nodules which may be missed on a CT or MRI examination.
    Tc-99m heat-damaged erythrocytes or In-111 labeled platelets are more sensitive than Tc-99m sulfur colloid in the detection of splenosis because of their better signal-to-background ratio, and their specificity for splenic tissue.
  - Tc-99m sulfur colloid scan:
    The diagnosis of splenosis can be confirmed with a Tc-99m sulfur colloid scan.
    This will show increased uptake as long as the splenosis nodule is 2 cm in diameter and larger.
    When Tc-99m sulfur colloid scan fails to confirm the presence of splenic tissue, a Tc-99m-tagged heat-damaged RBC scan (Tc-99m-DRBC) with autologous erythrocytes will help demonstrate and localize splenosis nodules.

• The diagnosis of splenosis must be confirmed pathologically. Tis is also important to exclude the possibility of malignancy.
  - A definitive diagnosis of splenosis is made histologically via an excisional biopsy.
  - CT guided biopsy will help rule out malignancy and demonstrate splenic tissue with abnormal architecture.
  - Laparoscopy as a minimally invasive technique is an alternative to CT guided biopsy.

6.4 Treatment

- Splenosis is a benign condition and does not require treatment.
- Surgery for splenosis is indicated in the following conditions:
  - Symptomatic patients
  - Patients with splenosis related complications
  - Post-splenectomy recurrent hematological diseases
  - When the diagnosis is uncertain and to avoid confusion with malignancy
- Asymptomatic splenosis should be treated conservatively
- For symptomatic splenosis, the treatment is surgical excision.
- It has been reported that splenosis may become functional and exert a protective immune response against bacterial infections in asplenic patients.

Further Reading

1. Imbriaco M, Camera L, Manciuria A, Salvatore M. A case of multiple intra-abdominal splenosis with computed tomography and magnetic resonance imaging correlative findings. World J Gastroenterol. 2008;14:1453–5.
2. Khosravi MR, Margulies DR, Alsabeh R, Nissen N, Phillips EH, Morgenstern L. Consider the diagnosis of splenosis for soft tissue masses long after any splenic injury. Am Surg. 2004;70:967–70.
3. Brancatelli G, Vilgrain V, Zappa M, Lagalla R. Case 80: splenosis. Radiology. 2005;234:728–32.
4. De Vuysere S, Van Steenbergen W, Aerts R, Van Hauwaert H, Van Beckevoort D, Van Hoe L. Intrahepatic splenosis: imaging features. Abdom Imaging. 2000;25:187–9.
5. Fleming CR, Dickson ER, Harrison EG Jr. Splenosis: autotransplantation of splenic tissue. Am J Med. 1976;61:414–9.
6. Berman AJ, Zahalsky MP, Okon SA, Wagner JR. Distinguishing splenosis from renal masses using ferumoxide-enhanced magnetic resonance imaging. Urology. 2003;62:748.
7. Yammine JN, Yatim A, Barbari A. Radionuclide imaging in thoracic splenosis and a review of the literature. Clin Nucl Med. 2003;28:121–3.
8. Feferman I, Cramer J. Splenosis: an unusual cause of intraabdominal hemorrhage. J Emerg Med. 1991;9:239–40.
9. Mortelé KJ, Mortelé B, Silverman SG. CT features of the accessory spleen. AJR Am J Roentgenol. 2004;183:1653–7.
10. Lin WC, Lee RC, Chiang JH, et al. MR features of abdominal splenosis. AJR Am J Roentgenol. 2003;180(2):493–6.
11. Brancatelli G, Vilgrain V, Zappa M, et al. Case 80: splenosis. Radiology. 2005;234(3):728–32.
12. Naylor MF, Karstaedt N, Finck SJ, et al. Noninvasive methods of diagnosing thoracic splenosis. Ann Thorac Surg. 1999;68(1):243–4.
13. Thourani VH, Sharma J, Duarte IG, et al. Intrathoracic splenosis. Ann Thorac Surg. 2005;80(5):1934–6.
14. Menth M, Herrmann K, Haug A, et al. Intra-hepatic splenosis as an unexpected cause of a focal liver lesion in a patient with hepatitis C and liver cirrhosis: a case report. Cases J. 2009;2(1):8335.
15. Mazurek A, Szaluś N, Stembrowicz-Nowakowska Z, et al. Detection of splenic tissue by 99mTc-labelled Sn-colloid SPECT/CT scintigraphy. Nucl Med Rev Cent East Eur. 2012;14(2):116–7.
16. Ehrlich CP, Papanicolaou N, Treves S, et al. Splenic scintigraphy using Tc-99m-labeled heat-denatured red blood cells in pediatric patients: concise communication. J Nucl Med. 1982;23(3):209–13.

17. Lake ST, Johnson PT, Kawamoto S, Hruban RH, Fishman EK. CT of splenosis: patterns and pitfalls. AJR Am J Roentgenol. 2012;199(6):686–93.
18. Tsitouridis I, Michaelides M, Sotiriadis C, Arvaniti M. CT and MRI of intraperitoneal splenosis. Diagn Interv Radiol. 2010;16(2):145–9.
19. Imbriaco M, Camera L, Manciuria A, Salvatore M. A case of multiple intraabdominal splenosis with computed tomography and magnetic resonance imaging correlative findings. World J Gastroenterol. 2008;14(9):1453–5.
20. Yildiz AE, Ariyurek MO, Karcaaltincaba M. Splenic anomalies of shape, size, and location: pictorial essay. Sci World J. 2013;2013:321810.
21. Lui EHY, Lau KKP. Intra-abdominal splenosis: How clinical history and imaging features averted an invasive procedure for tissue diagnosis. Australas Radiol. 2005;49(4):342–4.
22. Carr N, Turk E. The histological features of splenosis. Histopathology. 1992;21(6):549–53.
23. Ludtke FE, Mack SC, Schuff-Werner P, Voth E. Splenic function after splenectomy for trauma. Role of autotransplantation and splenosis. Acta Chir Scand. 1989;155(10):533–9.
24. Gunes I, Yilmazlar T, Sarikaya I, Akbunar T, Irgil C. Scintigraphic detection of splenosis: superiority of tomographic selective spleen scintigraphy. Clin Radiol. 1994;49(2):115–7.
25. Basile RM, Morales JM, Zupanec R. Splenosis, a cause of massive gastrointestinal hemorrhage. Arch Surg. 1989;124(9):1087–9.
26. Normand JP, Rioux M, Dumont M, Bouchard G. Ultrasonographic features of abdominal ectopic splenic tissue. Can Assoc Radiol J. 1993;44(3):179–84.
27. Sikov WM, Schiffman FJ, Weaver M, Dyckman J, Shulman R, Torgan P. Splenosis presenting as occult gastrointestinal bleeding. Am J Hematol. 2000;65(1):56–61.
28. Margari A, Amoruso M, D'Abbicco D, Notarnicola A, Epifania B. Massive gastrointestinal bleeding due to a splenotic nodule of the gastric wall. Chir Ital. 2008;60(6):863–5.
29. Wedemeyer J, Gratz KF, Soudah B, et al. Splenosis – important differential diagnosis in splenectomized patients presenting with abdominal masses of unknown origin. Z Gastroenterol. 2005;43(11):1225–9.
30. Balfanz JR, Nesbit ME Jr, Jarvis C, Krivit W. Overwhelming sepsis following splenectomy for trauma. J Pediatr. 1976;88(3):458–60.
31. Bresciani C, Ferreira N, Perez RO, Jacob CE, Zilberstein B, Cecconello I. Splenosis mimicking gastric GIST: case report and literature review. Arq Bras Cir Dig. 2011;24(2):183–5.
32. Alang N. Splenosis: an unusual cause of massive gastrointestinal bleeding. R I Med J. 2013;96(11):48–9.
33. Croskerry P. The importance of cognitive errors in diagnosis and strategies to minimize them. Acad Med. 2003;78(8):775–80.
34. Domellöf L, Enander LK, Nilsson F. Bleeding as a complication to endoscopic biopsies from the gastric remnant after ulcer surgery. Scand J Gastroenterol. 1983;18(7):951–4.
35. Lau JY, Sung JJ, Chan AC, et al. Stigmata of hemorrhage in bleeding peptic ulcers: an interobserver agreement study among international experts. Gastrointest Endosc. 1997;46(1):33–6.
36. Kiroff GK, Mangos A, Cohen R, Chatterton BE, Jamieson GG. Splenic regeneration following splenectomy for traumatic rupture. Aust N Z J Surg. 1983;53:431–4.
37. Fremont RD, Rice TW. Splenosis: a review. South Med J. 2007;100:589–93.
38. Brewster DC. Splenosis: report of two cases and review of the literature. Am J Surg. 1973;126:14–9.
39. Cotlar AM, Cerise EJ. Splenosis: the autotransplantation of splenic tissue following injury to the spleen. Report of two cases and review of the literature. Ann Surg. 1959;149:402–4.
40. Ksiadzyna D, Peña AS. Abdominal splenosis. Rev Esp Enferm Dig. 2011;103:421–6.
41. Storm BL, Abbitt PL, Allen DA, Ros PR. Splenosis: superparamagnetic iron oxide-enhanced MR imaging. AJR Am J Roentgenol. 1992;159:333–5.
42. Schiff RG, Leonidas J, Shende A, Lanzkowski P. The noninvasive diagnosis of intrathoracic splenosis using technetium-99m heat-damaged red blood cells. Clin Nucl Med. 1987;12:785–7.
43. Atkins HL, Goldman AG, Fairchild RG, et al. Splenic sequestration of Tc-99m labeled, heat treated red blood cells. Radiology. 1980;136:501–3.

7 Accessory Spleen

7.1 Introduction

- The spleen is considered the largest lymphoid organ in the body.
- It is located in the left upper quadrant of the abdomen, superior to the left kidney and posterior to the stomach. It is located below the left diaphragm and protected by the 9th–11th ribs.
- The spleen is a highly vascular organ.
- The spleen in made up of two important components:
  - The red pulp:

    This is made up of dense fibrovascular networks.

    Its main function is to filter the blood as it passes through this vascular network.
  - The white pulp:

    This is made up of lymphoid tissue surrounding the larger vessels.

    Its main function is immunological.
- The spleen has a relatively weak capsule and this can rupture more easily than other abdominal organs. This can lead to life-threatening hemorrhage when traumatized.
- There are several variants of spleen development which include:
  - Splenic clefts
  - Splenic notches
  - Splenic lobulations
  - An accessory spleen
  - Polysplenia
  - Wandering spleen
  - Asplenia
  - Splenogonadal fusion
- Asplenia:
  - This is a relatively rare condition.
  - Asplenia is often seen associated with:

A. H. Al-Salem, *The Spleen*, https://doi.org/10.1007/978-981-99-6191-7_7

    - Complex congenital heart disease
    - Bilateral "right-sidedness" such as bilateral three-lobed lungs
    - A right-sided stomach
    - A central liver
    - Intestinal malrotation
  - Patients with asplenia are at increased risk for overwhelming infections and should receive prophylactic antibiotics.
- Polysplenia:
  - Polysplenia is characterized by the presence of multiple accessory spleens rather than one normal spleen.
  - In polysplenia, there are multiple small spleens
  - Polysplenia often seen associated with biliary atresia
  - These are normal functional spleens and children with polysplenia have adequate splenic immune function.
  - Other conditions associated with polysplenia include:
    - Preduodenal portal vein
    - Situs inversus
    - Malrotation
    - Congenital heart disease
- Wandering Spleen:
  - The spleen has several ligamentous attachments which help hold the spleen in its normal anatomical position.
  - Wandering spleen is due to the lack of ligamentous attachments of the spleen to the diaphragm, colon, and retroperitoneum.
  - As a result of this the spleen will be mobile.
  - The spleen can move from the left upper quadrant of the abdomen to anywhere in the peritoneal cavity and sometimes can reach the right lower quadrant.
- Splenogonadal fusion:
  - Embryologically, the spleen develops near the urogenital ridge from which the gonads develop.
  - As a result of this close developmental relation, sometimes part of the developing spleen may attach to the gonad, and as they descend through the abdomen during development, they can produce either a continuous or a broken line of accessory splenic tissue.
  - This can result in one or more accessory spleens found along a path from the abdomen into the pelvis or scrotum.
  - Splenogonadal fusion develops as a result of attachment of an accessory spleen to the left testis.
  - Splenogonadal fusion occurs in two different forms:
    - Continuous splenogonadal fusion:
    - In this type there is a cordlike fibrous attachment between the spleen and gonad and accessory spleens are found within this cordlike structure.
    - Discontinuous splenogonadal fusion:

In this type there is a direct attachment of an accessory spleen to the gonad without a cordlike fibrous attachment.

- Ectopic splenic tissue is divided into two types:
  - Splenosis:
    This is due to auto transplantation of splenic tissue.
    This occurs following splenectomy and splenic trauma.
  - Accessory spleens:
    These are congenital healthy splenic nodules that are separate from the main body of the spleen.

CONGENITAL MALFORMATIONS OF THE SPLEEN

- **SPLENIC CLEFTS**
- **SPLENIC NOTCHES**
- **SPLENIC LOBULATIONS**
- **AN ACCESSORY SPLEEN**
- **POLYSPLENIA**
- **WANDERING SPLEEN**
- **ASPLENIA**
- **SPLENOGONADAL FUSION**

- An accessory spleen is also called supernumerary spleen, splenule, or splenunculus.
- An accessory spleen is defined as a small nodule of splenic tissue found separate from the main spleen. They resemble a small spleen but they can also resemble a lymph node.
- Accessory spleens are structurally identical to the normal spleen. Accessory spleens resemble normal spleen macroscopically and microscopically.
- Accessory spleen can be single or multiple.
- Accessory spleens are histologically and functionally similar to the normal spleen and if a patient fails therapeutic splenectomy, the possibility of an accessory spleen should be considered. This is specially so in children with idiopathic thrombocytopenic purpura.
- Accessory spleens arise from the failure of fusion of the splenic anlage located in the dorsal mesogastrium during the fifth week of fetal life. It is thought that accessory spleens result from inadequate fusion of primary lobules of the spleen during the second or third trimester of fetal development.
- An accessory spleen is a relatively common congenital abnormality. It is seen in up to 20% of healthy individuals and is found in up to 30% in patients with hematologic disorders.
- The presence of an accessory spleen has been reported in 10–30% of cases in postmortem studies and in 45–65% of patients after splenectomy.

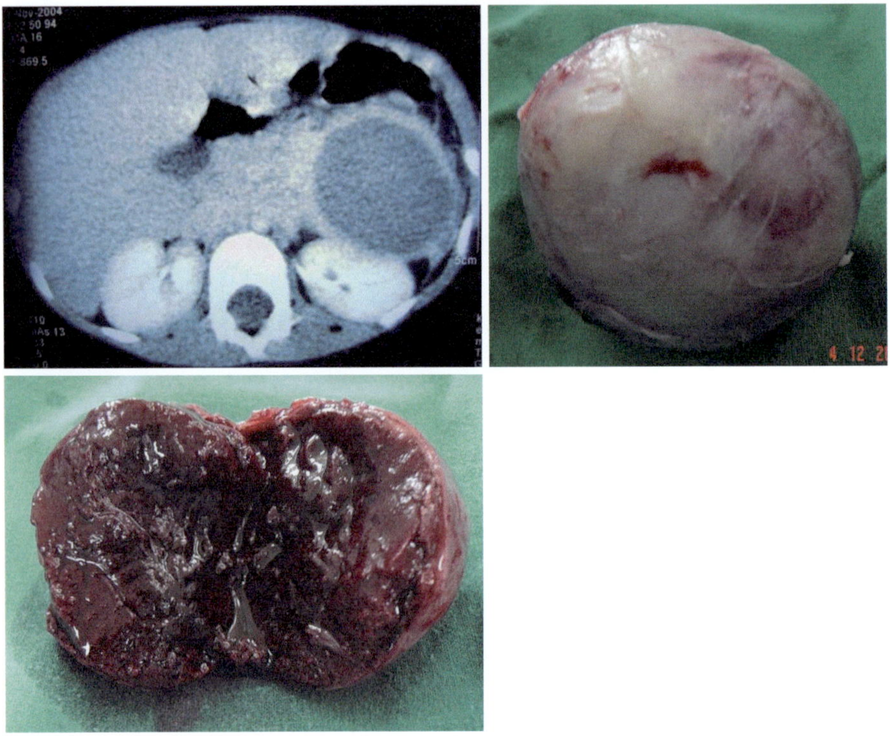

Figs. 7.1–7.3 Radiological CT-scan and clinical photographs showing an enlarged accessory spleen. Note the sie of the accessory spleen and the cut section showing similarity to the normal spleen

- Accessory spleens are found in approximately 10% of the population and in 20–30% of autopsies.
- The size of an accessory spleens is variable. Usually, accessory spleens are around one centimeter in diameter but may reach up to 4 cm or more under certain conditions (Figs. 7.1, 7.2, and 7.3).
- Accessory spleens are most commonly located:
  - Medial to the splenic hilum
  - Adjacent to or within the pancreatic tail
  - Below the spleen in the splenorenal ligament
  - Rarely may they be found elsewhere in the abdomen (Fig. 7.4).
- The diagnosis of an accessory spleen is made by ultrasound.
  - On ultrasound, accessory spleens appear as homogeneous, round, smaller than 2 cm in diameter and identical in echogenicity to the spleen.
  - The presence of a vascular pedicle and vascular supply arising from the main splenic vessels can frequently be seen with Doppler ultrasound and theses features are also helpful in diagnosing accessory spleens.

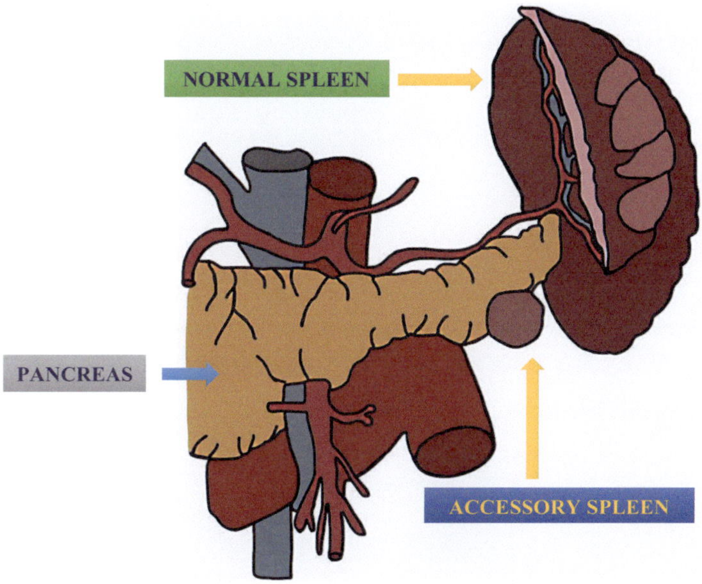

Fig. 7.4 Diagrammatic representation of an accessory spleen. Note its relation to the tail of the pancreas. Accessory spleens are most commonly located medial to the splenic hilum, adjacent to or within the pancreatic tail or below the spleen in the splenorenal ligament

- The diagnosis of an accessory spleen is usually simple but difficulties may arise in the following conditions:
  - Post splenectomy patients
  - When the accessory spleen is located in the pancreatic tail where it may mimic a pancreatic mass
  - When the accessory spleen is in an unusual location
  - In patients with a history of malignancy where accessory spleen may resemble tumor recurrence or lymphadenopathy
- Up to 19–20% of humans have accessory spleens. These accessory spleens will enlarge following splenectomy.
- If accessory spleens are not recognized and removed at the time of splenectomy for hematological or neoplastic disease, they can lead to relapse of the condition. This must be taken in consideration and looked for at the time of splenectomy especially for idiopathic thrombocytopenic purpura. The presence of an accessory spleen post splenectomy for idiopathic thrombocytopenic purpura may lead to recurrence.
- On the other hand, preservation of an accessory spleens at the time of splenectomy for trauma may preserve the immune function of the spleen (Figs. 7.5, 7.6, 7.7, and 7.8).

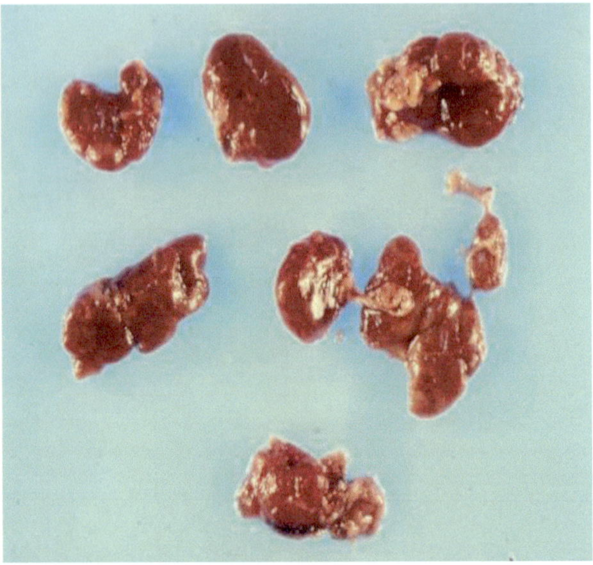

Fig. 7.5 A clinical photograph showing multiple spleniculi removed from a child at the time of splenectomy. Histological examination will confirm the diagnosis and differentiate these from lymph nodes which are also commonly seen in patients undergoing splenectomy

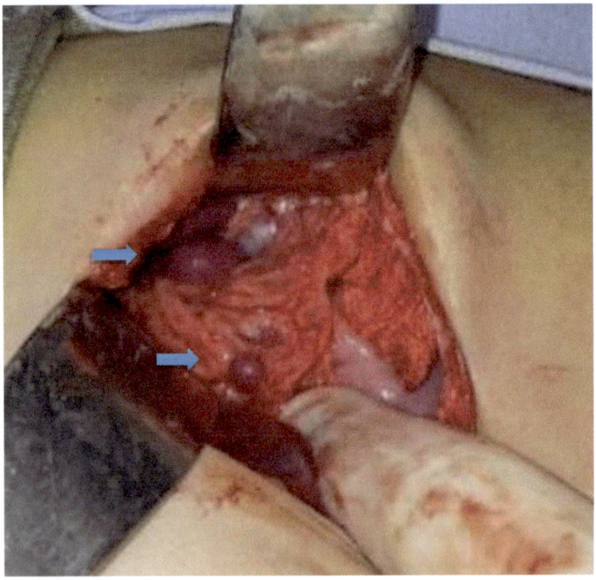

Fig. 7.6 Intraoperative photograph showing multiple accessory spleens. Note the shape and color of the accessory spleen which resembles a normal spleen. Note also the different sizes of accessory spleens

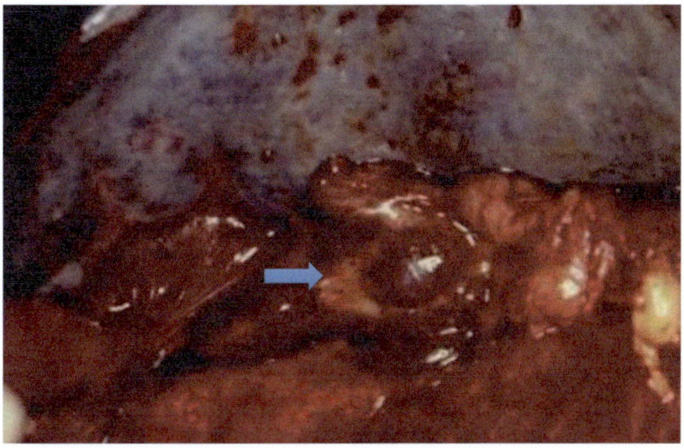

Fig. 7.7 Intraoperative photograph showing a splenculi. Note the position of the accessory spleen near the hilum of the spleen

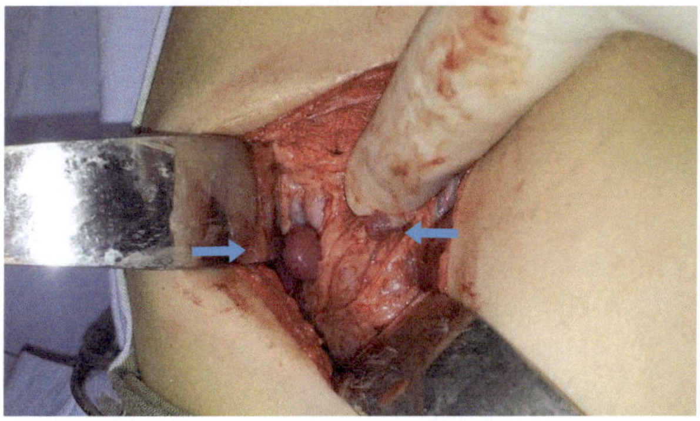

Fig. 7.8 Intraoperative photograph showing two splenculi. One is larger than the other

7.2 Embryology and Anatomy

- Embryologically, the spleen develops from the splenic primordium which is first recognized at the 8–10-mm stage.
- This appears as a mesenchymal bulge in the left dorsal mesogastrium between the stomach and pancreas.
- At around the 10–12-mm stage a true epithelium develops and sinusoids are present with communication to capillaries.
- At around 4 months of fetal development, the spleen starts producing red and white blood cells.
- This hematopoiesis function of the spleen is temporary that ceases later in gestation. The spleen is rarely the site of significant hematopoiesis in childhood except under certain conditions.
- The blood enters the spleen through splenic artery. The splenic artery divides into segmental vessels that branch into trabecular arteries; and, after further bifurcations, small arteries enter the white pulp of the spleen.
- Approximately 20% of the volume of the spleen is made up of white pulp, and the remainder is made up of the red pulp which consists of the endothelial cords of Billroth.

- The white pulp of the spleen consists of lymphocytes and macrophages which are arranged as a germinal center around the central artery. The microorganisms and particulate antigens reach the white pulp via branches of the central artery and this arrangement facilitate antibody formation in response to these antigens.
- Blood passing through the spleen traverses in either an open or a closed system.
  - Ten to 20% of blood passes through the closed system of the spleen directly from capillaries into venous sinusoids.
  - The remainder passes directly into the red pulp.
  - These cells and other particulate material must pass through the open system within the cords of Billroth and then migrate into the splenic sinusoids, which enter into trabecular veins.

7.3 Sites of Accessory Spleens

- Accessory spleens are developmental abnormalities formed during embryonic development of the spleen.
- They develop when some of the cells from the developing spleen become detached and deposited along the developmental path of the spleen.
- The most common locations for accessory spleens are:
  - The hilum of the spleen
  - Adjacent to the tail of the pancreas
- Accessory spleens may be found in the following sites (Fig. 7.9):
  - Along the splenic vessels
  - The gastrocolic ligament
  - The gastrosplenic ligament
  - The splenocolic ligament
  - The splenorenal ligament
  - The walls of the stomach mainly the greater curvature of the stomach
  - The walls of the small intestines
  - The tail of the pancreas (Intrapancreatic accessory spleen)
  - The greater omentum
  - The mesentery of small intestines
  - The female adnexa, mainly the left broad ligament
  - The left spermatic cord
  - The gonads and along their path of descent
- An accessory spleen derives its blood supply from branches of the splenic artery.
- The presence of an accessory spleen is important to look for and find in patients undergoing splenectomy for hematologic diseases especially patients with ITP (Idiopathic Thrombocytopenic Purpura).
- In patients with hematological diseases, accessory spleens can enlarge following splenectomy and lead to recurrence of their symptoms. This is of great importance in those with idiopathic thrombocytopenic purpura as an accessory spleen can lead to recurrence of symptoms if not excised (Figs. 7.10, 7.11, 7.12, 7.13 and 7.14).

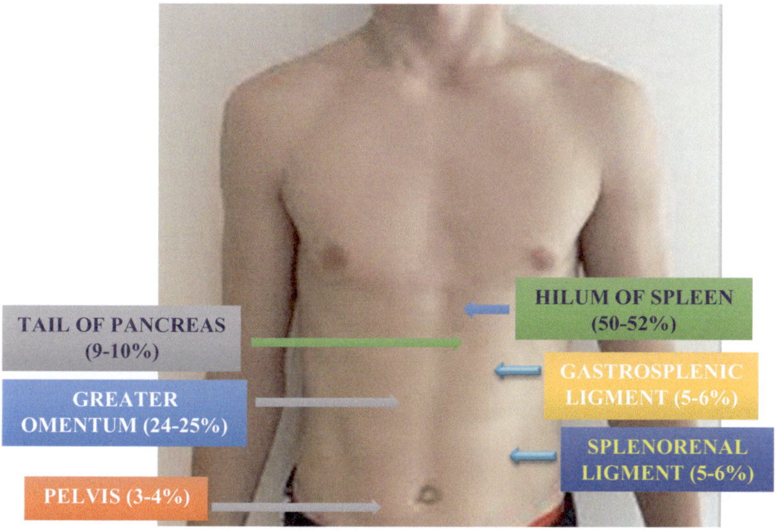

Fig. 7.9 Sites of accessory spleens. The hilum of the spleen is the commonest site followed by the tail of the pancreas and the greater omentum

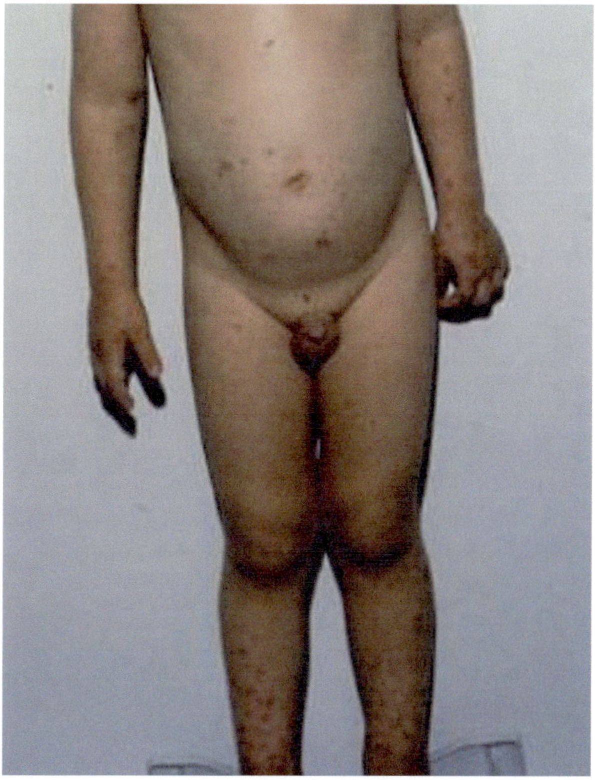

Fig. 7.10 A clinical photograph of a child with idiopathic thrombocytopenic purpura who will have splenectomy. It is important to look for splenculi in these patients to avoid recurrence

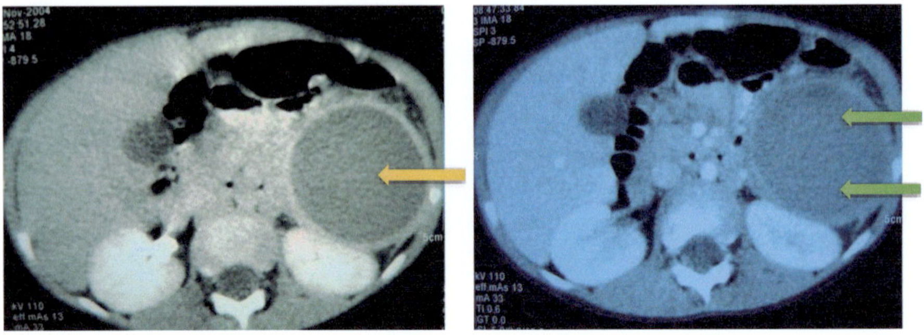

Figs. 7.11 and 7.12 CT-scan of the abdomen showing an enlarged spleniculi in a patient with congenital asplenia. This spleniculi enlarged and became ischemic causing abdominal pain. Note the site of the spleniculus which enlarged as a result of congenital asplenia

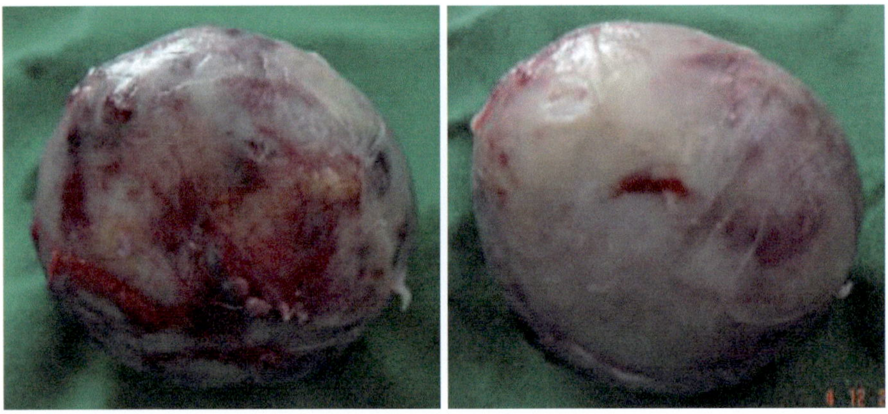

Figs. 7.13 and 7.14 Clinical photographs showing an enlarged spleniculi that was removed. Note the size of the removed spleniculi

7.4 Clinical Features

- An accessory spleen is a congenital anomaly consisting of normal splenic tissue.
- Accessory spleens are developmental abnormalities caused by failure of the splenic anlage to fuse during embryonic development.
- It is seen in approximately 10–15% of individuals.
- Accessory spleens are usually solitary.
- Approximately 10% of patients with one accessory spleen have a second one.
- More than 2 accessory spleens in the same patient are extremely rare.
- Nearly all accessory spleens are smaller than 2 cm in diameter, and 65% of them are 1 cm in diameter or less.
- Patients with accessory spleens are usually asymptomatic, and the lesions are usually discovered incidentally during investigations for other conditions.

- Preoperative diagnosis of accessory spleens is very difficult as they usually mimic lymphadenopathy or a tumor.
- Accessory spleens may become symptomatic because of:
  - Torsion on their vascular pedicle
  - Spontaneous rupture
  - Hemorrhage
  - Degeneration and or cyst formation
- The vascular pedicle of an accessory spleen may twist and cause torsion of the accessory spleen leading to rupture and torsion. As a result of this, these patients can present with abdominal pain.
- Accurate preoperative diagnosis is important but despite availability of multiple diagnostic investigations, it is difficult to make a preoperative diagnosis of accessory spleens.
- The majority of accessory spleens are located on the left side of the abdomen. The splenic hilum and tail of the pancreas are the most common sites for accessory spleens.
- When the main spleen is involved with hematologic or other systemic diseases including lymphoma, leukemia, thrombocytopenia, and hemosiderosis, ectopic spleens are also often affected by the primary disease.
- The definitive diagnosis of an accessory spleen is made by a histopathological examination of the excised nodule.
- Grossly, a typical accessory spleen appears as a solid mass with a smooth, round, ovoid, or minimally lobulated shape.
- Microscopically, accessory spleen resembles a normal spleen.
  - An accessory spleen usually has a well-defined fibrotic capsule.
  - A polymorphous population of hematopoietic cells can be seen, including lymphocytes, eosinophils, histiocytes, plasma cells, and red blood cells, admixed with a large number of small blood vessels representing splenic sinusoids.
  - Immunohistochemical staining for CD8 is important to confirm the diagnosis, as it specifically highlights the endothelial cells of the splenic sinus.
  - Lymphoproliferative disorder and hemosiderosis found in an accessory spleen will have an appearance similar to the same masses found within the main body of the spleen.
- The importance of accessory spleens needs to be emphasized and surgeons should be aware of the locations, gross appearance and imaging findings of an accessory. This will help in preoperative diagnosis and also intraoperative decisions.

7.5 Investigations

- Accessory spleens are structurally identical to the normal spleen.
- Accessory spleens are usually surrounded by a fibrotic capsule like normal spleens.

- Accessory spleens histology has red and white pulps similar to that of the normal spleen.
- The red pulp is made up of numerous vascular sinuses. Between these sinuses lie the lymphoid follicles and the cells of the reticuloendothelial system. These together make up the white pulp of the spleen.
- This spleen has a unique anatomy. The presence of the red pulp and whit pulp give different appearance on CT and MRI.
  - The red pulp of the spleen gives a low signal intensity (SI) on T2-weighted images as this part has the sinuses.
  - The white pulp gives a high SI on T2-weighted images. This is attributed to the higher water content of the lymphoid tissue.
- Physicians should be aware of accessory spleens and should consider them in the differential diagnosis of uncertain pancreatic and peripancreatic masses.
- The preoperative diagnosis of an accessory spleen is difficult, as it can be confused with other masses.
- Imaging studies such as ultrasonography (US), CT scan, and magnetic resonance imaging (MRI) are valuable but they are not specific.
- Ultrasonography:
  - On ultrasound, an accessory spleen appears as a homogeneous, round- or oval-shaped, well-demarcated mass that is isoechoic compared with the normal spleen.
  - This is nodule will appear hypoechoic to normal pancreatic parenchyma. Sometimes hyperechoic lesions have been described.
  - Ultrasonography may be helpful for the visualization of vascular hilum of an accessory spleen.
  - Accessory spleens on ultrasound are usually surrounded by a high-amplitude interface as a result of the covering capsule.
  - A color or Doppler US will reveal the presence of a vascular hilum entering the accessory spleen. This is reported to be a sensitive (90% of sensitivity) diagnostic feature of accessory spleen.
- Pancreatic tumors share signal characteristics with that of the spleen on MRI and so this accounts for the difficulty in differentiating intrapancreatic accessory spleen from pancreatic tumors.
- CT scan will usually detect an accessory spleen if it is large enough.
  - On an unenhanced CT scan, a typical accessory spleen will appear as a well-defined mass.
  - On a contrast-enhanced CT scan, an accessory spleen will enhance homogeneously as a normal spleen.
  - *Single photon emission computed tomography* (SPECT) is frequently used to diagnose and localize an accessory spleen.
- On magnetic resonance imaging, the spleen and accessory spleen have relatively long T1 and T2 relaxation times, which make them, appear dark on T1- and bright on T2-weighted images.
- MRI using superparamagnetic iron oxide as a negative contrast is also helpful for the diagnosing an accessory spleen.

  - In both T1-weighted and T2-weighted images, accessory spleen shows reduction of the signal intensity similar to that of the normal spleen.
  - Recently, MRI with diffusion-weighted imaging (DWI) has been increasingly used in abdominal imaging examination.
  - Accessory spleens show marked hyperintensity on high b-value DWI, similar to that of the normal spleen.
  - The combination of DWI and conventional MRI may improve the diagnosis of an accessory spleen.
- Tc-99m Heat-damaged Red Blood Cell Scintigraphy:
  - A radionuclide scanning with 99mTc (technetium)-labeled heat-damaged red blood cells may be helpful to confirm the diagnosis of an accessory spleen.
  - This is a very sensitive and specific test for differentiating accessory spleens from other tissues as about 90% of the injected erythrocytes are trapped by functional splenic tissue.
  - Scintigraphy with Tc-99m phytate may be the most helpful diagnostic method to establish the presence of a functioning accessory spleen.
  - The criterion used to diagnose an accessory spleen was the detection of a marked increase in uptake which should exceeds that of the cardiac blood pool and the major vessels at the site of suspected accessory spleen.
  - Autologous heat-damaged red blood cells are also used to detect an accessory spleen. These cells will be converted into rigid spherocytes and when injected intravenously they will be trapped by the accessory spleen.
  - Tc-99m heat-damaged red blood cells scintigraphy allows selective splenic visualization with an excellent spleen-to-liver ratio. This investigation my not be useful when there is minimal functioning splenic tissue.
- Intrapancreatic accessory spleen can be easily confused with other pancreatic tumors.
  - Nonfunctioning islet cell tumor can be differentiated from an accessory spleen based on the size of the lesion and the heterogeneity.
  - This is not the case with functioning islet cell tumor as these are difficult to differentiate from intrapancreatic accessory spleen.
- Epidermoid cysts of the spleen are relatively rare, accounting for only 10–20% of the non-parasitic splenic cysts.
- Epidermoid cysts are lined with stratified squamous epithelium, but they have no skin appendages such as sebaceous glands or sweat glands.
- It is interesting that epidermoid cysts occurring in accessory spleen are extremely rare and all the cases have been found in the pancreatic tail.
- A US-guided fine-needle aspiration biopsy is also an important tool for making a differential diagnosis of an accessory spleen. This will avoid unnecessary operation.
- If no definitive diagnosis is reached with noninvasive imaging tests, surgical resection remains the only option. This is diagnostic and curative.

7.6 Treatment

- The spleen is an important organ and has several functions.
- One of the important functions of the spleen is protection against bacterial infections. This was stressed in 1952 by King and Shumaker, who noted increased susceptibility of infants to infection after splenectomy.
- This prompted surgeons to adopt conservative nonoperative management of splenic injuries.
- Another significant development in the surgical management of splenic disorders was the development of laparoscopic splenectomy in 1991 by Delaitre and Maignien.
- This technique has quickly become the preferred technique for splenectomy both in adults and children.
- Treatment of an accessory spleen is controversial.
- It is usually asymptomatic.
- Asymptomatic accessory spleen should be treated conservativel.
- An accessory spleen can develop complications such as torsion of its pedicle and can result in infarction and rupture necissating surgical excision.
- Symptomatic accessory spleen can be excised laparoscopically. This is feasible and safe.
- Accessory spleens may mimic tumors, such as pancreatic tumor, adnexal tumor, abdominal tumor, retroperitoneal tumor, adrenal tumor, or testicular tumor.
- An accurate preoperative diagnosis will help avoid unnecessary surgery.
- It is important to diagnose accessory spleens preoperativel as they are usually asymptomatic and no treatment is necessary except in the following situations:
  - An accessory spleen may mimic lymphadenopathy or tumors in other abdominal organs, including the pancreas.
  - An accessory spleen may become symptomatic because of:
    Torsion
    Spontaneous rupture
    Hemorrhage
    Cyst formation
  - It is also important that all functional splenic tissue should be removed for the treatment of hematologic disorders such as idiopathic thrombocytopenic purpura (ITP). This is to avoid recurrence.

Further Reading

1. Halpert B, Gyorkey F. Lesions observed in accessory spleens of 311 patients. Am J Clin Pathol. 1959;32:165–8.
2. Wadham BM, Adams PB, Johnson MA. Incidence and location of accessory spleens. N Engl J Med. 1981;304:1111.
3. Dodds WJ, Taylor AJ, Erickson SJ, Stewart ET, Lawson TL. Radiologic imaging of splenic anomalies. AJR Am J Roentgenol. 1990;155:805–10.
4. Sica GT, Reed MF. Intrapancreatic accessory spleen. Radiology. 2000;217:134–7.

5. Halpert B, Alden ZA. Accessory spleens in or at the tail of the pancreas: a survey of 2700 additional necropsies. Arch Pathol. 1964;77:652–4.
6. Coote JM, Eyers PS, Walker A, Wells IP. Intra-abdominal bleeding caused by spontaneous rupture of an accessory spleen: the CT findings. Clin Radiol. 1999;54:689–91.
7. Appel MF, Bart JB. The surgical and hematologic significance of accessory spleens. Surg Gynecol Obstet. 1976;143:191–2.
8. Katz JF, Kane RA. Sonographic appearance of accessory spleen. J Clin Ultrasound. 1984;12:163–5.
9. Subramanyam BR, Balthazar EJ, Horii SC. Sonography of the accessory spleen. AJR Am J Roentgenol. 1984;143:47–9.
10. Hayward I, Mindelzun RE, Jeffrey RB. Intrapancreatic accessory spleen mimicking pancreatic mass on CT. J Comput Assist Tomogr. 1992;16:984–5.
11. Laeuffer JM, Baer HU, Maurer CA, et al. Intrapancreatic accessory spleen—a rare cause of a pancreatic mass. Int J Pancreatol. 1999;25:65–8.
12. Harris GN, Kase DJ, Bradnock H, McKinley MJ. Accessory spleen causing a mass in the tail of the pancreas: MR imaging findings. AJR Am J Roentgenol. 1994;163:1120–1.
13. Barawi M, Bekal P, Gress F. Accessory spleen: a potential cause of misdiagnosis at EUS. Gastrointest Endosc. 2000;52:769–72.
14. Bidet AC, Dreyfus-Schmidt G, Mas J, et al. Diagnosis of splenosis: the advantages of splenic scintiscanning with Tc 99m heat-damaged red blood cells. Eur J Nucl Med. 1986;12:357–8.
15. Ota T, Tei M, Yoshioka A, et al. Intrapancreatic accessory spleen diagnosed by technetium-99m heat damaged red blood cell SPECT. J Nucl Med. 1997;38:494–5.
16. Gunes I, Yilmazlar T, Sarikaya I, Akbunar T, Irgil C. Scintigraphic detection of splenosis: superiority of tomographic selective spleen scintigraphy. Clin Radiol. 1994;49:115–7.
17. Chung SY, Ryo Y, Pinsky S. Evaluation of a patient with splenosis by various imaging modalities. J Natl Med Assoc. 1986;78:458–63.
18. Gayer G, Zissin R, Apter S, Atar E, Portnoy O, Itzchak Y. CT findings in congenital anomalies of the spleen. Br J Radiol. 2001;74:767–72.
19. Mortele KJ, Mortele B, Silverman SG. CT features of the accessory spleen. AJR Am J Roentgenol. 2004;183:1653–7.
20. Velanovich V, Shurafa M. Laparoscopic excision of accessory spleen. Am J Surg. 2000;180:62–4.
21. Rosenthal CL, Bishop MC. Accessory spleen presenting as a retroperitoneal tumour. Eur Urol. 1981;7:314–6.
22. Stiris MG. Accessory spleen versus left adrenal tumor: computed tomographic and abdominal angiographic evaluation. J Comput Assist Tomogr. 1980;4:543–4.
23. Tsuchiyaa N, Satoa K, Shimodaa N, et al. An accessory spleen mimicking a nonfunctional adrenal tumor: a potential pitfall in the diagnosis of a left adrenal tumor. Urol Int. 2000;65:226–8.
24. Kapoor A, Jain A, Mahajan G, Singh A, Bajwa VS, Brar GS. Elusive retroperitoneal accessory spleen. Indian J Surg. 2004;66:298–9.
25. Campbell MF, Walsh PC, Retik AB, editors. Campbell's urology. 8th ed. New York: WB Saunders; 2002. p. 3523–5.

Asplenia (Congenital Absence of the Spleen)

8

8.1 Introduction

- Asplenia is defined as the absence of a spleen.
- Asplenia can develop as a result of:
  - Anatomic (Congenital) absence of the spleen
  - Functional asplenia secondary to a variety of diseases
  - Following splenectomy
- The spleen is the largest lymphoid organ in the body and it is located in the left upper quadrant of the abdomen under the left hemidiaphragm.
- Normally, the spleen is not palpable on physical exam and its size is variable.
- The spleen measures on average 10.6 cm × 5.2 cm.

CAUSES OF ASPLENIA
- **ANATOMIC (CONGENITAL) ABSENCE OF THE SPLEEN**
- **FUNCTIONAL ASPLENIA**
- **FOLLOWING SPLENECTOMY**

- The spleen has several important functions:
  - Filtration

    The spleen's main function is filtration of the blood and processing of old and abnormal blood cells, mainly red blood cells.
  - Immunological
- The spleen plays an important immunological role and helps protect the body against encapsulated microorganisms and respond to infectious pathogens.
- The spleen has two functionally and histologically distinct tissues:
  - The white pulp:

    The white pulp has a large mass of lymphoid tissue that produces antibodies against invading antigens.

A. H. Al-Salem, *The Spleen*, https://doi.org/10.1007/978-981-99-6191-7_8

- The red pulp:

  The red pulp has a tight network of sinusoids called the cord of Billroth which helps in blood filtration.

  It is the body largest filter of blood.

  It helps to remove old RBCs from the circulation and aid in the removal of blood-borne microorganisms.

- Asplenia is well known to be associated with increased risk of serious and life-threatening complications.
- One of these complications is the increased risk for serious infections caused by encapsulated microorganisms such as Neisseria meningitides or Streptococcal pneumoniae. These are serious and life-threatening if not recognized and untreated.
- It is estimated that patients with asplenia have a 200-fold greater risk of death from overwhelming sepsis compared to those with a normally functioning spleen.
- The spleen is an important organ and has several important immunological functions. This is attributed to the spleen immune cells and phagocytosis of circulating blood antigens and pathogens.
- This will help protect the body from invading pathogens especially capsulated bacteria.
- It is also important to assess the spleen function. This will help assess the risk of with diminished spleen function and their risk of developing overwhelming post-splenectomy sepsis.
- This will help plan their treatment and prevention of developing overwhelming sepsis with antibiotics and vaccines.
- Congenital asplenia is very rare.
- Congenital asplenia is usually associated with other congenital malformations including:
  - Cardiac malformations:

    These are seen in 80% of patients with congenital asplenia.

    This usually involves atrioventricular endocardial cushion and ventricular outflow tract.
  - Situs inversus
  - Anomalies of blood vessels, lung, and abdominal viscera.
- Asplenia can also be an acquired condition following splenectomy whether for medical or traumatic indications for splenectomy.
- Asplenia and splenic hypoplasia refer to the complete or partial lack of splenic tissue.
- Congenital asplenia
  - Congenital asplenia is a very rare life-threatening condition.
  - These patients often present with pneumococcal sepsis.
  - It may arise as part of situs abnormalities, or result from an unrelated specific defect of spleen development.
  - It is inherited usually as an autosomal dominant.
  - Rarely it is inherited as autosomal recessive
  - Sporadic cases of congenital asplenia have also been reported.

- Heterotaxy syndrome
  - This results from defects in the development of normal human left-right asymmetry.
  - The best-known heterotaxy syndrome is situs inversus.
  - Situs inversus is characterized by the presence of unpaired thoracic and abdominal organs which are positioned in a mirror-image fashion but are otherwise structurally normal.
  - Situs ambiguous or cardio splenic syndromes is the more common partial defects in left-right asymmetry.
  - Situs ambiguous or cardio splenic syndromes are divided into two distinct conditions:
    - Polysplenia syndrome/left isomerism
    - Asplenia syndrome/right isomerism
- Polysplenia syndrome, or left isomerism (bilateral left-sidedness, also known as Ivemark syndrome), is characterized by paired left-sided viscera.
  - Polysplenia syndrome is characterized by multiple spleens associated with cardiac anomalies which include:
    - Ventricular septal defect
    - Atrioventricular septal defects
    - Outflow tract abnormalities
    - Heart block
  - Patients with polysplenia syndrome also have:
    - Bilateral bilobed lungs
    - A malpositioned stomach
    - A midline liver
    - Interruption of the inferior vena cava with the venous return from the lower body occurring over the prominent azygos vein (right to the aorta) or hemiazygos vein (left to the aorta), running parallel to the descending aorta.
- Asplenia
  - Asplenia is defined as the loss of splenic function
  - Asplenia can develop as a result of:
    - Congenital absence of the spleen (Congenital asplenia). This is very rare.
    - Functional asplenia develops as a result of loss of splenic functions due to a variety of conditions including:
    - Sickle cell disease
    - Celiac disease
    - Alcohol-related liver disease
    - Functional asplenia can also occur after repeated splenic infarcts and splenic vein thrombosis
  - Surgical removal of the spleen (splenectomy)
    - This is seen in patients who require splenectomy after trauma or in patients with immunologic or hematologic diseases that require splenectomy such as:
    - Idiopathic thrombocytopenic purpura
    - Hypersplenism

    - Hereditary spherocytosis
    - Thalassemia
- Patients with asplenia are at increased risk to develop severe sepsis due to encapsulated microorganisms, primarily Streptococcus pneumoniae (pneumococcus), but also Haemophilus influenzae type b and Neisseria meningitidis (meningococcus).
- These patients should be educated and they should receive pneumococcal vaccine, meningococcal vaccine, and Haemophilus influenzae b vaccine.
- These patients should also receive influenza vaccine annually.
- Children with asplenia are also given prophylactic antibiotics such as penicillin or amoxicillin.

8.2 Etiology

- Asplenia is classified as:
  - Congenital
  - Acquired.
- Congenital asplenia is seen in patients who are born without a spleen.
- Acquired asplenia is seen:
  - Following splenectomy and this is the commonest cause of asplenia
    - Trauma is the commonest indication for splenectomy accounting for about 41% of the cases.
    - Other indications for splenectomy include hematologic malignancy (15%), and cytopenia (15%).
  - Splenectomy can be performed for a variety of benign and malignant hematologic disorders including:
    - Hereditary spherocytosis
    - Sickle cell disease
    - Thalassemia major
    - Thalassemia intermedia
    - Refractory immune thrombocytopenia purpura (ITP)
    - Myelofibrosis
    - Autoimmune hemolytic anemias
    - Lymphoproliferative disorders
    - Thrombotic thrombocytopenic purpura (TTP)
  - Functional asplenia:
    - This is seen in patients in whom the spleen is present but has no function or its function is reduced markedly.
- Congenital asplenia may occur as an isolated condition or as part of clinical syndromes such as Ivemark syndrome.
  - Ivemark syndrome:
    - This syndrome is classified under heterotaxy syndrome.

It is characterized by the presence of asplenia or hypoplasia of the spleen, congenital malformations of the heart, and abnormal arrangements of organs of the chest and abdomen.

- Other diseases known to be associated with asplenia or hypofunction of the spleen include the followings:
  - Gastrointestinal (GI) disorders such as:
    Celiac disease
    Whipple disease
    Inflammatory bowel diseases (IBD)
  - Infectious diseases and acquired immune deficiency syndromes including HIV.
  - Hepatic disorders such as:
    Alcoholic liver disease
    Cirrhosis
    Portal hypertension
    Hepatorenal syndrome
  - Rheumatologic conditions such as systemic lupus erythematosus (SLE) have been associated with reduced spleen function.
- Functional asplenia can be caused by repeated attacks of sequestration of red blood cells (RBCs) in the spleen leading to their entrapment as seen in patients with sickle cell disease. This entrapment leads to splenic enlargement and then later splenic atrophy and loss of function.
- Atrophy of the spleen is also known as auto splenectomy.
- Auto splenectomy:
  - This is commonly seen in patients with sickle cell anemia
  - Patients with sickle cell anemia commonly develop small, multiple and repeated attacks of splenic infarcts.
  - These infarcts will ultimately lead to fibrosis and atrophy of the spleen (Auto splenectomy).
  - This atrophic spleen may be seen as a small calcified nodule (Figs. 8.1, 8.2, and 8.3).

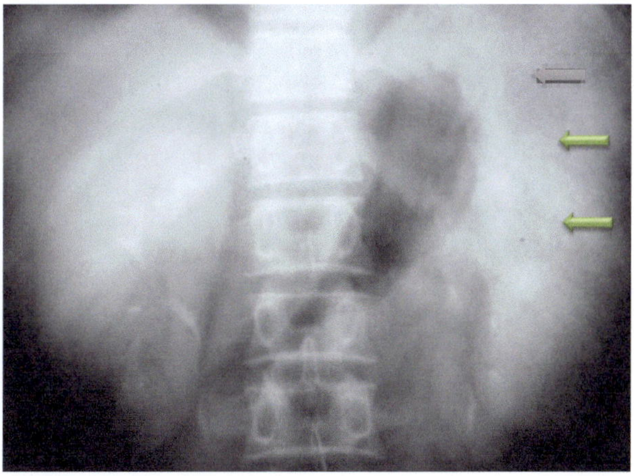

Fig. 8.1 Plain abdominal X-ray showing an enlarged calcified spleen in a patient with sickle cell anemia

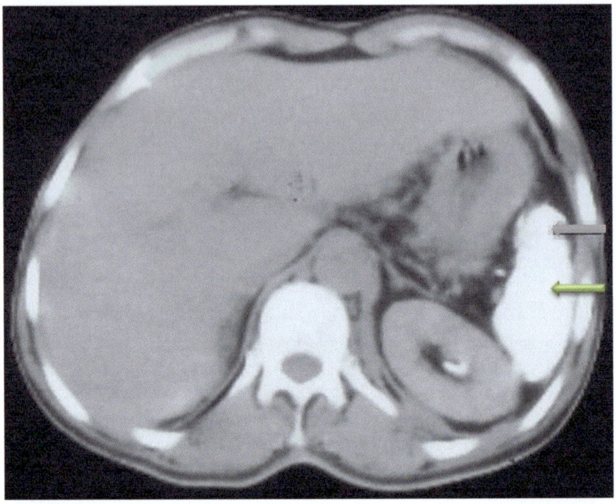

Fig. 8.2 Abdominal CT-scan showing a small calcified spleen in a patient with sickle cell anemia

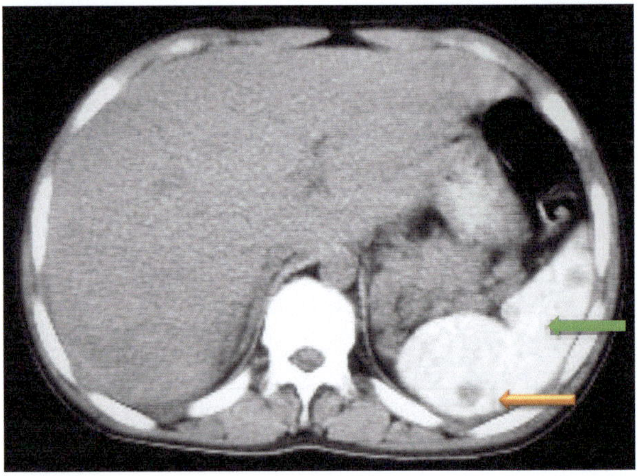

Fig. 8.3 Abdominal CT-scan showing a calcified spleen in a patient with sickle cell anemia

ETIOLOGY OF ASPLENIA

- **CONGENITAL ASPLENIA: THIS IS A DEVELOPMENTAL ABNORMALITY OF THE SPLEEN**
  - **ISOLATED CONGENITAL ASPLENIA**
  - **PART OF THE HETERTAXY SYNDROME**
- **AQUIRED ASPLENIA**
  - **FOLLOWING SPLENECTOMY**
  - **FUNCTIONAL ASPLENIA:**

    (1) **GASTROINTESTINAL DISESES**

    CELIAC DISEASE

    WHIPPLE DISEASE

    INFLAMMATORY BOWEL DISEASE

(2) **INFECTIOUS DISEASES AND AQUIRED IMMUNODEFICIENCY SYNDROME**
(3) **HEPATIC DISORDERS**
  ALCHOLIC LIVER DISEASE
  LIVER CIRRHOSIS
  PORTAL HYPERTENSION
  HEPATORENAL SYNDROME
(4) **RHEUMATOID CONDITIONS SUCH AS SYSTEMIC LUPUS ERYTHROMATOSUS**
- **AUTOSPLENECTOMY AS SEEN IN PATIENTS WITH SICKLE CELL ANEMIA**

- In hematological malignancies, functional asplenia may develop as a result from direct infiltration of the malignant cells into the parenchyma of the spleen.
- Many gastrointestinal disorders can lead to functional asplenia.
  - In cases of hepatic dysfunction or failure, asplenia can develop due to the disruption of the normal hepatic circulation, such as in portal hypertension.
  - Alcohol consumption has been known to cause a direct toxic effect on splenic tissue.
  - In Celiac disease, asplenia is caused by excessive loss of lymphocytes through inflamed enteric mucosa, leading to reticuloendothelial atrophy of the spleen.

8.3 Pathophysiology

- Asplenia is defined as the absence of the spleen and/or its functions.
- Absence of splenic tissue can be total as seen in patients with congenital asplenia or partial in those with hypoplastic spleen.
- Loss of splenic tissue and function develops following splenectomy for trauma or other conditions such as hemoglobinopathies, massive enlargement of the spleen and malignancy.
- Auto splenectomy is the loss of splenic function due to multiple and repeated splenic infarcts as seen in patients with sickle cell anemia.
- Embryologically, the spleen begins to develop as early as 12 days of gestation, along with the splanchnic mesodermal plate. This is one of the processes involved with formation of the asymmetrical left-right axis.
- The spleen is also important as the site for early hematopoietic development, particularly the development of red blood cells during the first 4 months of gestation.
- After birth, the spleen has several important functions.
- The spleen is made up of two main components:
  - The white pulp of the spleen:
    This contains the germinal centers, with lymphocytes, plasma cells, and macrophages.
    These play an important role in the immune response

It is important in both innate and adaptive immunity of the body.
The spleen plays an important role in the production of immunoglobulin (IgM) antibodies and complement. These are important to opsonize bacteria which is essential in fightening invading microorganisms and protecting the body.
Once these microorganisms have been tagged, the spleen with its phagocytosis function plays an important role in the destruction of these bacteria.
The spleen is also important for the maturation of antibodies and act as a reservoir for both B and T lymphocytes.
The spleen is also an important reservoir for T-cells and the number of total T cells (CD3) and T-helper cells (CD4).
The spleen plays an important role in granulocyte homeostasis by eliminating senescent cells and regulating the production of granulocyte in the bone marrow.

- The red pulp of the spleen:
  The red pulp main function id filtration of the blood as it passes through.
  The spleen plays an important role as a scavenger. This is important for the destruction of senescenced red blood cells, white blood cells, and platelets.
  The spleen is rich with macrophages and these play an important role in the recycle of iron which is produced from the destruction of red blood cells in the spleen.
  The spleen plays an important role in removal of abnormally shaped red blood cells such as sickled cells, spherocytes, and poikilocytoses.
  The spleen also plays an important role in removal of intracellular inclusions such as Heinz bodies and Howell Jolly bodies.
  These functions are known as culling and pitting, respectively. Loss of these functions results in the persistence of abnormal red cells and inclusions in the peripheral blood smear in patients with anatomical or functional asplenia.

- Heinz bodies (Fig. 8.4):
  - Heinz bodies (also referred to as “Heinz-Ehrlich bodies”) are inclusions found within the red blood cells and composed of denatured hemoglobin.
  - Heinz bodies are indicative of oxidative injury to the red blood cells.
  - They are inclusions of irreversibly denatured hemoglobin attached to the red blood cells membrane. They are formed from the breakdown of hemoglobin in red blood cells.
  - They develop as a result of oxidative damage to the red blood cells from (Fig. 8.4):
    Toxins
    Medications
    Underlying G6PD deficiency
    Underlying thalassemia
- Howell Jolly bodies (Figs. 8.5 and 8.6):
  - Howell-Jolly bodies are remnants of RBC nuclei that are normally removed by the spleen.

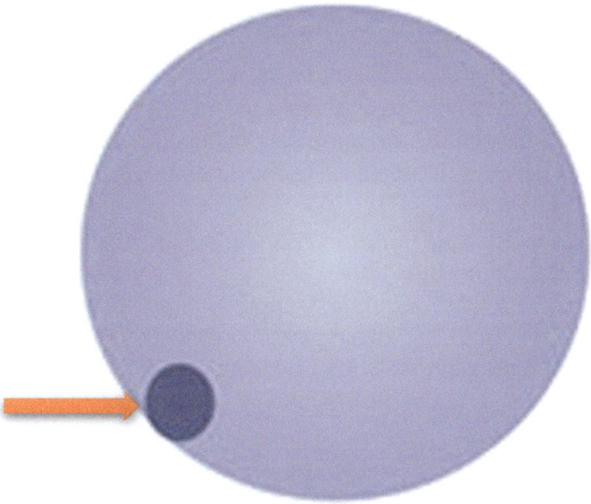

Fig. 8.4 A microscopic picture showing Heniz body. Hese are inclusions of irreversibly denatured hemoglobin attached to the red blood cells membrane

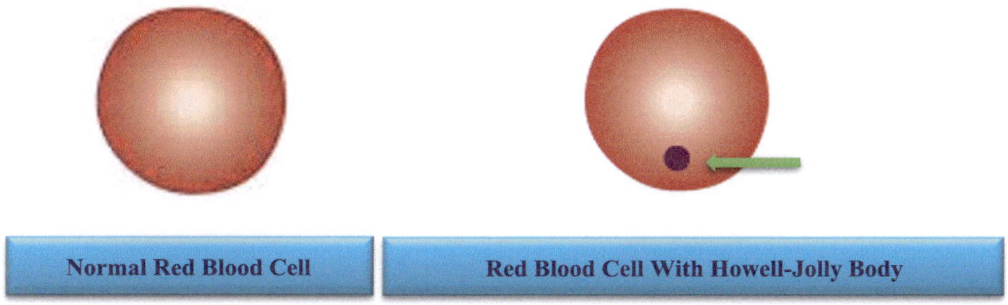

Figs. 8.5 and 8.6 Normal red blood cell and red blood cell with Howell-Jolly body. Howell-Jolly bodies are remnants of RBC nuclei

  - They are seen in patients who had splenectomy or in those with functional asplenia.
  - A Howell–Jolly body is made up of basophilic nuclear remnants (clusters of DNA) in circulating erythrocytes.
- In patients with asplenia several factors contribute to the increased susceptibility to serious and often fatal bacterial infections. These include:
  - Impaired clearance of opsonized organisms
  - Decreased production of immunoglobulin IgM
  - Poor antibody production
- In patients with asplenia, the most common and serious infections are due to gram positive encapsulated organisms.
  - Streptococcus pneumonia is most common
  - Haemophilus influenzae type b
  - Neisseria meningitides

  - Staphylococcus aureus
  - Salmonella species
  - Pseudomonas aeruginosa
- In infants younger than 6 months, gram-negative enteric organisms such as Klebsiella species and Escherichia coli are the most common pathogens.
- Endocarditis due to Bordetella holmesii was also reported in a patient with asplenia.
- The younger the patient at the time of splenic function loss, the higher the risk for serious infection.
- One of the common complications following splenectomy is persistent and sometimes significant thrombocytosis. This may lead to the development of thromboembolic complications.
- Isolated congenital asplenia is extremely rare and inherited as autosomal dominant.
- Most cases of congenital asplenia or polysplenia are seen in association with other abnormalities. These are seen most commonly as a result from interference in the establishment of normal right-left symmetry during embryogenesis.
- This is called heterotaxy syndrome, laterality sequences.
  - Congenital asplenia may be viewed as bilateral right-sidedness and is associated with dextrocardia in approximately one third of the cases.
  - Polysplenia may be regarded as bilateral left-sidedness and may be associated with left atrial isomerism.
  - Both asplenia and polysplenia are associated with congenital cardiac anomalies.
    These congenital cardiac anomalies are more common, severe, and generally complex in patients with asplenia.
    Cyanotic congenital heart defects tend to be more commonly seen in patients with asplenia.
    Acyanotic congenital heart defects are more commonly seen in patients with polysplenia.
  - These cardiac anomalies include:
    Endocardial cushion defects
    Atrioventricular canal defects
    Pulmonary atresia
    Pulmonary stenosis
    Transposition of the great vessels
    Total anomalous pulmonary venous return
    Double-outlet right ventricle
- Congenital asplenia is most often found in association with other developmental anomalies.
- The most common is Ivemark syndrome:
  - This is also referred to as asplenia syndrome
  - It is characterized by visceral heterotaxy with bilateral right-sidedness.
  - The body right-sided organs are usually duplicated

  - The body organs that are normally present on the left side are absent
  - Infants with Ivemark syndrome usually present during the neonatal period with cyanosis and respiratory distress.
  - It more common in males than in females.
  - The majority of these patients (79%) die in their first year of life due to cardiovascular complications.
  - Ivemark syndrome is usually associated with complex congenital cardiac anomalies. These anomalies include:
    Transposition of the great arteries
    Pulmonary stenosis
    Pulmonary atresia
    Total anomalous venous drainage
  - Other anomalies associated with Ivmark syndrome include:
    The liver tends to be symmetrical and transverse in position
    The stomach may be in the midline and hypoplastic
- Pearson syndrome (pancreatic insufficiency, sideroblastic anemia) is a mitochondrial disorder associated with splenic atrophy.
- Asplenia is also present in Stormorken syndrome (thrombocytopenia and miosis).
- Asplenia may be present in Smith-Fineman-Myers syndrome:
  - Mental retardation
  - Short stature
  - Cryptorchidism
- Asplenia can be seen in patients with ATR-X syndrome (α thalassemia and mental retardation).
- Asplenia may be associated with caudal deficiency or cystic disease of the liver, kidney, and pancreas.
- Asplenia has also been reported in association with Fanconi aplastic anemia.
- Asplenia can be found in patients with autoimmune polyendocrine syndrome type-1.
- Horseshoe adrenal glands have also been associated with asplenia syndrome.
- Asplenia was also seen in Cat eye syndrome.
- Functional asplenia:
  - This is known to be associated with hemoglobinopathies and develop as a result of repeated splenic infarctions.
  - Most these children start losing the splenic function in the first two years of life and become anatomically asplenic secondary to splenic infarction which ultimately leads to splenic atrophy by the second decade of life.
  - This is called auto splenectomy and it is seen in patients with:
    Homozygous sickle cell disease
    Sickle cell disease hemoglobin C disease
    Sickle cell hemoglobin (Hb S) β thalassemia
- Other clinical conditions known to be associated with splenic hypofunction include:
  - Rheumatoid arthritis

  - Systemic lupus erythematous
  - Inflammatory bowel disease
  - Graft versus host disease
  - Nephrotic syndrome

8.4 Clinical Features

- Patients with asplenia can be asymptomatic
- They can also present with the followings:
  - Malaise
  - Fatigue
  - Fever
  - Encephalopathy
  - In patients with functional asplenia (auto splenectomy), the spleen will usually not be palpable due to fibrosis and atrophy.
- Patients with functional asplenia will have symptoms of the original disease such as those with hemoglobinopathies.
- Patients with asplenia may also present with signs or symptoms of underlying infection.
  - Severe infection due to Streptococcus pneumoniae, Haemophilus influenzae, and Neisseria meningitides are the most common.
  - Pneumococcal meningitis is more fatal in patients with asplenia when compared to other forms of bacterial meningitis.

8.5 Investigations

- Analysis of the peripheral blood smear may demonstrate the presence of Howell-Jolly bodies.
- The presence of Howell-Jolly bodies and pitted red blood cells seen on a peripheral blood smear are suggestive of asplenia or reduced function of the spleen.
- The peripheral blood smear in patients with asplenia shows Howell-Jolly (HJ) bodies in RBCs. These are nuclear remnants seen as small, round, densely stained inclusions in RBCs.
- The presence of Howell-Jolly bodies in the peripheral blood smear of a patient older than 7 days should suggest splenic dysfunction.
- In patients with asplenia there is an increased presence in peripheral blood smear of:
  - Target cell
  - Heinz bodies
  - Pappenheimer bodies: These are small basophilic inclusions that contain non-heme iron
  - Reticulocytes
  - Spherocytes
- A complete blood count will reveal thrombocytosis and leukocytosis.

Fig. 8.7 Pitted (pocked) red blood cell

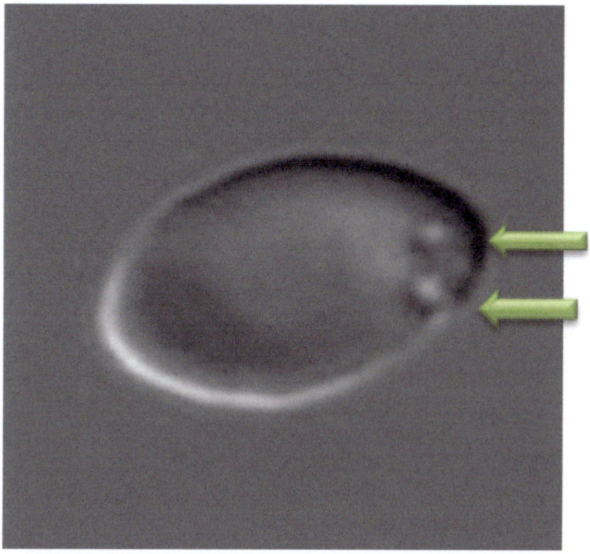

- In patients with asplenia, a useful investigation is the pit count (autophagic vacuoles) which count the number of pocked erythrocytes (Fig. 8.7).
  - This technique requires the use of Nomarski optics
  - These optics are not always readily available.
  - Normally, less than 2% of red cells have these pocks or pits.
  - A pocked erythrocyte count of more than 3.5% suggests functional hyposplenia.
  - A pocked erythrocyte count of more than 12% is indicative of asplenia.
- Other non-specific cytologic findings of asplenia include:
  - Monocytosis
  - Lymphocytosis
  - Thrombocytosis
- Various imaging techniques can be used to assess splenic function and measurement, including CT scanning, MRI, ultrasonography, and technetium-99m scintigraphy.
- Abdominal ultrasound (US), CT-scan, MRI, or spleen scintigraphy can be used to confirm the presence or absence of the spleen. They are also useful in measuring the size of the spleen if present.
- Abdominal MRI or CT scan may reveal absence or hypoplasia of the spleen.
- Spleen scintigraphy using a 99-Tc labeled radio colloid is specific for splenic tissue. This is an important investigation and failure of uptake by the spleen is indicative of asplenia.
- The presence of an anatomical spleen does not ensure it is adequately functioning.
- A color Doppler ultrasonography can be used to assess splenic function. The presence of a small spleen with no parenchymal vascularization on color Doppler ultrasonography has been associated with functional asplenia.
- Asplenia whether anatomic or functional can be confirmed with a technetium-99m radionuclide scan.

8.6 Treatment

- Patients with asplenia are at a significantly increased risk of developing severe bacterial infection, especially from encapsulated microorganisms.
- Every effort should be made to prevent these infections as they are known to be associated with increased morbidity and mortality.
- It is also important to diagnose and manage these infections as early as possible.
- The most common bacterial pathogen leading to infection in patients with asplenia is Streptococcus pneumoniae.

Up to 87% of patients with asplenia were found to have been infected with Streptococcus pneumoniae.

- One way to avoid these serious infections is that all patients with asplenia should be vaccinated against these encapsulated bacteria.
- They should be vaccinated before splenectomy and in cases of emergency after the splenectomy and prior to their discharge from the hospital.
- Patients with functional asplenia or auto splenectomy should also be vaccinated against these encapsulated bacteria.
- It is recommended that patients undergoing elective splenectomy should receive:

The pneumococcal conjugate vaccine (PCV-13) 4–8 weeks prior to splenectomy
The pneumococcal polysaccharide vaccine (PPSV-23)
Haemophilus influenzae type B vaccine (Hib)
The quadrivalent meningococcal conjugate vaccine 14 days before splenectomy.

- These patients should also receive a booster dose of the PPSV-23 vaccine 5 years after splenectomy.
- It is recommended that all asplenic patients should receive both the conjugated 13-valent pneumococcal vaccine and the 23-valent polysaccharide vaccine. Four doses at two, four, six and 12–15 months of age should be given.
- The pneumococcal polysaccharide vaccine against 23 serotypes should be given as soon as possible after 24 months of age for supplemental protection.
- If an asplenic patient has previously received only PPV23, then he should receive one dose of PCV13 one year after receipt of the PPV23 vaccine.
- The meningococcal conjugate vaccine should be repeated every 5 years.
- Patients with asplenia should be encouraged to receive annual vaccination against the common strains of influenza.
- It has been recommended that prophylactic penicillin should be given to children with asplenia until the age of 5, although this prophylaxis is not recommended in adults.
- It was shown that overwhelming post-splenectomy sepsis is more common in children younger than 16 years old.
- Patients with asplenia who present with signs of infection should be admitted to the hospital and immediately covered with broad-spectrum antibiotics until the causative organism is isolated and identified from cultures.
- Patients with asplenia are also prone to deteriorate quickly and develop septic shock. They should be hydrated well and they may also require vasopressors to maintain their blood pressure. In certain circumstances, mechanical ventilation may be necessary if patients develop respiratory failure.

- It is also recommended that these patients should receive vaccination for Hib. A series of three doses should be given at two, four and six months of age, with a booster dose at 18 months of age.
- All patients with asplenia ≥5 years of age who have never received Hib immunization or have missed one or more doses should receive one dose.
- Yearly seasonal influenza vaccine is also recommended, starting at six months of age.
- To obviate the risk of Salmonella infection, all patients with asplenia travelling to less developed areas of the world should be immunized for S typhi.
- The best response to the vaccination occurs when vaccines are given at least two weeks before surgery. This is important to consider when planning an elective splenectomy.
- Patients undergoing emergency splenectomy should be immunized at least two weeks post splenectomy or before discharge from hospital.
- Immunizations do not fully protect against infections with encapsulated bacteria.
- Prophylactic antibiotics is important and should be given to young children who had splenectomy.
- S pneumoniae is the most common cause of severe infections in children with asplenia or hyposplenia and patients younger than five years of age should all receive antibiotic prophylaxis (Table 8.1).
- There are several controversial issues with respect to antibiotic prophylaxis. These include:
  - The degree of patient compliance
  - The optimal duration of prophylaxis antibiotics
  - The effect of prophylaxis on the emergence of penicillin-resistant pneumococci.
  - It is recommended that prophylaxis antibiotics should be given until the child is five years of age and a minimum of one year of prophylaxis for children older than five years of age post splenectomy. This is provided the child has received all the appropriate pneumococcal vaccinations.
  - Since most post splenectomy sepsis occurs within the first two to three years after splenectomy, it is recommended that antibiotic prophylaxis should be given for a minimum of two years post splenectomy and for all children <5 years of age.

Table 8.1 Antibiotic prophylaxis recommended for children with asplenia

Age	Oral prophylaxis
Birth - 3 months	Amoxicillin/clavulanate 10 mg/kg/dose two times per day, with penicillin VK 125 mg two times per day OR amoxicillin 10 mg/kg/dose two times per day being an alternative if not tolerated
> 3 months- 5 years	Penicillin VK 125 mg per dose two times per day OR amoxicillin 10 mg/kg/dose two times per day
> 5 years	Penicillin V 250 mg or 300 mg per dose two times per day OR amoxicillin 250 mg per dose two times per day

- It is important for parents to understand that children with asplenia must be evaluated by a physician immediately for every febrile illness.
- Sepsis in patients with asplenia or hyposplenia is considered as a medical emergency. These children can die within several hours of fever onset despite appearing well initially.
  - A blood culture should be performed.
  - All children should be admitted to the hospital and covered with intravenous broad-spectrum antibiotics.
  - Antibiotics should be modified once blood culture results become available.
  - These children should be monitored closely as clinical deterioration can be rapid even after antibiotic administration.
  - These children should be shifted to the intensive care unit if any sign of deterioration appears.
- There are several controversies with respect to antibiotic prophylaxis. These include:
  - The degree of patient compliance
  - The optimal duration of prophylaxis antibiotics
  - The effect of prophylaxis on the emergence of penicillin-resistant pneumococci.

ANTIBIOTIC PROPHYLAXIS RECOMMENDATIONS FOR CHILDREN WITH ASPLENIA

8.7 Antibiotic Prophylaxis

- Antibiotic prophylaxis should be initiated immediately upon the diagnosis of asplenia because these patients are at significant risk of pneumococcal infections.
  - For children younger than 2 years, oral penicillin V may be given twice a day.
  - Amoxicillin has also been recommended as an appropriate prophylactic antibiotic.
  - Erythromycin is an alternative choice for patients who are allergic to penicillin.
  - In general, antimicrobial prophylaxis should be considered for all children with asplenia or splenic dysfunction until age 5 years and for at least 1 year after splenectomy.

8.8 Immunization

- Patients with asplenia should receive all standard childhood and adolescent immunizations at the recommended age.
- In addition, patients with asplenia should receive the following vaccinations:

  - Pneumococcal conjugate and/or polysaccharide
  - H influenzae type b conjugate
  - Meningococcal conjugate and/or polysaccharide vaccines
- The conjugate pneumococcal vaccine (PCV13) covers most of the 13 pneumococci serotypes.
- These pneumococci serotypes include types 1, 3, 4, 5, 6A, 6B, 7F, 9V, 14, 18C, 19A, 19F, and 23F.
- These are the commonest serotypes causing invasive pneumococcal infections in patients with asplenia.
- The immunization schedule for pneumococcal conjugate vaccine (PCV13) consists of a primary series of 4 doses (0.5 mL each) at age 2, 4, 6, and 12-15 months.
- The pneumococcal polysaccharide vaccine against 23 serotypes (PPV23) should be given after age 24 months.
- A booster dose of PPV23 vaccine is given 3-5 years after the first dose.
- Children 6-18 years old with functional or anatomic asplenia should receive PCV13 regardless of whether they received the previously available 7-valent pneumococcal conjugate vaccine (PCV7) or the 23-valent pneumococcal polysaccharide vaccine (PPSV23).
- These patients should also receive the quadrivalent meningococcal vaccine.
  - Four meningococcal vaccines against serotypes A, C, Y, and W-135 are available.
  - Two of these vaccines are meningococcal conjugate vaccines (MCV4) and are recommended for infants and toddlers under the age of 2 years with asplenia.
  - MCV4 vaccine is recommended for all children 11-12 years old with asplenia.
  - A booster dose of MCV4 vaccine should be given 5 years after the initial vaccine.
  - Patients who were previously vaccinated at age 7 years or older should be revaccinated 5 years after their previous meningococcal vaccine.
  - Patients who were previously vaccinated at age 2-6 years should be revaccinated 3 years after their previous meningococcal vaccine.
  - It is recommended that patients with asplenia should continue to be revaccinated with MCV4 vaccine at 5-year intervals.
- Patients with asplenia should be vaccinated against H influenza. The recommended vaccination schedule for H influenzae type b is:
  - A primary series of 3 doses of H influenza vaccine is given at age 2, 4, and 6 months or only 2 doses are given at age 2 and 4 months, depending on the particular conjugate vaccine product administered.
  - It is also recommended to give these children a booster dose of H influenza vaccine at age 12 months.
  - Children who are scheduled for splenectomy and completed their primary series of vaccination are advised to receive a booster dose of conjugate vaccine at least 7-10 days before splenectomy.
  - Patients 5 years or older who never received Hib immunization should receive one dose of vaccination.

- H influenza type b conjugate vaccine may provide long-term protection to patients with asplenia and it should be given to patients with asplenia regardless of their previous vaccinations and time from splenectomy.
- Yearly influenza vaccine is also recommended for patients with asplenia to minimize the likelihood of secondary bacterial infections.

- Children 2 years of age or older undergoing elective splenectomy should receive one or both pneumococcal vaccines and the meningococcal vaccine at least 2 weeks prior to splenectomy.
- Children younger than 2 years of age should receive PCV13 prior to elective splenectomy.
- It is important to remember that the risk of serious bacterial infection is always present in patients with asplenia. Many of these patients present initially with mild trivial symptoms and rapidly develop fulminant sepsis and death within hours.
- All patients with asplenia with suspected infection must be urgently and promptly evaluated including investigations and treatment. These should include:
  - A complete blood count and differential
  - Blood, urine, and, if indicated, cerebrospinal fluid (CSF) cultures.
  - Start broad-spectrum intravenous antibiotics which are specifically effective against S pneumoniae, H influenzae type b, and N meningitidis.
  - Intravenous antibiotics should be started immediately and before bacteriological results are available. Second-generation or third-generation cephalosporins may be the initial choices. If multiple-drug resistance is a concern, vancomycin should be added to the regimen.
  - Supportive care with intravenous fluids, volume expanders, and when necessary pressor support also.
- Patients with asplenia are at an increased risk of sepsis, shock, and meningitis secondary to Capnocytophaga canimorsus resulting from dog, cat, or rodent bites. Penicillin is the treatment of choice in these cases, but cephalosporins, clindamycin, and erythromycin may also be used.
- Elective splenectomy for children with hemoglobinopathies should be discouraged before age 6 years and should be delayed as long as possible.
- Alternatives to total splenectomy should be considered when appropriate. These include:
  - Conservative management of splenic trauma, and auto transplantation
  - Splenorrhaphy in trauma cases
  - Partial splenectomy in trauma cases
  - Partial splenectomy for selective cases of hemoglobinopathies
  - Splenic artery embolization

8.9 Complications

- The most common complication in patients with asplenia is the development of an overwhelming post-splenectomy infection (OPSI).

- Asplenic patients are at higher risk of developing infection with encapsulated microorganisms' mainly streptococcal pneumonia.
- Most cases of fulminant sepsis in patients with asplenia is due to encapsulated bacteria.
  - Streptococcus pneumoniae is the most common organism causing sepsis in patients with asplenia and it is isolated in at least 50% of cases.
  - Haemophilus influenzae type b (Hib), Neisseria meningitidis, Salmonella species and Escherichia coli are also common.
  - Less commonly encountered causative organisms include Pseudomonas, Klebsiella, streptococci and staphylococci.
- OPSI results in massive bacteremia caused by encapsulated microorganisms. In these patients, there is a short prodromal phase, followed by septic shock which may be associated with disseminated intravascular coagulopathy (DIC).
- OPSI is known to be associated with 60–70% mortality if not diagnosed early and appropriately treated.
- Vascular complications were also observed following splenectomy. These are mostly seen in patients with hematologic diseases like beta-thalassemia.
  - In these patients, there have been reports of increased incidences of:
    Stroke
    Myocardial infarction
    Coronary arterial disease
  - These vascular complications could be due to chronic inflammation and dysfunction of the endothelium. The destruction of the endothelial lining could lead to platelet activation and thrombosis.
- Patients with asplenia are at increased risk of pulmonary hypertension.
  - The incidence of pulmonary hypertension in patients with asplenia is 8–11.5%.
  - This is even higher in patients with hemoglobinopathies following splenectomy.
  - In patients with sickle cell anemia or beta-thalassemia, the incidence of pulmonary hypertension could be as high as 30%.
- Patients with asplenia are more prone to develop adrenal hemorrhage and Waterhouse-Friedrichsen syndrome and purpura fulminans.

8.10 Prognosis

- The prognosis of patients with asplenia is poor if left unrecognized and untreated.
- These patients are at increased risk of developing serious overwhelming infections.
- These patients and their families should be aware of this risk. They should also report to the hospital if they develop febrile illnesses.
- It is important for these patients to receive proper vaccination.
- Antibiotic prophylaxis and vaccinations are the recommended and should be given to patients with asplenia.

- Patients with asplenia who do not receive immunizations are at a high risk of developing severe bacterial infection, sepsis, septic shock, and death.
- In patients with asplenia, the main aim of management should be directed at preventive measures to avoid overwhelming sepsis.
- Patients with asplenia should be educated thoroughly and they need to understand the risks and complications of asplenia, especially the risk of serious overwhelming infections which can be fatal.
- Vaccinations against pneumococcus, meningitis, Haemophilus, and influenza and the use of prophylactic antibiotics namely amoxicillin and penicillin need to be emphasized to these patients and their families.
- Although immunization and prophylactic antibiotics are effective, they do not provide complete protection.
- Children with asplenia and their families must be educated about the risk of sepsis and instructed to seek medical attention promptly when the child is ill or has a fever.
- The importance of using prophylactic antibiotics and vaccines should be emphasized repeatedly.
- It is also recommended that patients with asplenia should wear a Medic Alert bracelet. This will help alert the treating physicians and help them initiate treatment promptly.
- Every effort should be made to preserve the spleen during emergency operations for trauma.
- Non-surgical management of splenic trauma should be attempted whenever possible.
- It is important to preserve the spleen as this will promote a healthy immune system, and avoid the infectious and thromboembolic complications often seen in patients with asplenia.
- These procedures include:
  - Repair of traumatized spleen (Splenorrhaphy)
  - Partial splenectomy
  - Angiographic embolization
- Asplenia syndrome:
  - This is a syndrome of right isomerism or attempted bilateral right sidedness.
  - Asplenia syndrome is associated with:
    Complex congenital heart disease
    Situs anomalies of other thoraco-abdominal organs
  - The reported incidence of asplenia syndrome is about 1% of newborns with symptomatic congenital heart defects.
  - The most common associated congenital heart defects seen in patients with asplenia syndrome include the followings:
    Pulmonary stenosis
    Transposition of the great arteries
    Total Anomalous Pulmonary Venous Connection (TAPVC)
  - Heterotaxia is the discordant alignment of different organs is considered a characteristic feature of asplenia syndrome.

- Apart from complete forms of laterality defects, such as the asplenia syndrome, partial situs inversus defects have been described which probably represent the other end of a spectrum of laterality defects.
- Other associated anomalies include:
  - Intestinal malrotation
  - Biliary atresia
  - Splenic abnormalities
  - Defective gastric suspension mechanisms
  - Displacement of abdominal viscera
  - Aberrant vascular structures and vascular connections
- Although most cases of heterotaxia/asplenia syndrome are sporadic, many cases are familial, and some of them are X linked.
- Asplenia syndrome is more common in males (60%) than in females.
- Once the diagnosis of asplenia is made, antibiotics prophylaxis and vaccination with pneumococcal polysaccharide, Haemophilus influenzae type b and meningococcal conjugate vaccine are the most important steps to prevent serious infections.
- Associated congenital anomalies require a multidisciplinary approach and these anomalies are also responsible for increased morbidity and mortality in these patients.

- Asplenia (and polysplenia) may be sporadic or familial (autosomal dominant).
- Neonates with congenital asplenia have high morbidity and mortality rates usually caused by associated severe cardiovascular anomalies.
- Infants, who have asplenia as part of heterotaxy syndromes, often have increased mortality and morbidity. This mostly due to the associated congenital abnormalities in neonatal period rather than asplenia. These patients if they survive past age 1 month, they have a higher risk of dying from sepsis than from the associated cardiac anomalies.
- Asplenia syndrome (Ivemark syndrome) is more common in males with male to female ratio of 2:1.
- Polysplenia syndrome is more common in females, whereas asplenia is more common in males.
- All patients with congenital or acquired asplenia or splenic dysfunction are at significant risk of fulminant bacteremia, especially from encapsulated bacteria. This must be kept in mind all the time.
- The primary care physician plays an important role in the diagnosis and long-term follow up and treatment of patients with asplenia.
- Serious infections must be identified as early as possible and managed promptly and aggressively.
- The most important component in the treatment of pediatric patients with asplenia is parent education. They should be instructed to seek medical advice at the first sign of illness.
- "Asplenia" or "No Spleen." Medic Alert bracelet should be worn by asplenic children.
- Written instructions on a card should be given to the parents in a form that they can keep with them.

Further Reading

1. Paterson A, Frush DP, Donnelly LF, Foss JN, O'Hara SM, Bisset GS 3rd. A pattern-oriented approach to splenic imaging in infants and children. Radiographics. 1999;19(6):1465–85.
2. Mikoluc B, Michalkiewicz J, Motkowski R, Smolka D, Pietrucha B, Piotrowska-Jastrzebska J, et al. Neutrophil phenotypic characteristics in children with congenital asplenia and splenectomized for hereditary spherocytosis. Immunol Investig. 2012;41(1):61–74.
3. Mikoluc B, Motkowski R, Käyhty H, Heropolitanska-Pliszka E, Pietrucha B, Bernatowska E. Antibody response to Haemophilus influenzae type-b conjugate vaccine in children and young adults with congenital asplenia or after undergoing splenectomy. Eur J Clin Microbiol Infect Dis. 2012;31(5):805–9.
4. Motkowski R, Michalkiewicz J, Mikoluc B, Smolka-Afifi D, Pietrucha B, Kubiszewska I, et al. Peripheral blood T lymphocyte subsets in children with congenital asplenia. Hum Immunol. 2012;73(11):1091–7. Morgan TL, Tomich EB. Overwhelming post-splenectomy infection (OPSI): a case report and review of the literature. J Emerg Med. 2012 Oct 43(4):758-63.
5. Uchida Y, Matsubara K, Wada T, Oishi K, Morio T, Takada H, et al. Recurrent bacterial meningitis by three different pathogens in an isolated asplenic child. J Infect Chemother. 2012;18(4):576–80.
6. Konda S, Zell D, Milikowski C, Alonso-Llamazares J. Purpura fulminans associated with Streptococcus pneumoniae septicemia in an asplenic pediatric patient. Actas Dermosifiliogr. 2013;104(7):623–7.
7. Committee on Infectious Diseases. Meningococcal conjugate vaccines policy update: booster dose recommendations. Pediatrics. 2011;128(6):1213–8.
8. Angelski CL, McKay E, Blackie B. A case of functional asplenia and pneumococcal sepsis. Pediatr Emerg Care. 2011;27(7):639–41.
9. Casey B, Devoto M, Jones KL, Ballabio A. Mapping a gene for familial situs abnormalities to human chromosome Xq24-q27.1. Nat Genet. 1993;5(4):403–7.
10. Chellapandian D, Schneider A. Anatomical asplenia in cat eye syndrome: an expansion of the disease spectrum. Case Rep Pediatr. 2013;2013:218124.
11. Committee on Infectious Diseases American Academy of Pediatrics. Red book: 2006 report of the committee on infectious diseases. 27th ed.; 2006.
12. Couturier AP, Dahl K. Bordetella holmesii endocarditis: case report and review of literature. Pediatr Infect Dis J. 2014;33(6):661–4.
13. Elsayes KM, Narra VR, Mukundan G, et al. MR imaging of the spleen: spectrum of abnormalities. Radiographics. 2005;25(4):967–82.
14. Escobar-Diaz MC, Friedman K, Salem Y, Marx GR, Kalish BT, Lafranchi T, et al. Perinatal and infant outcomes of prenatal diagnosis of heterotaxy syndrome (asplenia and polysplenia). Am J Cardiol. 2014;114(4):612–7.
15. Fremont RD, Rice TW. Splenosis: a review. South Med J. 2007;100(6):589–93.
16. Garson A, Bricker JT, McNamara DG. The science and practice of pediatric cardiology, vol. 3. Baltimore: Williams & Wilkins; 1990. p. 1288–97.
17. Gaschignard J, Levy C, Chrabieh M, Boisson B, Bost-Bru C, Dauger S, et al. Invasive pneumococcal disease in children can reveal a primary immunodeficiency. Clin Infect Dis. 2014;59(2):244–51.
18. Gill DG, Kara M, Moore L. Septicaemia and adrenal haemorrhage in congenital asplenia. Arch Dis Child. 1991;66:1366.
19. Gonzalez P, Humeres P, Arroyo A. Liver uptake with phytate colloid in normal subjects and patients with cirrhosis: a possible alternative to extrahepatic measurements. J Nucl Med. 1991;32(7):1467–8.
20. Gorg C, Eickhorn M, Zugmaier G. The small spleen: sonographic patterns of functional hyposplenia or asplenia. J Clin Ultrasound. 2003;31(3):152–5.
21. Harrod VL, Howard TA, Zimmerman SA, Dertinger SD, Ware RE. Quantitative analysis of Howell-Jolly bodies in children with sickle cell disease. Exp Hematol. 2007;35(2):179–83.

22. Katcher AL. Familial asplenia, other malformations, and sudden death. Pediatrics. 1980;65(3):633–5.
23. Kevy SV, Tefft M, Vawier GF, Rosen FS. Hereditary splenic hypoplasia. Pediatrics. 1968;42(5):752–7.
24. Kim SJ. Heterotaxy syndrome. Korean Circ J. 2011;41(5):227–32.
25. Lane PA, O'Connell JL, Lear JL, et al. Functional asplenia in hemoglobin SC disease. Blood. 1995;85(8):2238–44.
26. Leahy RT, Philip RK, Gibbons RJ, et al. Asplenia in ATR-X syndrome: a second report. Am J Med Genet A. 2005;139(1):37–9.
27. Lin XZ, Chang TM, Tsai HM, et al. Liver, spleen and tumor volume measured by personal computer. Hepatogastroenterology. 1999;46(26):838–42.
28. Lindor NM, Smithson WA, Ahumada CA, et al. Asplenia in two father-son pairs. Am J Med Genet. 1995;56(1):10–1.
29. Long WA. Ivemark syndrome (cardiosplenic or heterotaxy syndrome). Fetal Neonatal Cardiol. 1990:616–9.
30. Mahlaoui N, Minard-Colin V, Picard C, Bolze A, Ku CL, Tournilhac O, et al. Isolated congenital asplenia: a French nationwide retrospective survey of 20 cases. J Pediatr. 2011;158(1):142–8, 148.e1.
31. Mebius RE, Kraal G. Structure and function of the spleen. Nat Rev Immunol. 2005;5(8):606–16.
32. Melles DC, de Marie S. Prevention of infections in hyposplenic and asplenic patients: an update. Neth J Med. 2004;62(2):45–52.
33. Moller JH, Neal WA. Congenital cardiac anomalies. Fetal Neonatal Inf Cardiac Dis. 1990:767–71.
34. Morgan MS. Prophylaxis should be considered even for trivial animal bites. BMJ. 1997;314(7091):1413.
35. Panagopoulos MI, Saint Jean M, Brun D, Guiso N, Bekal S, Ovetchkine P, et al. Bordetella holmesii bacteremia in asplenic children: report of four cases initially misidentified as Acinetobacter lwoffii. J Clin Microbiol. 2010;48(10):3762–4.
36. Pearson HA. The spleen and disturbances of splenic function. Hematology. 1993;2:1058–9.
37. Pollak U, Bar-Sever Z, Hoffer V, Marcus N, Scheuerman O, Garty BZ. Asplenia and functional hyposplenism in autoimmune polyglandular syndrome type 1. Eur J Pediatr. 2009;168(2):233–5.
38. Price VE, Blanchette VS, Ford-Jones EL. The prevention and management of infections in children with asplenia or hyposplenia. Infect Dis Clin N Am. 2007;21(3):697–710, viii-ix.
39. Price VE, Dutta S, Blanchette VS, et al. The prevention and treatment of bacterial infections in children with asplenia or hyposplenia: practice considerations at the Hospital for Sick Children, Toronto. Pediatr Blood Cancer. 2006;46(5):597–603.
40. Rose V, Izukawa T, Moes CA. Syndromes of asplenia and polysplenia. A review of cardiac and non- cardiac malformations in 60 cases with special reference to diagnosis and prognosis. Br Heart J. 1975;37(8):840–52.
41. Stiehm ER. Immunodeficiency disorders. In: Immunologic disorders in infants and children. 4th ed. Philadelphia: W.B. Saunders; 1996. p. 326–7.
42. Strouse PJ, Haller JO, Berdon WE, Rosovsky MA, Bellah RD, Barr M Jr. Horseshoe adrenal gland in association with asplenia: presentation of six new cases and review of the literature. Pediatr Radiol. 2002;32(11):778–82.
43. Tham KT, Teague MW, Howard CA, Chen SY. A simple splenic reticuloendothelial function test: counting erythrocytes with argyrophilic inclusions. Am J Clin Pathol. 1996;105(5):548–52.
44. Tribioli C, Lufkin T. The murine Bapx1 homeobox gene plays a critical role in embryonic development of the axial skeleton and spleen. Development. 1999;126(24):5699–711.
45. Updated recommendation from the Advisory Committee on Immunization Practices (ACIP) for revaccination of persons at prolonged increased risk for meningococcal disease. MMWR Morb Mortal Wkly Rep. 2009;58(37):1042–3.
46. Updated recommendations on the use of meningococcal vaccines. Pediatrics. 2014;134(2):400–3.

47. US Preventive Services Task Force. Screening for sickle cell disease in newborns: recommendation statement. Am Fam Physician. 2008;77(9):1300–2.
48. Use of 13-valent pneumococcal conjugate vaccine and 23-valent pneumococcal polysaccharide vaccine among children aged 6-18 years with immunocompromising conditions: recommendations of the Advisory Committee on Immunization Practices (ACIP). MMWR Morb Mortal Wkly Rep. 2013;62(25):521–4.
49. Waldman JD, Rosenthal A, Smith AL, et al. Sepsis and congenital asplenia. J Pediatr. 1977;90(4):555–9.
50. Wang JK, Hsieh KH. Immunologic study of the asplenia syndrome. Pediatr Infect Dis J. 1991;10(11):819–22.
51. Wolla CD, Hlavacek AM, Schoepf UJ, Bucher AM, Chowdhury S. Cardiovascular manifestations of heterotaxy and related situs abnormalities assessed with CT angiography. J Cardiovasc Comput Tomogr. 2013;7(6):408–16.
52. Wong T, Yeung J, Hildebrand KJ, Junker AK, Turvey SE. Human primary immunodeficiencies causing defects in innate immunity. Curr Opin Allergy Clin Immunol. 2013;13(6):607–13.
53. Woods CR. Prevention of meningococcal infections in the first 2 years of life. Pediatr Ann. 2013 Aug;42(8):164–71.
54. Yamamura K, Joo K, Ohga S, Nagata H, Ikeda K, Muneuchi J, et al. Thrombocytosis in asplenia syndrome with congenital heart disease: a previously unrecognized risk factor for thromboembolism. Int J Cardiol. 2013;167(5):2259–63.
55. Salvadori MI, Price VE, Canadian Paediatric Society, Infectious Diseases and Immunization Committee. Preventing and treating infections in children with asplenia or hyposplenia. Paediatr Child Health. 2014;19(5):271–8.
56. Singer DB. Postsplenectomy sepsis. Perspect Pediatr Pathol. 1973;1:285–311.
57. Holdwoth RJ, Irving AD, Cuschieri A. Postsplenectomy sepsis and its mortality rate: actual versus perceived risks. Br J Surg. 1991;78(9):1031–8.
58. Bisharat N, Omari H, Lavi I, Raz R. Risk of infection and death among post-splenectomy patients. J Infect. 2001;43(3):182–6.
59. Lion C, Escande F, Burdin JC. Capnocytophaga canimorsus infections in humans: review of the literature and cases report. Eur J Epidemiol. 1996;12(5):521–33.
60. Davies JM, Barnes R, Milligan D, British Committee for Standards in Haematology, Working Party of the Haematology/Oncology Task Force. Update of guidelines for the prevention and treatment of infection in patients with an absent or dysfunctional spleen. Clin Med. 2002;2(5):440–3.
61. Krause PJ, Gewurz BE, Hill D, et al. Persistent and relapsing babesiosis in immunocompromised patients. Clin Infect Dis. 2008;46(3):370–6.
62. O'Brien KL, Hochman M, Goldblatt D. Combined schedules of pneumococcal conjugate and polysaccharide vaccines: is hyporesponsiveness an issue? Lancet Infect Dis. 2007;7(9):597–606.
63. Shatz DV, Schinsky MF, Pais LB, Romero-Steiner S, Kirton OC, Carlone GM. Immune responses of splenectomized trauma patients to the 23-valent pneumococcal polysaccharide vaccine at 1 versus 7 versus 14 days after splenectomy. J Trauma. 1998;44(5):760–5.
64. American Academy of Pediatrics. Immunization in special clinical circumstances. In: Pickering LK, Kimberlin DW, Long SS, editors. Red book: 2012 report of the committee on infectious diseases. 29th ed. Elk Grove Village, IL: AAP; 2012.
65. Bach O, Baier M, Pullwitt A, et al. Falciparum malaria after splenectomy: a prospective controlled study of 33 previously splenectomized Malawian adults. Trans R Soc Trop Med Hyg. 2005;99(11):861–7.
66. Public Health Agency of Canada. Canadian recommendations for the prevention and treatment of malaria among international travellers – 2009: supplement. Can Commun Dis Rep. 2009;35(Suppl 1):1–8.
67. O'Neal HR, Niven AS, Karam GH. Critical illness in patients with asplenia. Chest. 2016;150(6):1394–402.

68. Swirski FK, Nahrendorf M, Etzrodt M, Wildgruber M, Cortez-Retamozo V, Panizzi P, Figueiredo JL, Kohler RH, Chudnovskiy A, Waterman P, Aikawa E, Mempel TR, Libby P, Weissleder R, Pittet MJ. Identification of splenic reservoir monocytes and their deployment to inflammatory sites. Science. 2009;325(5940):612–6.
69. Taher A, Isma'eel H, Mehio G, Bignamini D, Kattamis A, Rachmilewitz EA, Cappellini MD. Prevalence of thromboembolic events among 8,860 patients with thalassaemia major and intermedia in the Mediterranean area and Iran. Thromb Haemost. 2006;96(4):488–91.
70. Schilling RF. Spherocytosis, splenectomy, strokes, and heat attacks. Lancet. 1997;350(9092):1677–8.
71. Phrommintikul A, Sukonthasarn A, Kanjanavanit R, Nawarawong W. Splenectomy: a strong risk factor for pulmonary hypertension in patients with thalassaemia. Heart. 2006;92(10):1467–72.
72. Hale AJ, LaSalvia M, Kirby JE, Kimball A, Baden R. Fatal purpura fulminans and Waterhouse-Friderichsen syndrome from fulminant Streptococcus pneumoniae sepsis in an asplenic young adult. IDCases. 2016;6:1–4.
73. Katz SC, Pachter HL. Indications for splenectomy. Am Surg. 2006;72(7):565–80.
74. Rosati C, Ata A, Siskin GP, Megna D, Bonville DJ, Stain SC. Management of splenic trauma: a single institution's 8-year experience. Am J Surg. 2015;209(2):308–14.
75. Hosey RG, Mattacola CG, Kriss V, Armsey T, Quarles JD, Jagger J. Ultrasound assessment of spleen size in collegiate athletes. Br J Sports Med. 2006;40(3):251–4; discussion 251-4.
76. Suttie AW. Histopathology of the spleen. Toxicol Pathol. 2006;34(5):466–503.
77. Crary SE, Buchanan GR. Vascular complications after splenectomy for hematologic disorders. Blood. 2009;114(14):2861–8.
78. Erdem SB, Genel F, Erdur B, Ozbek E, Gulez N, Mese T. Asplenia in children with congenital heart disease as a cause of poor outcome. Cent Eur J Immunol. 2015;40(2):266–9.
79. Huebner ML, Milota KA. Asplenia and fever. Proc (Bayl Univ Med Cent). 2015;28(3):340–1.
80. DeFrances CJ, Cullen KA, Kozak LJ. National Hospital Discharge Survey: 2005 annual summary with detailed diagnosis and procedure data. Vital Health Stat 13. 2007;(165):1–209.
81. Rose AT, Newman MI, Debelak J, Pinson CW, Morris JA, Harley DD, Chapman WC. The incidence of splenectomy is decreasing: lessons learned from trauma experience. Am Surg. 2000;66(5):481–6.
82. Masiwal P, Chenthil KS, Priyadarsini B, Gnanaprakasam J, Srihari I. Ivemark syndrome. J Assoc Physicians India. 2016;64(5):73–5.
83. Kirkineska L, Perifanis V, Vasiliadis T. Functional hyposplenism. Hippokratia. 2014;18(1):7–11.
84. William BM, Corazza GR. Hyposplenism: a comprehensive review. Part I: basic concepts and causes. Hematology. 2007;12(1):1–13.
85. Waghorn DJ. Overwhelming infection in asplenic patients: current best practice preventive measures are not being followed. J Clin Pathol. 2001;54(3):214–8.
86. Rubin LG, Schaffner W. Clinical practice. Care of the asplenic patient. N Engl J Med. 2014;371(4):349–56.
87. Rhodes A, Evans LE, Alhazzani W, Levy MM, Antonelli M, Ferrer R, Kumar A, Sevransky JE, Sprung CL, Nunnally ME, Rochwerg B, Rubenfeld GD, Angus DC, Annane D, Beale RJ, Bellinghan GJ, Bernard GR, Chiche JD, Coopersmith C, De Backer DP, French CJ, Fujishima S, Gerlach H, Hidalgo JL, Hollenberg SM, Jones AE, Karnad DR, Kleinpell RM, Koh Y, Lisboa TC, Machado FR, Marini JJ, Marshall JC, Mazuski JE, McIntyre LA, McLean AS, Mehta S, Moreno RP, Myburgh J, Navalesi P, Nishida O, Osborn TM, Perner A, Plunkett CM, Ranieri M, Schorr CA, Seckel MA, Seymour CW, Shieh L, Shukri KA, Simpson SQ, Singer M, Thompson BT, Townsend SR, Van der Poll T, Vincent JL, Wiersinga WJ, Zimmerman JL, Dellinger RP. Surviving sepsis campaign: international guidelines for management of sepsis and septic shock: 2016. Intensive Care Med. 2017;43(3):304–77.

9 Polysplenia

9.1 Introduction

- Polysplenia is the presence of multiple spleens.
- This is a rare condition (Fig. 9.1).
- It is commonly seen associated with extrahepatic biliary atresia.
- Some of the spleens in polysplenia are large in size while others are small in size.
- It is a rare congenital anomaly which was initially described by Helwig in 1929.
- Since then, few cases were described in the literature with an incidence of approximately 1/250,000 live births.
- There are known associated anomalies with polysplenia and these include:
  - Intestinal malrotation

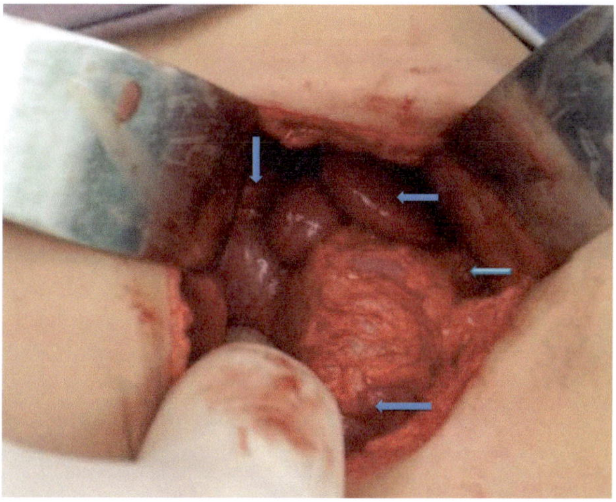

Fig. 9.1 Intraoperative photograph in a patient with polysplenia. Note the variable size of these spleniculi. This patient was a case of biliary atresia

A. H. Al-Salem, *The Spleen*, https://doi.org/10.1007/978-981-99-6191-7_9

  - Situs inversus
  - Biliary atresia
  - Congenital cardiac malformations
- Heterotaxia is generally classified into 2 major syndromes, polysplenia and asplenia syndromes.
  - Polysplenia syndrome is a rare congenital subtype of heterotaxia syndrome and it is associated with various visceral and vascular abnormalities.
  - It is characterized by presence of 2 or more spleens and anomalies of other asymmetric organs.
  - Polysplenia syndrome refers to the association of 2 or more multiple spleens with multiple congenital abnormalities in abdomen and chest.
  - However, some cases of polysplenia syndrome have been described where there is a single bilobed spleen or a single normal spleen.
  - In all cases of polysplenia, the spleens are located on the same side of the stomach along the greater curvature of the stomach. This is related to the embryonic development of the spleen in the posterior mesogastrium.

9.2 Polysplenia

- Polysplenia is a very rare condition that was initially described by Helwig in 1929.
- Polysplenia is characterized by the presence of more than one spleen in the abdominal cavity.
- Polysplenia is characterized by the presence of multiple small accessory spleens, rather than a single, full-sized, normal spleen.
- Polysplenia is seen more commonly in female patients.
- The estimated incidence of polysplenia is approximately 1–1.5/10,000 live births.
- Polysplenia is more common in Asia than in North America and Europe.
- It is also more common in children born to Black or Hispanic mothers than in children born to white mothers.
- Polysplenia Syndrome is a very rare congenital anomaly characterized by abdominal, cardiac, vascular and thoracic abnormalities.
- Polysplenia Syndrome is a very rare congenital malformation. The estimated incidence is approximately 2.5:100,000 live births.
- It is generally diagnosed in early childhood due to associated and severe cardiac abnormalities.
- The exact cause of polysplenia has not been clearly defined.
- In asymptomatic patients, polysplenia may be found incidentally during abdominal surgery for other unrelated conditions or during radiological examination.
- Polysplenia sometimes occurs alone, but it is often accompanied by other developmental abnormalities.
- Abnormalities associated with polysplenia include:
  - Gastrointestinal abnormalities, such as intestinal malrotation or biliary atresia.
  - Cardiac abnormalities, such as dextrocardia.

- Polysplenia syndrome:
  - In polysplenia syndrome there are frequently associated congenital anomalies which are related to abnormalities that are related to deviations in the development of anatomical asymmetries in early embryonic stages.
  - The associated abdominal anomalies include:
    - Heterotaxy syndrome
    - Intestinal malrotation
    - Situs inversus
    - Biliary atresia
  - Cardiac malformations include:
    - Dextrocardia
    - Atrial situs ambiguous
    - Ventricular inversion
    - VA concordance with left posterior aorta
- Polysplenia is also called heterotaxy syndrome or bilateral left-sidedness and it is characterized by the absence of the main spleen and the presence of multiple small accessory spleens.
- The associated severe cardiovascular malformations contribute to the early mortality of these patients. Most patients with polysplenia syndrome die by the age of 5 years.
  - 50% of children with polysplenia syndrome die by 4 months of age.
  - 75% of children with polysplenia syndrome die before 5 years of age.
  - Patients with polysplenia syndrome often present with atrioventricular septal defects.
  - Approximately 5–10% of patients with polysplenia syndrome have normal hearts or only minor cardiac anomalies. In these patients polysplenia may first be discovered in adulthood.
  - Polysplenia syndrome in these patients generally causes no symptoms until adulthood. These patients are often diagnosed incidentally as a result of radiological investigations performed for other reasons.
  - Polysplenia syndrome is known to have a high mortality. This is related to the associated cardiac anomalies, biliary atresia or intestinal atresia.
    - 50% of these patients may survive till they are 4-months old.
    - 25% of polysplenia syndrome patients may survive until they are 5 years of age.
    - Patients with minor cardiac anomalies or those without any cardiac anomalies comprise 10% of all patients with polysplenia syndrome. These patients represent the group of patients with polysplenia syndrome that will survive till their adulthood.
  - Severe cardiovascular anomalies contribute to mortality and these include:
    - Interruption of the suprarenal inferior vena cava
    - Atrioventricular septal defects
    - Ipsilateral pulmonary venous drainage
    - Ventricular outflow tract obstruction
    - Dextrocardia
    - Abnormal great vessel relationships

- The most frequent abnormality detected in polysplenia syndrome is the presence of multiple spleens.
  - The spleens are variable in size ranging from 1-6 cm in dimension
  - They can be located at the right side or left side of the abdomen but they are most commonly found on the left side of the abdominal cavity.
  - In relation to the stomach, generally they are adjacent to the greater curvature of the stomach.
- The second frequent abnormality detected in polysplenia syndrome is vascular anomalies.
  - The continuation of the inferior vena cava, azygos vein or hemiazygos vein and the total absence of the hepatic segment are the most frequently encountered vascular anomalies.
  - They can be seen in 50-60% of the cases.
  - The main problem is the absence of the vena cava inferior suprarenal segment, which drains into the superior vena cava instead of the right atrium as it shows a continuation with the expanded azygos or hemiazygos veins located at infrarenal vena cava.
- In Polysplenia cases where a left isomerism is seen the splenic tissue is always located alongside the greater curvature of the stomach.
- In these patients, a short pancreas forms due to developmental failure of the tail and corpus of the pancreas as a result of an undeveloped dorsal bud during the development stage of the pancreas.
- Situs ambiguous or heterotaxy is characterized by abnormal arrangement of organs and vessels as opposed to the orderly arrangement typically seen in situs solitus and situs inversus.

9.3 Heterotaxia syndrome (Figs. 9.2, 9.3 and 9.4)

- The term "heterotaxy" is from the Greek words "heteros," meaning "other than," and "taxis," meaning "arrangement."
- Heterotaxy syndrome results from complex birth defects affecting the heart, lungs, liver, spleen, intestines, and other organs.
- Heterotaxy is defined as a congenital abnormality where the internal thoraco-abdominal organs are abnormally arranged across the left- right axis of the body.
- Heterotaxy syndrome can affect the structure of the heart and the attachment of the large blood vessels.
- There are three types of splenic anomalies found in patients with heterotaxy syndrome:
  - Congenital asplenia where the spleen is absent.
  - Polysplenia where there are a number of small spleens or a large spleen may be found together with several smaller spleens, or the spleen may be made up of more than one lobe (multilobed spleen).

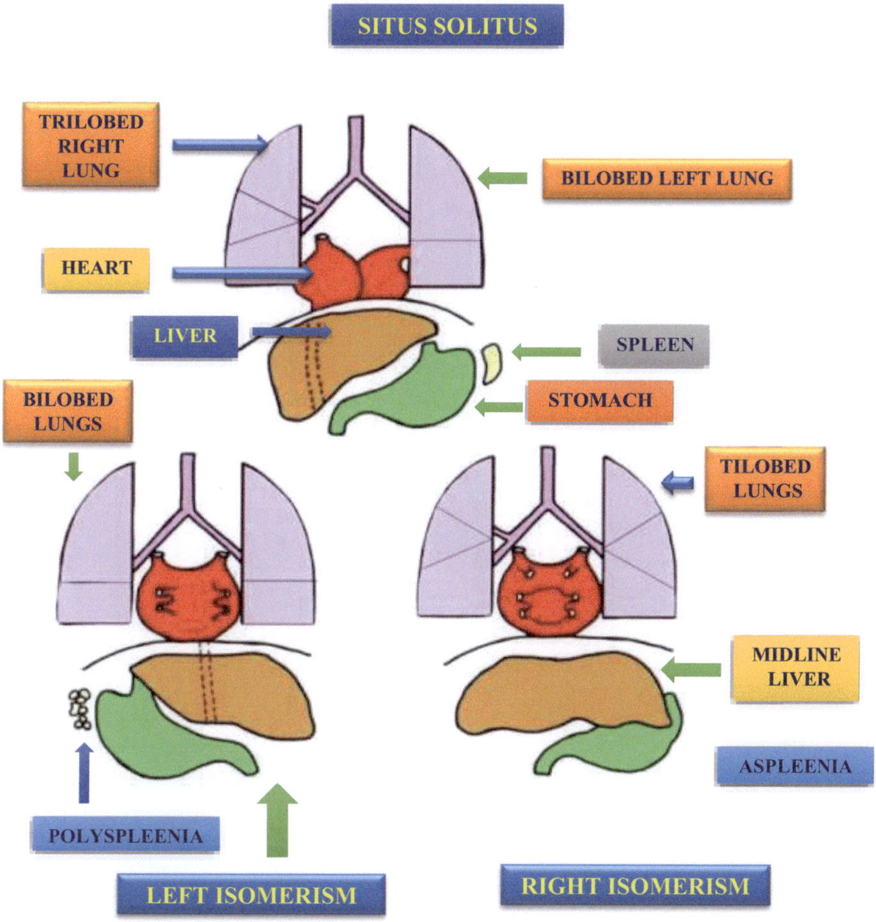

Fig. 9.2 Diagrammatic representation of situs solitus, right isomerism and left isomerism. **Situs solitus** refers to the normal position of the thoracic and abdominal organs in the body. This will include a left-sided heart, also known as levocardia. Anatomically, this means that the heart is on the left with the pulmonary atrium on the right and the systemic atrium on the left along with the cardiac apex. Right-sided organs are the liver, the gall bladder and a trilobed lung as well as the inferior vena cava, while left-sided organs are the stomach, single spleen, a bilobed lung, and the aorta. Right isomerism is a severe complex congenital heart defect resulting from embryonic disruption of proper left-right axis determination. Right isomerism is usually characterized by complex atrioventricular septal defect with a common atrium and univentricular AV connection, total anomalous pulmonary drainage, and transposition or malposition of the great arteries. Other associated abnormalities include bilateral trilobed lungs, midline liver, and asplenia, as well as situs inversus affecting other organs. Left isomerism, also called polysplenia, is associated with paired left-sidedness viscera (left atrial appendages, bilobed lungs and long bronchi) and multiple small spleens. Those patients with isomeric left atrial appendages frequently have bilaterally bilobed lungs, each with a long bronchus, and multiple spleens

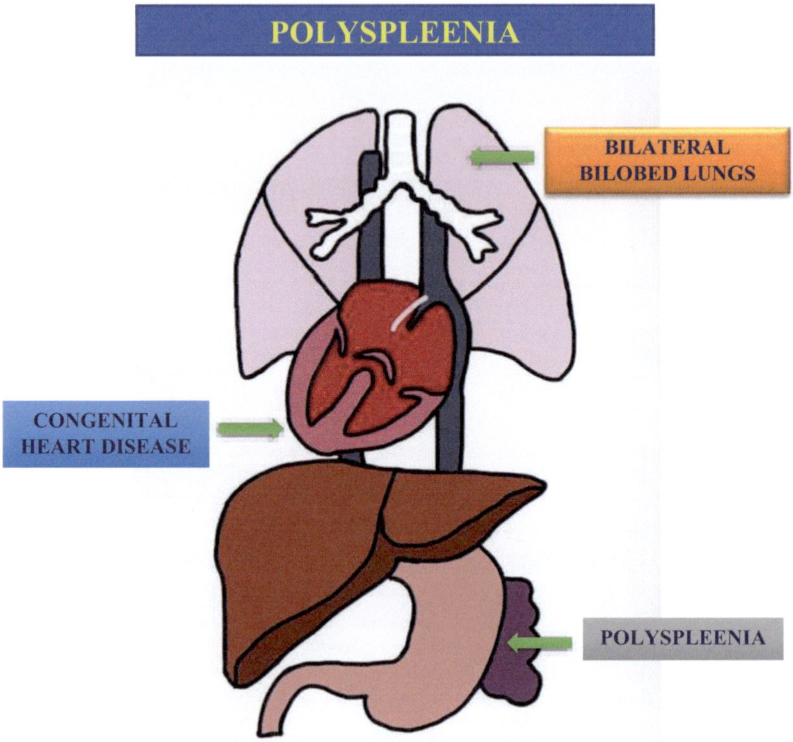

Fig. 9.3 Diagrammatic representation of polysplenia. Polysplenia is characterized by bilateral bilobed lungs and bilateral hyparterial bronchi and multiple small rounded spleens. Congenital heart anomalies occur in 75% of these patients and is usually mild

  - The spleen may be of normal size but located in the right upper quadrant of the abdomen rather than the normal location in the left upper quadrant of the abdomen.
- The prognosis of patients with complex cardiac lesions and heterotaxy is poor.
  - The 1-year mortality is >85% for patients with asplenia.
  - The 1-year mortality is >50% for patients with polysplenia.
- Heterotaxy is generally classified into 2 major syndromes:
  - Polysplenia syndrome (Left Isomerism)
  - Asplenia syndromes (Right Isomerism)
- Long-term outcome of polysplenia heterotaxy syndrome is determined by the severity of the associated cardiac anomalies.
- Polysplenia syndrome is defined as a congenital malformation characterized by an abnormal morphology and position of the thoraco-abdominal organs. These organs do not coexist with the normal arrangement of body organs with the nor-

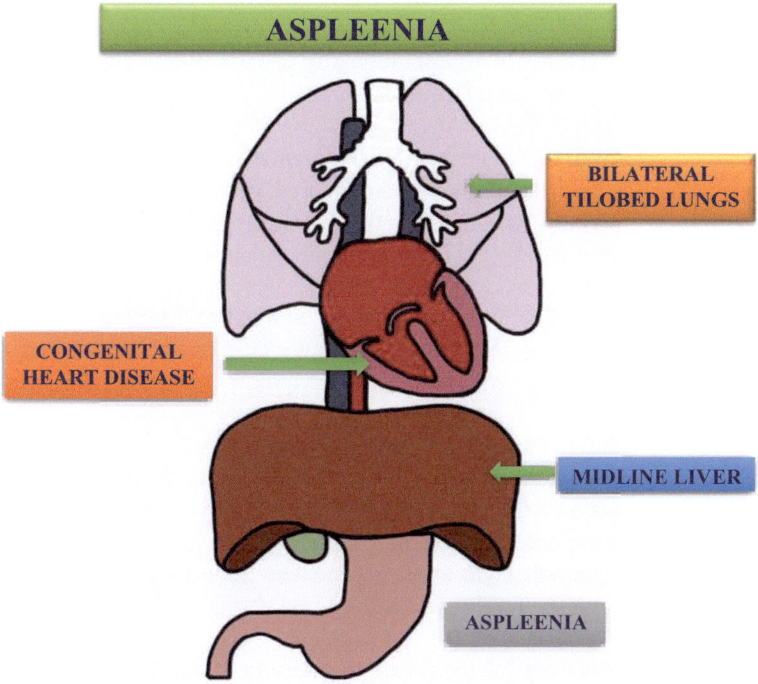

Fig. 9.4 Diagrammatic representation of asplenia. Asplenia is characterized by bilateral trilobed lungs and bilateral eparterial bronchi. Liver is often midline, spleen is absent, and position of stomach is variable. Congenital heart disease is nearly universal. Associated cardiac defects are generally severe and include common atrium and single ventricle with pulmonary stenosis or pulmonary atresia

mal left-right asymmetry (situs solitus) or the reversed/mirrored arrangement of the abdominal and thoracic organs (situs inversus).

- Patients with right isomerism (asplenia) have a higher incidence of congenital heart anomalies when compared with polysplenia. These anomalies include:
  - Univentricular circulation
  - Complete atrioventricular septal defect
  - Pulmonary atresia
  - Total anomalous pulmonary venous return
- Patients with left isomerism (polysplenia) have a high association of congenital gastrointestinal abnormalities. These anomalies include:
  - Intestinal malrotation
  - Biliary atresia
  - Intestinal volvulus
- The word isomerism is also derived from the Greek: iso-meaning equal and meros-meaning part.

- Isomerism refers to structures that are themselves mirror-imaged.
- The term "isomerism" has been used to describe the situation in which morphologically right structures or morphologically left structures are found on both sides of the body in the same individual.
- Left isomerism:
  - Left isomerism in the context of the congenitally malformed heart is defined as a subset of heterotaxy where some paired structures on opposite sides of the left-right axis of the body are symmetrical mirror images of each other, and have the morphology of the normal left-sided structures.
  - Patients with isomeric left atrial appendages frequently have bilaterally bilobed lungs, each with a long bronchus, and multiple spleens.
  - Patients with isomeric left appendages have pulmonary veins connecting to both atrial chambers as if both atriums were morphologically left atriums.
- Right isomerism:
  - Right isomerism in the context of the congenitally malformed heart is defined as a subset of heterotaxy where some paired structures on opposite sides of the left-right axis of the body are symmetrical mirror images of each other, and have the morphology of the normal right-sided structures.
  - Patients with isomeric right atrial appendages frequently have bilaterally trilobed lungs, each with a short bronchus, and absence of the spleen.
- Situs Ambiguous:
  - Situs ambiguous is defined as an abnormality in which there are components of situs solitus and situs inversus in the same person.
  - Situs ambiguous is present when the thoracic and abdominal organs are positioned with respect to each other in such a way that they are not clearly lateralized and thus have neither the usual, or normal, nor the mirror-imaged arrangements.
- All patients with heterotaxy should also be considered to have "situs ambiguous" and all patients with "situs ambiguous" should also have heterotaxy syndrome.
- Heterotaxy syndrome accounts for approximately 3 percent of all congenital heart anomalies.
- Situs solitus:
  - This is the normal arrangement of the thoracic and abdominal organs of the body.
  - The heart, spleen, stomach and pancreas are located on the left side of the body.
  - The liver is located on the right side of the body.
- Situs inversus:
  - In this condition, the orientation of the internal thoracic and abdominal organs is completely flipped from right to left.
  - The heart, spleen stomach and pancreas are located on the right side of the body.
  - The liver is located on the left side of the body.
- Situs ambiguous:
  - This is also called Heterotaxy syndrome.

– In this condition, the arrangement of internal thoracic and abdominal organs of the body is somewhere between situs solitus and situs inversus.
– The abnormal arrangement of body organs in heterotaxy syndrome often causes serious medical problems.

9.4 Associated Anomalies

- Polysplenia syndrome is also known as left isomerism, bilateral left-sidedness or Ivemark syndrome (Figs. 9.5, 9.6, 9.7 and 9.8).
- It is considered a type of heterotaxy syndrome.
- It is characterized by the presence of 2 or more multiple spleens as part of left-sided isomerism associated with various congenital visceral and vascular malformations.
- The following associated anomalies are known to be associated with polysplenia syndrome:
  – Multiple spleens (Polysplenia)
  – Cardiac anomalies:
    Ventricular septal defects
    Atrioventricular septal defects
    Outflow tract abnormalities
    Heart block

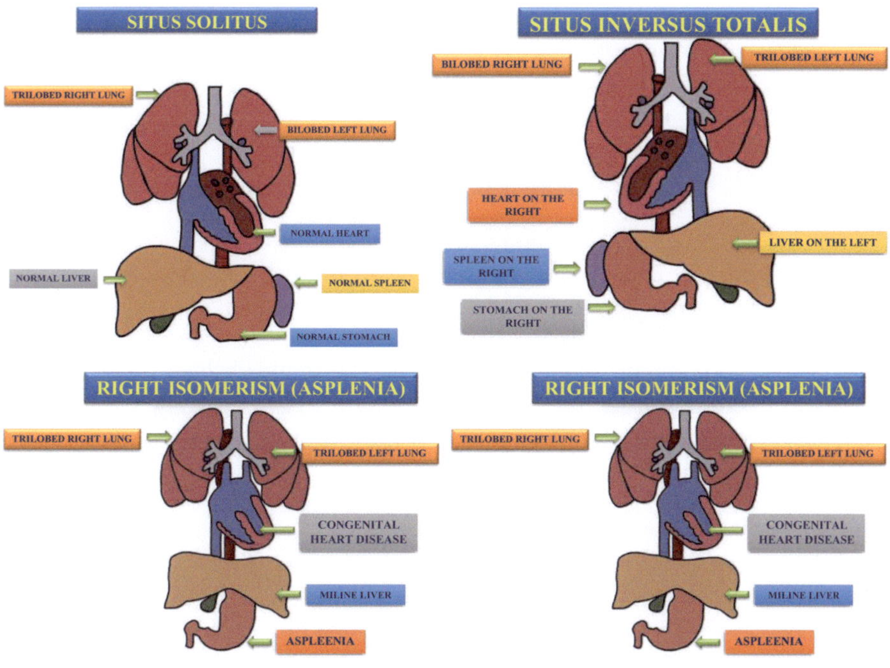

Figs. 9.5–9.8 Diagrammatic representations of situs solitus, situs inversus, polysplenia and asplenia

  - Bilateral bilobed lungs
  - A malpositioned stomach
  - A midline liver
  - Interruption of the inferior vena cava with the venous return from the lower body via the prominent azygos vein or hemiazygos vein.
- Congenital heart anomalies present in >50% of patients and include:
  - Atrial septal defect
  - Endocardial cushion defect
  - Bilateral left atrium
  - Ventricular septal defects
  - Double outlet right ventricle
  - Pulmonary stenosis
  - Outflow tract abnormalities
  - Heart block
  - Interruption of hepatic segment of inferior vena cava
  - Abnormal pulmonary venous return (70%, partial anomalous pulmonary venous return /total anomalous pulmonary venous return)
  - Dextrocardia (37%)
- Gastrointestinal anomalies:
  - Semi-annular pancreas/congenitally short pancreas
  - Midgut malrotation (80%)
  - Gallbladder agenesis (50%)
  - Biliary atresia
  - Mobile cecum
  - Tracheo-esophageal fistula
- Genitourinary anomalies:
  - Renal cyst
  - Renal agenesis
  - ovarian cyst
- Vascular anomalies:
  - Inferior vena cava (IVC) anomalies
  - Intrahepatic IVC interruption with azygos/hemiazygos continuation
  - Portal vein anomalies
  - Preduodenal portal vein
- Asplenia syndrome or right isomerism (bilateral right-sidedness), is characterized by the followings:
  - Agenesis of the spleen (Asplenia)
  - Paired right-sided viscera
  - Right atrial isomerism
  - Bilateral superior vena cava
  - Anomalous pulmonary venous connections
  - Atrioventricular septal defects
  - Outflow tract abnormalities
  - Bilateral trilobed lungs
  - Midline liver

9.5 Diagnosis

- The clinical diagnosis of polysplenia can be difficult and challenging because detection of the small spleens can be difficult by abdominal ultrasound.
- The presence of abnormal splenic tissue or multiple spleens can be detected along the greater curvature of the stomach with contrast enhancement abdominal CT scan.
- Polysplenia is highly variable and this may appear as multiple small nodules, a few small nodules, or a large lobulated spleen.
- Lack of any splenic tissue and the absence of a splenic artery and vein are suggestive of congenital asplenia.
- Bilobed spleens and accessory spleens can be seen in 10% of normal healthy people.
- In many patients with situs ambiguous, plain radiographs of the chest and abdomen provide the initial clues to the diagnosis, which should be suspected in patients with:
  - A midline liver
  - A right-sided or midline stomach
  - Mesocardia or dextrocardia
- Radiological investigations including ultrasonography, abdominal computed tomography (CT) scan, and abdominal magnetic resonance imaging (MRI) are valuable in the assessment and diagnosis of the various abnormalities in patients with heterotaxy.
- Abdominal CT scan is important to detect the location and numbers of spleens and to detect other anomalies in cases of heterotaxy syndrome.
- Definitive confirmation of polysplenia can be made by a radiocolloid scan using 99mTc-Sulfur-colloid which is taken by the ectopic splenic tissue.
- An echocardiogram or an electrocardiogram may also be useful to diagnose polysplenia.
- Plain chest radiography and chest ultrasonography can be used in patients with suspected congenital heart disease and to evaluate the mediastinal anatomy.
- Abdominal ultrasonography is valuable in determining the position of the stomach and this may also detect splenic tissue along the greater curvature of the stomach. These will appear as multiple, widely spaced spherules above the left kidney.
- MRI is particularly useful:
  - It can delineate cardiac defects.
  - It can detect the presence of spleen tissue.
  - It can roll out associated congenital abnormalities, such as biliary atresia and intestinal malrotation.
- Abdominal ultrasonography is also valuable in evaluating:
  - The position of the liver in the abdomen
  - The presence and location of the gallbladder
  - The hepatic venous anatomy
  - The course of the portal vein and whether it is preduodenal or retro duodenal.

  - The superior mesenteric vein extends anterior to the pancreas and duodenum and courses ventrally into the liver; on transverse images, its cross-section is seen anterior to the pancreas.
  - A dilated duodenum in cases of duodenal stenosis and/or atresia
  - Midline horseshoe adrenal gland
  - Renal anomalies
- Situs ambiguous is associated with major congenital anomalies which include:
  - Intestinal malrotation
  - Biliary atresia
  - Splenic abnormalities
  - Abnormal gastric suspension mechanisms
  - Displacement of abdominal viscera
  - Aberrant vascular structures and vascular connections
- Many infants with situs ambiguous present with severe congenital cardiac anomalies. In these patients it is important to know if an associated interruption of the inferior vena cava is present before performing cardiac catheterization.
- In patients with polysplenia, an upper gastrointestinal study demonstrates intestinal nonrotation and a right-sided stomach.
- A right-sided aortic arch is most frequently seen in patients with situs solitus.
- Rotational anomalies of the midgut are most often seen in patients with a solitus condition.
- Normally the liver and spleen are clearly identified with the use of technetium-99 m sulfur-colloid (^{99m}Tc-SC) scintigraphy.
  - When spherules are close together, the splenic mass appears as a single spleen, which is difficult to differentiate from situs solitus or situs inversus totals.
  - If the hepatic mass extends to both upper quadrants, identification of the splenic tissue may be difficult and necessitate selective splenic scintigraphy with radiolabeled, heat-damaged red blood cells and/or hepatobiliary scintigraphy.
  - In some patients, splenic activity may be clearly distinct from that of the liver if the spherules are clearly separated from the liver and from each other.
- Before the development of cross-sectional imaging techniques, angiography was used to diagnose heterotaxy. It is also useful to define cardiac anatomy, and demonstrate any abdominal viscerovascular heterotaxy.
- Currently, MRI and echocardiography are extensively used for the evaluation of cardiac malformations in these patients.

9.6 Treatment

- The treatment of polysplenia is organ specific and depends on which organ is affected and how severe the specific organ is affected.
- The treatment is also influenced by the nature and severity of the associated cardiac and non-cardiac anomalies.

- The associated cardiac anomalies may require surgery early to correct any heart defects.
- A Ladd's procedure may be necessary to correct an associated intestinal malrotation.
- A multidisciplinary team approach is recommended to follow, diagnose and treat patients diagnosed with heterotaxy.
- Patients with right isomerism (asplenia) have a higher incidence of univentricular circulation, complete atrioventricular septal defect, pulmonary atresia, and total anomalous pulmonary venous return compared to patients with left isomerism (polysplenia).
- Long-term outcome of heterotaxy syndrome is determined by the severity of the associated cardiac anomalies.
- Patients with heterotaxy syndrome are known to have a high association of congenital gastrointestinal abnormalities including:
  - Malrotation of the bowel
  - Biliary atresia
  - Intestinal volvulus
  - Splenic anomalies
- These anomalies are known to significantly affect the long-term survival in infants with the heterotaxy syndrome.
- Patients with polysplenia are known to have a variety of associated anomalies including:
  - Intestinal malrotation anomalies ranging from nonrotation to reversed rotation and faulty peritoneal attachments.
  - Intestinal rotation abnormalities include:

    Nonrotation: The small bowel is entirely on the right of the spine and the colon on the left.

    Incomplete rotation: The appearance is between normal and nonrotation.

    Reversed complete or incomplete rotation.

    In nonrotation and incomplete rotation, the mesenteric attachment is very narrow which predisposes these patients to the risk of midgut volvulus.
  - Duodenal atresia or stenosis
  - Biliary atresia
  - A short pancreas: This results from maldevelopment or agenesis of dorsal pancreatic bud.
  - A midline adrenal gland
  - Renal anomalies
  - Interruption of inferior vena cava has been reported in as many as 50-60% of patients with polysplenia.
  - In these patients, the venous return occurs via the right or left sided azygous systems.
  - In patients with intact inferior vena cava, both aorta and inferior vena cava can lie to one side of midline or on opposite sides.

  - Inferior vena cava (IVC) interruption with azygous continuation is the second most common abnormality observed in patients with polysplenia. The first abnormality is the presence of multiple spleens.
  - The portal vein is preduodenal.
  - A preduodenal portal vein is one of the anomalies seen in association with polysplenia.
  - Common cardiac anomalies in patients of polysplenia include atrioventricular septal defects.
  - A small number of patients with situs ambiguous also have central nervous system anomalies including:
    Holoprosencephaly
    Neural tube defects
    Caudal regression syndrome
- Heterotaxy syndromes are known to be associated with increased incidence of complex cardiac anomalies.
- Cardiac anomalies are generally less commonly seen in polysplenia syndrome than in patients with asplenia.
- These anomalies include:
  - Atrial septal defect
  - Ventricular septal defect
  - Bilateral superior vena cava
  - Right-sided aortic arch
  - Partial anomalous pulmonary venous return
  - Transposition of the great arteries
  - Pulmonary valvular stenosis
  - Subaortic stenosis

Further Reading

1. Lagrotta G, Moises M. Heterotaxy polysplenia syndrome in adulthood: focused review and a case report. Cureus. 2020;12(1):e6822.
2. Cupers S, Linthout CV, Desprechins B, Rausin L, Demarche M, Seghaye MC. Heterotaxy syndrome with intestinal malrotation, polysplenia and azygos continuity. Clin Pract. 2018;8(1):1004.
3. Ryerson LM, Pharis S, Pockett C, Soni R, Fruitman D, Guleserian KJ, et al. Heterotaxy syndrome and intestinal rotation abnormalities. Pediatrics. 2018;142(2):e20174267. https://doi.org/10.1542/peds.2017-4267.
4. Pauvels LSP, Langer FW, Dos Santos D, Haygert CJP. Heterotaxy syndrome. Radiol Bras. 2018;51(5):344–5.
5. Yoneyama H, Kondo C, Yamasaki A, Nakanishi T, Sakai S. Comparison of situs ambiguous patterns between heterotaxy syndromes with polysplenia and asplenia. Eur J Radiol. 2015;84(11):2301–6.
6. Dong SZ, Zhu M, Li F. Preliminary experience with cardiovascular magnetic resonance in evaluation of fetal cardiovascular anomalies. J Cardiovasc Magn Reson. 2013;15:40.
7. Loomba R, Shah PH, Anderson RH. Fetal magnetic resonance imaging of malformations associated with heterotaxy. Cureus. 2015;7(5):e269.

8. Esposito F, Vitale V, Noviello D, Di Serafino M, Vallone G, Salvatore M, et al. Ultrasonographic diagnosis of midgut volvulus with malrotation in children. J Pediatr Gastroenterol Nutr. 2014;59(6):786–8.
9. Görg C. The forgotten organ: contrast enhanced sonography of the spleen. Eur J Radiol. 2007;64(2):189–201.
10. Gayer G, Zissin R, Apter S, et al. CT findings in congenital anomalies of the spleen. Br J Radiol. 2001;74(884):767–72.
11. Buca DIP, Khalil A, Rizzo G, Familiari A, Di Giovanni S, Liberati M, et al. Outcome of prenatally diagnosed fetal heterotaxy: systematic review and meta-analysis. Ultrasound Obstet Gynecol. 2018;51(3):323–30.
12. Rodríguez Vargas D, Parada Blázquez MJ, Vargas SB. Diagnostic imaging of abnormalities in the number and location of the spleen. Radiologia. 2019;61(1):26–34.
13. Ditchfield MR, Hutson JM. Intestinal rotational abnormalities in polysplenia and asplenia syndromes. Pediatr Radiol. 1998;28(5):303–6.
14. Hernanz-Schulman M, Ambrosino MM, Genieser NB, et al. Pictorial essay. Current evaluation of the patient with abnormal visceroatrial situs. AJR Am J Roentgenol. 1990;154(4):797–802.
15. Gayer G, Apter S, Jonas T, et al. Polysplenia syndrome detected in adulthood: report of eight cases and review of the literature. Abdom Imaging. 1999;24(2):178–84.
16. Wang JK, Li YW, Chiu IS, et al. Usefulness of magnetic resonance imaging in the assessment of venoatrial connections, atrial morphology, bronchial situs, and other anomalies in right atrial isomerism. Am J Cardiol. 1994;74(7):701–4.
17. Oates E, Austin JM, Becker JL. Technetium-99m-sulfur colloid SPECT imaging in infants with suspected heterotaxy syndrome. J Nucl Med. 1995;36(8):1368–71.
18. Geva T, Vick GW 3rd, Wendt RE, Rokey R. Role of spin echo and cine magnetic resonance imaging in presurgical planning of heterotaxy syndrome. Comparison with echocardiography and catheterization. Circulation. 1994;90(1):348–56.
19. Jelinek JS, Stuart PL, Done SL, Ghaed N, Rudd SA. MRI of polysplenia syndrome. Magn Reson Imaging. 1989;7(6):681–6.
20. Nemec SF, Brugger PC, Nemec U, Bettelheim D, Kasprian G, Amann G, et al. Situs anomalies on prenatal MRI. Eur J Radiol. 2012;81(4):e495–501.
21. Martínez-Frías ML. Primary midline developmental field. I. Clinical and epidemiological characteristics. Am J Med Genet. 1995;56(4):374–81.
22. Martínez-Frías ML, Urioste M, Bermejo E, et al. Primary midline developmental field. II. Clinical/epidemiological analysis of alteration of laterality (normal body symmetry and asymmetry). Am J Med Genet. 1995;56(4):382–8.
23. Opitz JM, Gilbert EF. CNS anomalies and the midline as a "developmental field". Am J Med Genet. 1982;12(4):443–55.
24. Maldjian PD, Saric M. Approach to dextrocardia in adults: review. AJR Am J Roentgenol. 2007;188(Suppl. 6):S39–49. quiz S35-8
25. Jacobs JP, Anderson RH, Weinberg PM, et al. The nomenclature, definition and classification of cardiac structures in the setting of heterotaxy. Cardiol Young. 2007;17(Suppl. 2):1–28.
26. Uemura H, Ho SY, Devine WA, Anderson RH. Analysis of visceral heterotaxy according to splenic status, appendage morphology, or both. Am J Cardiol. 1995;76(11):846–9.
27. Horwich A, Brueckner M. Left, right and without a cue. Nat Genet. 1993;5(4):321–2.
28. Cohen MS, Anderson RH, Cohen MI, Atz AM, Fogel M, Gruber PJ, et al. Controversies, genetics, diagnostic assessment, and outcomes relating to the heterotaxy syndrome. Cardiol Young. 2007;17(Suppl. 2):29–43.
29. Herman TE, Siegel MJ. Polysplenia syndrome with congenital short pancreas. AJR Am J Roentgenol. 1991;156(4):799–800.
30. Varga I, Babala J, Kachlik D. Anatomic variations of the spleen: current state of terminology, classification, and embryological background. Surg Radiol Anat. 2018;40(1):21–9.
31. Ruscazio M, Van Praagh S, Marrass AR, et al. Interrupted inferior vena cava in asplenia syndrome and a review of the hereditary patterns of visceral situs abnormalities. Am J Cardiol. 1998;81(1):111–6.

32. Merran S, Karila-Cohen P, Servois V. CT anatomy of the normal spleen: variants and pitfalls. J Radiol. 2007;88(4):549–58.
33. Kouwenhoven JW, Bartels LW, Vincken KL, et al. The relation between organ anatomy and pre-existent vertebral rotation in the normal spine: magnetic resonance imaging study in humans with situs inversus totalis. Spine. 2007;32(10):1123–8.
34. Wang P, Jing H, Li F, Wang Z, Huo L. 99mTc-Labeled native RBC scintigraphy in distinguishing polysplenia from abdominal masses in a patient with situs inversus totalis. Clin Nucl Med. 2019;44(12):998–1000.
35. Kesim S, Inanir S. 99m-Technetium-labeled red blood cells' single-photon emission computed tomography/computed tomography in situs ambiguous with polysplenia. Indian J Nucl Med. 2020;35(1):84–6.
36. Izpisúa Belmonte JC. How the body tells left from right. Sci Am. 1999;280(6):46–51.
37. Casey B. Two rights make a wrong: human left-right malformations. Hum Mol Genet. 1998;7(10):1565–71.
38. Fujinaga M. Development of sidedness of asymmetric body structures in vertebrates. Int J Dev Biol. 1997;41(2):153–86.
39. Srivastava D. Left, right ... which way to turn? Nat Genet. 1997;17(3):252–4.
40. Danias PG, Manning WJ. Is this right? (...or is it left?). Circulation. 1999;100(2):209–10.
41. Applegate KE, Goske MJ, Pierce G, Murphy D. Situs revisited: imaging of the heterotaxy syndrome. Radiographics. 1999;19(4):837–52. discussion 853–4
42. Fulcher AS, Turner MA. Abdominal manifestations of situs anomalies in adults. Radiographics. 2002;22(6):1439–56.
43. Chen SJ, Li YW, Chang CI, Hsieh SC, Su CT, Hsu JCY. Usefulness of electron beam computed tomography in children with heterotaxy syndrome. Am J Cardiol. 1998;81(2):188–94.
44. Rasool F, Mirza B. Polysplenia syndrome associated with situs inversus abdominus and type I jejunal atresia. APSP J Case Rep. 2011;2(2):18.
45. Durmaz MS, Cengiz A, Arslan S, Erdogan H, Tolu I, Cengiz A. An incidental finding of polysplenia syndrome in an adult patient with multiple anomalies. Clin Med Rev Case Rep. 2016;3:148.
46. Maier M, Wiesner W, Mengiardi B. Annular pancreas and agenesis of the dorsal pancreas in a patient with polysplenia syndrome. Am J Roentgenol. 2007;188(2):W150–3.
47. Kandpal H, Sharma R, Gamangatti S, Srivastava DN, Vashisht S. Imaging the inferior vena cava: a road less traveled. Radiographics. 2008;28(3):669–89.
48. Bass JE, Redwine MD, Kramer LA, Huynh PT, Harris JH Jr. Spectrum of congenital anomalies of the inferior vena cava: cross-sectional imaging findings 1: (CME available in print version and on RSNA link). Radiographics. 2000;20(3):639–52.
49. Pickhardt PJ, Bhalla S. Intestinal malrotation in adolescents and adults: spectrum of clinical and imaging features. Am J Roentgenol. 2002;179(6):1429–35.
50. Strouse PJ. Disorders of intestinal rotation and fixation ("malrotation"). Pediatr Radiol. 2004;34(11):837–51.
51. De Wailly P, Metzler P, Sautot-Vial N, Olivier D, Meunier B, Faure JP. Pre-duodenal portal vein in polysplenia syndrome: clinical effects and surgical application. Surg Radiol Anat. 2011;33(5):451–4.
52. Low JP, Williams D, Chaganti JR. Polysplenia syndrome with agenesis of the dorsal pancreas and preduodenal portal vein presenting with obstructive jaundice—a case report and literature review. Br J Radiol. 2011;84(1007):e219–22.
53. Kapa S, Gleeson FC, Vege SS. Dorsal pancreas agenesis and polysplenia/heterotaxy syndrome: a novel association with aortic coarctation and a review of the literature. JOP. 2007;8(8):433–7.
54. Rose V, Izukawa T, Moes CA. Syndromes of asplenia and polysplenia. A review of cardiac and non-cardiac malformations in 60 cases withspecial reference to diagnosis and prognosis. Heart. 1975;37(8):840–52.
55. Fukuda K, Onda T, Kimura Y, Miura S, Matsumori R, Masaki Y. An adult case of polysplenia syndrome associated with sinus node dysfunction, dextrocardia, and systemic venous anomalies. Internal Med. 2015;54(9):1071–4.

56. Peoples WM, Moller JH, Edwards JE. Polysplenia: a review of 146 cases. Pediatr Cardiol. 1983;4(2):129–37.
57. Bartram U, Wirbelauer J, Speer CP. Heterotaxy syndrome—asplenia and polysplenia as indicators of visceral malposition and complex congenital heart disease. Biol Neonate. 2005;88(4):278–90.
58. Shiraishi I, Ichikawa H. Human heterotaxy syndrome from molecular genetics to clinical features, management, and prognosis. Circ J. 2012;76:2066–75.
59. Lin J-H, Chang C-I, Wang J-K, et al. Intrauterine diagnosis of heterotaxy syndrome. Am Heart J. 2002;143:1002–8.
60. Rasool F, Mirza B. Polysplenia syndrome associated with situs inversus abdominus and type I jejunal atresia. APSPS J Case Rep. 2011;2(2):18.
61. Chinya A, Naranje K, Mandeliaa A. Situs Inversus abdominalis, polysplenia, complex jejunal atresia and malrotation in a neonate: a rare association. Int J Surg Case Rep. 2019;56:93–5.
62. Garcia-Rodriguez E, Garcia-Garcia E, Perez-Sanchez A, Pavon-Delgado A. A New observation of 13q deletion syndrome: severe undescribed features. Genet Couns. 2015;26(2):213–7.
63. Marx MV, Van Allan RSIR. film panel case: heterotaxia with polysplenia. J Vasc Interv Radiol. 2005;16:1055–9.
64. Escobar-Diaz M, Friedman K, Salem Y, et al. Perinatal and infant outcomes of prenatal diagnosis of heterotaxy syndrome (asplenia and polysplenia). Am J Cardiol. 2014;114(4):612–7.

Hepatolienal Fusion

10

- Embryologically, the spleen is derived from a mass of mesenchymal cells which are located between the layers of the dorsal mesogastrium.
- The spleen begins to develop during the fifth week of intrauterine life.
- The proliferating mesenchymal cells invade the underlying angiogenetic mesenchyme, which becomes condensed and vascularized.
- This process occurs simultaneously in several closely related areas. These adjoining areas fuse to form a lobulated fetal spleen.
- The lobules of the spleen normally disappear before birth.
- Accessory spleens may be formed during embryonic development as heterotopic splenic tissue.
  - Accessory spleen is common and reported in 4–15% of the population.
  - The most common sites for an accessory spleen are:
    - The hilum of the main spleen
    - The great omentum
    - The gastrosplenic ligament
    - The spleno-renal ligament
    - The pelvis
  - Spleno-visceral fusion:

    This is a developmental abnormality where a developmentally separate splenic nodule is directly connected to another organ.
- Embryologically, parts of splenic tissue can get separated and detached from the main spleen and develop as a congenital abnormality.
  - Accessory spleens or spleniculli

    If these nodules of splenic tissue become separate and isolated from the main spleen, they are called accessory spleens or spleniculli.
  - Spleno-visceral fusions

    Nodules of splenic tissue can separate from the main spleen and become adherent or merge with other organs during embryonic development, they are called spleno-visceral fusions.

A. H. Al-Salem, *The Spleen*, https://doi.org/10.1007/978-981-99-6191-7_10

Most often, they merge with the tail of the pancreas forming spleno-pancreatic fusion or an intrapancreatic accessory spleen.
When they merge with the gonads, they form spleno-gonadal fusion.
When they merge with the kidney, they form spleno-renal fusion.
When they merge with the adrenal gland, they form spleno-adrenal fusion

- Embryologically, spleno-visceral fusions develop as a result of migration of splenic tissue along the dorsal mesentery to the urogenital ridge, together with primitive germ cells. This occurs towards the end of the fifth week and during the sixth week of embryonic development.

SPLENO-VISCERAL FUSIONS

- **SPLENO-PANCREATIC FUSION**
- **INTRAPANCREATIC ACCESSORY SPLEEN**
- **SPLENO-GONADAL FUSION**
- **SPLENO-RENAL FUSION**
- **SPLENO-ADRENAL FUSION**
- **HEPATO-LIENAL FUSION**

- Spleno-visceral fusions are usually asymptomatic conditions.
- Their main clinical significance lies in the confusion they cause as they may be confused with tumors of other organs or metastasis from other organs.
- Rarely Spleno-visceral fusions cause complications, such as:
  - Torsion of a wandering accessory spleen.
  - Bleeding caused by spontaneous rupture of an accessory spleen.
- Fusion of organs of the body is a rare developmental abnormality.
- This has been described and most frequently encountered in the following conditions:
  - Horseshoe kidney
  - Splenogonadal fusion
  - Hepatorenal fusion
  - Adrenal fusion
  - Female genital tract
  - Urinary tract
- In 1978, Cotelingam emphasized that: "the presence of capsular fusion, the absence of inflammatory reaction or stigmata of previous injury and the normal embryologic relationship strongly support a developmental genesis".
- Developmental abnormalities of the liver are unusual, including hepatolienal fusion.
- Hepatolienal fusion is an extremely rare abnormality in which there is fusion of liver and spleen.
- The first case of hepatolienal fusion was reported by Cotelingam in 1978 and the second case was reported by Wang in 2017.

- Eagle syndrome is a more frequent condition that is caused by elongation of the styloid process or calcification of the stylohyoid ligament, which was first described by Eagle in 1937.
- Patients with Eagle syndrome present with a variety of symptoms including:
  - Throat and neck pain
  - Sensation of foreign body in the throat
  - Dysphagia
- Hepatolienal fusion was also reported combined with Eagle syndrome.

Further Reading

1. Varga I, Galfiova P, Adamkov M, Danišovič L, Polak S, Kubikova E, Galbavy S. Congenital anomalies of the spleen from an embryological point of view. Med Sci Monit. 2009;15:RA269–76.
2. Cotelingam JD, Saito R. Hepatolienal fusion: case report of an unusual lesion. Hum Pathol. 1978;9(2):234–6.
3. Wang K, Shen D. Hepatolienal fusion: a rare case report (Poster No. 88). In: Abstracts and case studies from the College of American Pathologists; 2017.
4. Eagle WW. Elongated styloid processes: report of two cases. Arch Otolaryngol. 1937;25:584–7.
5. Raina D, Gothi R, Rajan S. Eagle syndrome. Indian J Radiol Imaging. 2009;19(2):107–8.
6. Zeckler SR, Betancur GA, Yaniv G. The eagle is landing: Eagle syndrome-an important differential diagnosis. Br J Gen Pract. 2012;62(602):501–2.
7. Clement R, Barrios L. Eagle syndrome and sudden and unexpected death: forensic point of view about one case. J Foren Res. 2014;5(3):1–4.
8. Kumar P, Raymane AP, Subbaramaiah M. Sudden death due to Eagle syndrome. Am J Foren Pathol. 2013;34:231–3.

Spleno-gonadal Fusion

11

11.1 Introduction

- Splenogonadal fusion is a very rare congenital malformation of the spleen.
- It is defined as fusion of the ectopic spleen and gonad.
- The first case of splenogonadal fusion was described by Bostroem in 1883.
- Eugen Woldemar Bostroem (13 October 1850–24 May 1928) was a German pathologist. He received his medical degree in 1876. In 1890 Bostroem managed to isolate the causative organism of actinomycosis from a culture of grain, grasses, and soil. In 1883 Bostroem was the first to describe splenogonadal fusion (Wikipedia, the free encyclopedia).
- The first detailed review of splenogonadal fusion was published by Putschar and Manion in 1956. They also put the classification of splenogonadal fusion (Fig. 11.1).
- There are two types of splenogonadal fusion:
  - Continuous type
  - Discontinuous type
- The continuous type:
  - This is he more common type
  - It is characterized by a cord that connects the normal spleen and ectopic spleen.
  - This cord may also be beaded with small splenunculi.
- The discontinuous type:
  - This is rare when compared to the continuous type
  - In this type, there is no cord present connecting the normal spleen to the ectopic spleen.
- Splenogonadal fusion is typically diagnosed incidentally during surgery for inguinal hernia and/or cryptorchidism (Fig. 11.2).
- It is important to note that in certain cases, splenogonadal fusion may closely resembles a primary testicular neoplasm, and the patient may undergo an unnecessary orchiectomy.

A. H. Al-Salem, *The Spleen*, https://doi.org/10.1007/978-981-99-6191-7_11

Fig. 11.1 Walter G. J. Putschar (1904–1987). Walter G. J. Putschar was working in the Department of Pathology of Massachusetts General Hospital (MGH) from 1959 to 1984. The first detailed review of splenogonadal fusion was published by Putschar and Manion in 1956 wherein a splenogonadal fusion classification system was also established (Wikipedia, the free encyclopedia)

- Splenogonadal fusion on ultrasound resembles a mass within the scrotal sac of low reflectivity in comparison to the normal testicular parenchyma. This is a differentiating point but the mass may not be seen separate from the testis. This makes splenogonadal fusion easily confused with a primary testicular tumor.
- On cooler Doppler ultrasound flow in the abnormal tissue assumes a pattern similar to that seen in the central aspect of the normal testis or that seen in splenic tissue.
- A 99mTc-sulphur colloid scan is the best investigation to demonstrate uptake within the ectopic splenic tissue.

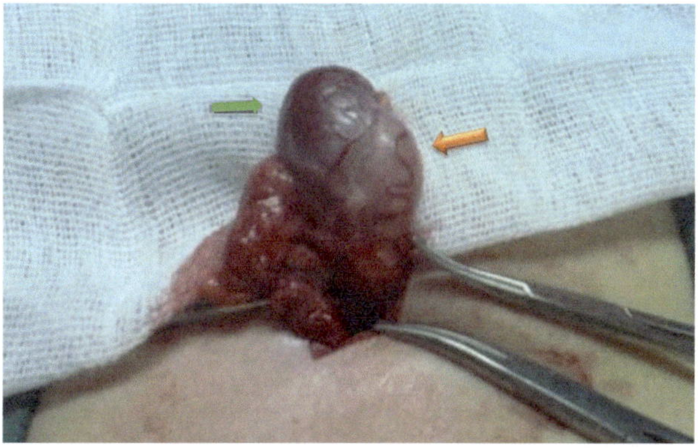

Fig. 11.2 Intraoperative photograph showing discontinuous Splenogonadal fusion

11.2 Associated Anomalies

- Splenogonadal fusion is known to be associated with other congenital anomalies or discovered during the evaluation of these associated anomalies.
- These are more seen in those with the continuous type of splenogonadal fusion.
- Continuous type of splenogonadal fusion is known to be associated with other associated anomalies in as much as 30% of the cases.
- There is also an association between splenogonadal fusion and testicular malignant tumors.
- It is interesting to note that the reported cases of splenogonadal fusion in association with testicular malignancy, the malignant tumors developed in adults how had undescended testes or in those following orchidopexy for undescended testes.
- This may represent an association rather than an increased risk in this group of patients. Patients with undescended testes have a well-known increased risk of malignancy.
- There are several anomalies associated with splenogonadal fusion.
- Inguinal hernia and cryptorchidism are the most common.
- These associated anomalies include:
  - Inguinal hernia
  - Cryptorchidism
  - Micrognathia
  - Peromelia which is categorized as a separate syndrome (Splenogonadal fusion limb syndrome)
  - Cleft palate
  - Congenital heart disease
  - Moebius syndrome
  - Hypospadias
  - Osteogenesis imperfecta
  - Persistent Mullerian duct syndrome
  - Potter syndrome
  - Gastrointestinal malrotation

- Anal stenosis
- Transverse testicular ectopia
- Hip dysplasia
- Hypoplastic left heart
- Phenotypical sex reversal
- Hypoglossia
- Craniosynostosis
- Spina bifida
- Microgastria
- Congenital diaphragmatic hernia
- Anorectal malformations

11.3 Etiology and Embryology

- The exact etiology of splenogonadal fusion is not known.
- Embryologically, splenogonadal fusion occurs between the fifth and eighth weeks of gestation.
- During the fifth and sixth weeks of gestation, the spleen starts to develops from the splenic anlage in the left dorsal mesogastrium.
- At about the same time, the gonadal ridge starts to develop between the mesonephros and dorsal mesogastrium.
- Embryologically, the testis starts to descend from its initial embryological position between the dorsal mesogastrium and the mesonephros at around the eighth week of intrauterine life. This coincides with the time of splenic development.
- At around 5 weeks of intrauterine life the embryonic gut starts to rotate and the dorsal mesogastrium rotates to the left. As a result of this, the developing spleen will become into close proximity with the left urogenital fold which contains the gonadal mesoderm.
- This close proximity of the developing spleen and developing gonad remains until the gonads start to descend and the mesonephros involute during the eighth week of gestation.
- Splenogonadal fusion is thought to result from partial fusion of splenic and gonadal tissues during the fourth-eighth weeks of intra-uterine life.
- The subsequent descent of the gonad during the eighth-tenth weeks of gestation and in some patients and because of this close proximity of the developing spleen and developing gonad may results in descent of a part of the developing spleen together with the descending gonad.
- In the discontinuous type, there is complete detachment from the normal spleen and the ectopic spleen is completely separate from the main spleen.
- In the continuous type there is an attachment to the normal spleen by a cord-like structure. This cord can be made up of splenic tissue or totally fibrotic.
- Occasionally, there are multiple nodules along this cord which represent foci of splenic tissue that got detached from the main spleen and developed separately.
- This however does not fully explain the occasional occurrence of right sided Splenogonadal fusion.

- Several mechanisms have been proposed to explain this abnormal fusion between the splenic and gonadal tissues.
  - Embryologically, the close proximity between the developing left gonad and spleen during embryogenesis explains the fact that splenogonadal fusion almost always occurs on the left side.
  - During embryonic development, the descent of the testis can sometimes carry with it part of the developing spleen but this remains attached to the main spleen by a long band of tissues (Continuous splenogonadal fusion) or only carry a portion of the developing spleen down as it descends without any attachment to the main spleen (Discontinuous splenogonadal fusion).

11.4 Classification

- According to Putschar and Manion, there are two types of splenogonadal fusion:
  - Continuous type
  - Discontinuous type
- Continuous Splenogonadal fusion (55% of cases) is the commonest.
- The continuous type develops when the normally located spleen is attached to the gonad by a discrete long band or cord that maybe:
  - Totally made up of splenic tissue.
  - Made up of multiple connected beads or small nodules of splenic tissue (splenic rosary bead).
  - A cord made up of fibrous tissue.
- In discontinuous splenogonadal fusion (45% of cases), there is no connection between the normal spleen and the ectopic accessory spleen which is fused with the gonad. This type is completely separated from the normal spleen.
- The discontinuous splenogonadal fusion is considered a rare variant of an accessory spleen.
- Both continuous and discontinuous types of splenogonadal fusion occur with almost equal frequency and the discontinuous type may be discovered incidentally during:
  - Herniotomy
  - Orchidopexy
  - Or present as a scrotal swelling
- It is estimated that approximately 37% of the reported patients with splenogonadal fusion had orchidectomy because of suspicion of a testicular tumor.

11.5 Clinical Features

- Splenogonadal fusion is considered a rare condition where an accessory spleen is fused to the gonads.
- Commonly, splenogonadal fusion is asymptomatic and discovered incidentally during routine herniotomy or orchidopexy.
- Many of these cases however go unnoticed or discovered at autopsy.

- In the pediatric age group:
  - Splenogonadal fusions are asymptomatic and discovered incidentally during routine herniotomy or orchidopexy.
  - These patients commonly present with a scrotal swelling and mass.
  - They rarely present with an acute scrotal pain as a result of:
    Torsion of splenic tissue
    Involvement of splenic tissue with other pathological conditions such as mumps, malaria, leukemia, trauma and infectious mononucleosis.
- The majority of splenogonadal fusion cases occur on the left side.
- The left side is far more commonly affected than the right in 98% of the cases.
- It is seen more commonly in males than females with an M: F of 16:1.
- This however may not be totally true.
  - The reported male dominance is most likely due to underestimation of Splenogonadal fusion occurrence in females as females' gonads are hidden inside the pelvis of the body and therefore are not noticeable or accessible for clinical examination than the male gonads.
  - Add to this the fact that there are fewer complications associated with female gonads than with male gonads.
- The age of presentation of splenogonadal fusion is less than 10 years in half of the reported cases.
- About 82% of the reported cases occur in patients younger than 30 years.
- The clinical presentation of splenogonadal fusion is not specific and the final diagnosis is often made on histological examination of the resected specimen.
- These patients usually present with an inguinal hernia, undescended testis or as a scrotal mass.
- Splenogonadal fusion is often confused with and misdiagnosed as testicular tumor. This will lead to unnecessary orchidectomy.
- Awareness of this clinical entity is important as this could avoid unnecessary orchidectomies. The testis is usually separable from the fused splenic tissue and can be preserved while the lesion is excised.
- Splenogonadal fusion is typically diagnosed incidentally while exploring the inguinal region for cryptorchidism, an inguinal hernia, or a hydrocele.
- Splenogonadal fusion can also present as an asymptomatic scrotal mass. The discontinuous form of splenogonadal fusion usually presents as a hard scrotal nodule, mimicking a testicular tumor.
- The continuous form of splenogonadal fusion can also rarely present as intestinal obstruction from the band of tissue attaching it to the main spleen.

11.6 Investigations

- It is estimated that about 20–30% of patients with splenogonadal fusion underwent unnecessary orchidectomy because of suspicion of malignancy.
- This is important and calls for increasing awareness of this condition among treating physicians and try to use better diagnostic investigations.

- Splenogonadal fusion should be considered in the differential diagnosis of any patient presenting with a testicular mass.
- The radiologic diagnosis of splenogonadal fusion is important to prevent unnecessary orchidectomy.
- The following investigations are helpful to diagnose splenogonadal fusion:
  - B-mode ultrasound
  - Computed tomography (CT)
  - MRI
  - Technicium-99 m sulfur colloid liver-spleen scan
- Doppler ultrasonography will show a hyper vascular mass on the upper pole of testis. This mass will appear similar to the normal spleen.
- Ultrasound is a valuable diagnostic investigation and can be used for scrotal imaging.
  - Ultrasound could help to diagnose continuous splenogonadal fusion when a cord connecting the spleen to the testicle is visualized.
  - On ultrasound, the splenic tissue appears as a well-encapsulated, extra-testicular homogenous hypoechoic or isoechoic mass.
  - Doppler ultrasound could monitor blood flow to the testes.
- CT is helpful to diagnose the abnormal vasculature of the spleen.
- 99 m Tc-Sulphur colloid liver spleen scan is a very reliable investigation for detecting accessory spleen and help to diagnose splenogonadal fusion.
- MRI is valuable in detecting the position and shape of the testes and to exclude other associated congenital anomalies.
- MRI is reliable and accurate in detecting a testicular or scrotal mass.
- MRI is also helpful in differentiating intratesticular and extra testicular lesions.
- An intra-operative biopsy with frozen section should be considered when the diagnosis is in doubt to determine the nature of the mass.
- Diagnostic laparoscopy when necessary is recommended as it is safe, reliable and very accurate in diagnosing and treating continuous splenogonadal fusion.

11.7 Treatment

- The diagnosis of Splenogonadal fusion should be considered in the differential diagnosis of scrotal swelling and if suspected preoperatively, the diagnosis can be confirmed by a ^{99m}Tc-sulphur colloid scan.
- When a definite diagnosis of splenogonadal fusion is made, it is important to completely excise the splenic tissue and preserve the testis.
- The treatment of splenogonadal fusion is surgical excision and every attempt should be made to preserve the gonad at the time of excision. This should not be difficult since true fusion with the gonad is rare and a demarcation line between the two is present (Figs. 11.3 and 11.4).
- For some patients, surgery may not even be necessary.
- Unnecessary orchidectomy should be avoided.

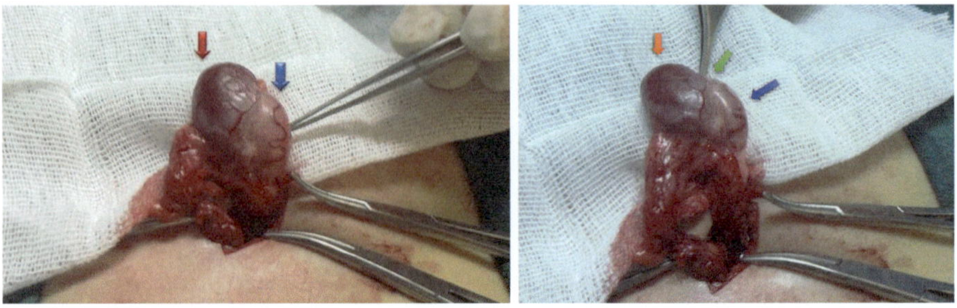

Figs. 11.3 and 11.4 Intraoperative photographs showing discontinuous splenogonadal fusion. Note the line of demarcation between the ectopic spleen and gonad

Further Reading

1. Lin CS, Lazarowicz JL, Allan RW, Maclennan GT. Splenogonadal fusion. J Urol. 2010;184:332–3.
2. Abokrecha A, Almatrfi A. Discontinued splenogonadal fusion and bilateral empty scrotum in an 18-month-old boy. Eur J Pediatr Surg Rep. 2017;5:e1–3.
3. Carragher AM. One hundred years of splenogonadal fusion. Urology. 1990;35:471–5.
4. Putschar WG, Manion WC. Splenicgonadal fusion. Am J Pathol. 1956;32:15–33.
5. Malik RD, Liu DB. Splenogonadal fusion: an unusual case of an acute scrotum. Rev Urol. 2013;15:197–201.
6. Lopes RI, de Medeiros MT, Arap MA, Cocuzza M, Srougi M, Hallak J. Splenogonadal fusion and testicular cancer: case report and review of the literature. Einstein (Sao Paulo). 2012;10:92–5.
7. Xiang X, Jiang Y, Liu JX, Qiu L. A rare content of congenital inguinal hernia: a case report of splenogonadal fusion. BMC Pediatr. 2019;19:422.
8. Speare R, Roberts J, Cohen M, Wales J. Splenogonadal fusion and sex reversal. J Pediatr Endocrinol Metab. 2012;25:541–2.
9. Duhli N, Venkatramani V, Panda A, Manojkumar R. Splenogonadal fusion: pathological features of a rare scrotal mass. Indian J Pathol Microbiol. 2013;56:474–6.
10. Ferrón SA, Arce JD. Discontinuous splenogonadal fusion: new sonographic findings. Pediatr Radiol. 2013;43:1652–5.
11. Celik A, Tiryaki S, Darcan S, Ergun O. Splenogonadal fusion-limb deformity syndrome: a rare but important cause of undescended testis. World J Pediatr. 2016;12:246–8.
12. Li WF, Luan MX, Ma Z, Chen YJ. Splenogonadal fusion: report of four cases and review of the literature. Exp Ther Med. 2013;6:816–8.
13. Liu W, Wu R, Guo Z. The diagnosis and management of continuous splenogonadal fusion in a 6-year-old boy. Int Urol Nephrol. 2013;45:21–4.
14. Croxford WC, Pfistermuller KL, Scott F, Pope AJ. Splenogonadal fusion presenting clinically and radiologically as a seminoma. Urol Case Rep. 2015;3:204–5.
15. Bal K, Ermete M, Balcı U, Dinçel Ç. Splenogonadal fusion: a very rare congenital anomaly in the differential diagnosis of a testicular mass. Turk J Urol. 2014;40:62–4.
16. Harris AM. Splenogonadal fusion: a rare benign testicular mass in a 55-Year-old male. Urol Case Rep. 2016;9:41–2.
17. Karray O, Oueslati A, Chakroun M, Ayed H, Bouzouita A, Cherif M, Ben Slama MR, Derouiche A, Chebil M. Splenogonadal fusion—a rare cause of scrotal swelling: a case report. J Med Case Rep. 2018;12:172.

18. Kocher NJ, Tomaszewski JJ, Parsons RB, Cronson BR, Altman H, Kutikov A, Smith RE. Splenogonadal fusion: a rare etiology of solid testicular mass. Urology. 2014;83:e1–2.
19. Kumar S, Jayant K, Agrawal S, Parmar KM, Singh SK. A rare case of continuous type splenogonadal fusion in a young male with primary infertility. Case Rep Urol. 2014;2014:796761.
20. Lakshmanan PM, Reddy AK, Nutakki A. A surprising content of congenital hernia: complete splenogonadal fusion band. BMJ Case Rep 2014; 2014. pii: bcr2014203640.
21. Sountoulides P, Neri F, Bellocci R, Schips L, Cindolo L. Splenogonadal fusion mimicking a testis tumor. J Postgrad Med. 2014;60:202–4.
22. Zhang Y, Tong Z, Zhang Y. X-ray computed tomography and sonography features of continuous splenogonadal fusion. J Xray Sci Technol. 2013;21:303–8.
23. Shadpour P, Rezaimehr B. "Rosary of testes": splenogonadal fusion in association with bilateral abdominal testes presenting as polyorchidism. Case Rep Urol. 2015;2015:317189.
24. Trottmann M, Marcon J, Mai V, D'Anastasi M, Becker A, Stief C, Reiser M, Clevert DA. Characterization of splenogonadal fusion by contrast-enhanced ultrasound (CEUS) and elastography. Ultraschall Med. 2015;36:97–100.
25. Jayasundara JA, Vithana VH, Lamahewage AK. A case of continuous-type splenogonadal fusion. Singap Med J. 2013;54:e123–4.
26. Bosnalı O, Cici İ, Moralıoğlu S, Cerrah-Celayir A. Continuous-type splenogonadal fusion: report of a rare case. Turk J Pediatr. 2014;56:680–3.
27. Jakkani R, Alhajri FA, Alteriki A, Almuteri MF, Athyal RP, Hashem KZ. Discontinuous splenogonadal fusion diagnosed on computed tomography. Indian J Radiol Imaging. 2016;26:506–9.
28. Uglialoro AD, Goltzman ME, Niazi M, Lehman D, Silletti J, Bjurlin MA. Splenogonadal fusion presenting as an asymptomatic testicular mass. Urology. 2016;97:1–4.
29. Akama Y, Shimanuki K, Asahi S, Watanabe Y, Ko K, Takano R, Amano H, Kawaguchi T, Uchida E. Incidentally detected splenogonadal fusion in a laparoscopic transabdominal preperitoneal hernia repair operation: a case report. Int J Surg Case Rep. 2017;35:29–32.
30. Huang G, Huang Y, Zeng L, Yuan M, Wu Y, Huang L. Continuous-type splenogonadal fusion: a case report. Exp Ther Med. 2017;13:2019–21.
31. Li X, Ye J, Jiang G. Sonographic diagnosis of splenogonadal fusion in a 2-year-old boy. J Clin Ultrasound. 2017;45:179–82.
32. Preece J, Phillips S, Sorokin V, Herz D. Splenogonadal fusion in an 18-month-old. J Pediatr Urol. 2017;13:214–5.
33. Chiaramonte C, Siracusa F, Li VG. Splenogonadal fusion: a genetic disorder? Report of a case and review of the literature. Urol Case Rep. 2014;2:67–9.
34. Srinivasa Rao RC, Radhakrishna V, Rao N, Rakshit S. Torsion of a Splenule in a case of Splenogonadal fusion mimicking a strangulated inguinal hernia. J Indian Assoc Pediatr Surg. 2018;23:100–2.
35. Shakeri A, Shakeri A, Rasolmali R, Shakeri S. A case of splenogonadal fusion accompanied by accessory spleen in a 4-year-old boy. Urol Ann. 2018;10:406–8.
36. Zhou L, Muthucumaru M, Stunden R, Lenghaus D. Splenogonadal fusion: a rare scrotal mass in a 9-year-old boy. ANZ J Surg. 2018;88:E81–2.
37. Chen CJ, Kavoussi N, Jacobs MA. Splenogonadal fusion: a rare finding during routine orchiopexy. Urol Case Rep. 2019;27:100904.
38. Grosu S, Rübenthaler J, Knösel T, Trottmann M, Marcon J, Clevert DA. Splenogonadal fusion evaluation using contrast enhanced ultrasound and elastography. A case report. Med Ultraso. 2019;21:356–8.
39. Mann JA, Ritchie E. Splenogonadal fusion: a case of two lesions. Urol Case Rep. 2019;24:100878.
40. Patre V, Netam S, Luka P, Mandle H. Discontinuous type of splenogonadal fusion syndrome with limb defects. Indian J Urol. 2012;28:94–5.
41. Seager MJ, Alexander S, Muneer A, Walkden M. Splenogonadal fusion: a rare paratesticular lesion and how to recognise it on ultrasound. Ultrasound. 2020;28:54–7.
42. Riley DS, Barber MS, Kienle GS, Aronson JK, von Schoen-Angerer T, Tugwell P, Kiene H, Helfand M, Altman DG, Sox H, Werthmann PG, Moher D, Rison RA, Shamseer L, Koch CA,

Sun GH, Hanaway P, Sudak NL, Kaszkin-Bettag M, Carpenter JE, Gagnier JJ. CARE guidelines for case reports: explanation and elaboration document. J Clin Epidemiol. 2017;89:218–35.
43. Ahmed SF, Khwaja O, Hughes IA. The role of a clinical score in the assessment of ambiguous genitalia. BJU Int. 2000;85:120–4.
44. Herenger Y, Stoetzel C, Schaefer E, Scheidecker S, Manière MC, Pelletier V, Alembik Y, Christmann D, Clavert JM, Terzic J, Fischbach M, De Saint MA, Dollfus H. Long term follow up of two independent patients with Schinzel-Giedion carrying SETBP1 mutations. Eur J Med Genet. 2015;58:479–87.
45. Alam A, Delto JC, Blachman-Braun R, Wayne G, Mittal AG, Castellan M, Kozakowski K, Labbie A, Gosalbez R. Staged Fowler-Stephens and single-stage laparoscopic orchiopexy for intra-abdominal testes: Is there a difference? A single institution experience. Urology. 2017;101:104–10.
46. Mohrs OK, Thoms H, Egner T, Brunier A, Eiers M, Kauczor HU, Hallscheidt P. MRI of patients with suspected scrotal or testicular lesions: diagnostic value in daily practice. AJR Am J Roentgenol. 2012;199:609–15.
47. Elzeneini WM, Mostafa MS, Dahab MM, Youssef AA, AbouZeid AA. How far can one-stage laparoscopic Fowler-Stephens orchiopexy be implemented in intra-abdominal testes with short spermatic vessels? J Pediatr Urol. 2020;16:197.
48. Lopes RI, de Medeiros MT, Arap MA, Cocuzza M, Srougi M, Hallak J. Splenogonadal fusion and testicular cancer: case report and review of the literature. Einstein (Sao Paulo). 2012;101:92–5.
49. Khairat AB, Ismail AM. Splenogonadal fusion: case presentation and literature review. J Pediatr Surg. 2005;40:1357–60.
50. Varma DR, Sirineni GR, Rao MV, Pottala KM, Mallipudi BV. Sonographic and CT features of splenogonadal fusion. Pediatr Radiol. 2007;37:916–9.
51. Gouw AS, Elema JD, Bink-Boelkens MT, de Jongh HJ, ten Kate LP. The spectrum of splenogonadal fusion. Case report and review of 84 reported cases. Eur J Pediatr. 1985;144:316–23.
52. Cortes D, Thorup JM, Visfeldt J. The pathogenesis of cryptorchidism and splenogonadal fusion: a new hypothesis. Br J Urol. 1996;77:285–90.
53. Karaman MI, Gonzales ET. Jr Splenogonadal fusion: report of 2 cases and review of the literature. J Urol. 1996;155:309–11.

Splenorenal Fusion 12

12.1 Introduction

- Developmental abnormalities of the spleen are common.
- The most common abnormality is an accessory spleen which is commonly seen in the area of the splenic hilum.
- An accessory spleen is estimated to occur in 10–30% of patients in autopsy series.
- A much less common developmental abnormality is splenic fusion.
- Splenic fusion is well-described in gonads, liver, pancreas but rarely in kidney.
- There are different types of splenic fusion and these include:
  - Splenogonadal fusion
  - Splenopancreatic fusion
  - Hepatolienal fusion
  - Splenorenal fusion
  - Splenic fusion is a rare abnormality.
- It may result from disturbances during embryogenesis of the spleen because both organs develop from the dorsal mesogastrium. Add to this their close proximity to each other which may result in their fusion.
- Gonzalez-Crussi et al. first reported splenorenal fusion in 1977.
- A review of the literature revealed only nine reported cases of splenorenal fusion, six were reported on the left side and three on the right side. Seven of these cases were reported in adults and two cases were in children.
- Awareness of this condition and proper evaluation and diagnosis is important to avoid unnecessary nephrectomy.

TYPES OF SPLENIC FUSION
(1) **SPLENOGONADAL FUSION**
(2) **SPLENOPANCREATIC FUSION**
(3) **HEPATOLIENAL FUSION**
(4) **SPLENORENAL FUSION**

A. H. Al-Salem, *The Spleen*, https://doi.org/10.1007/978-981-99-6191-7_12

- The presentation of splenorenal fusion is variable.
- They may present as a renal mass which cannot be reliably distinguished from clear cell carcinoma, Wilms tumor or other renal neoplasms.
- Preoperative diagnosis of splenorenal fusion is important.
- CT scan and MRI cannot however distinguish splenorenal fusion from other renal neoplasms.
- The diagnosis of splenorenal fusion can be confirmed with 99 m Tc-sulfur-colloid scan or 99 m Tc–labeled, heat-damaged red blood cell scan.
- Fine-needle aspiration (FNA) or percutaneous biopsy under ultrasound or CT-scan guidance can be used to confirm the diagnosis.

12.2 Embryology

- Embryologically, the spleen develops from mesenchymal cells that migrate between the leaves of the dorsal mesogastrium. This occurs during the fifth embryonic week.
- These mesenchymal cells subsequently fuse together to form the splenic tissue.
- Failure of fusion of the mesenchymal cells results in the development of accessory spleens.
- As a result of the rotation of the embryonic gut, the splenic primordium lies on the left side close to the left mesonephric ridge.
- There are two theories explaining the development of spleno-visceral fusion.
  - The first one is the continuous theory:

    This theory describes a band of tissue that adhere the spleen with the organ between the seventh and eighth week of gestation.
  - The second one is the discontinuous theory:

    This theory postulates migration of the splenic cells caudally to reach the developing mesonephric ridge and retroperitoneum, where fusion takes place.
- Two types of ectopic splenic tissue (splenules) can be found:
  - Accessory spleens
  - Splenosis
- Accessory spleens:
  - These are congenital developmental abnormality and arise from the left side of the dorsal mesogastrium during the embryological development of the spleen.
  - Accessory spleens are also called supernumerary spleens, splenunculi, or splenules.
  - Accessory spleens have been reported within the pancreas, kidney, and scrotum and as an adnexal mass.
  - An accessory spleen is defined as ectopic splenic tissue that develops as a result of failure of fusion of cells during embryonic development of the spleen as they migrate from their midline location to the left upper quadrant.

- The most common locations for accessory spleens are the hilum of the spleen (75%) and adjacent to the tail of the pancreas (25%).
- Accessory spleens may occur anywhere in the abdomen including:
  - The gastrosplenic ligament
  - The splenorenal ligament
  - The greater omentum
  - Stomach
  - Mesentery
  - Adrenal glands
  - Gonads
- Accessory spleen may resemble and confused with many tumours according to its location such as:
  - Pancreatic tumor
  - Adnexal tumor
  - Abdominal tumor
  - Retroperitoneal tumor
  - Adrenal tumor
  - Testicular tumor
- Accessory spleens are fairly common and found in approximately 10-30% of the population in autopsy series and 16% of patients undergoing contrast enhanced abdominal CT.
- They are typically around 1 cm in diameter but may reach up to 20 cm.
- Spleno-organ fusion is considered as a subtype of accessory spleens.

• Splenosis:
  - This refers to splenic tissue that has been auto transplanted in a heterotopic location following splenic trauma or splenectomy.
  - Splenosis can be found in any location. The most common locations are:
    - The peritoneum
    - Omentum
    - Mesentery
  - Splenosis has been also reported in:
    - The pericardium
    - Subcutaneous tissue
    - The brain
    - The liver

• Most patients with splenic heterotopia are usually asymptomatic and found incidentally.

12.3 Clinical Features

• Splenorenal fusion is very rare.
• Splenorenal fusion is a very rare benign condition characterized by the presence of splenic tissue within the renal capsule.

  - When it occurs in patients who had splenectomy it has been classified as splenosis.
  - When it is found in patients with intact spleens it is called splenorenal fusion. This is a developmental anomaly resulting in fusion of splenic and renal tissue.
- Splenorenal fusion is more commonly seen in the left kidney.
- Splenorenal fusion is usually asymptomatic discovered during the evaluation of other unrelated conditions.
- Splenorenal fusion can present as a renal mass which can be confused with primary or secondary renal neoplasms on radiological investigations.
- Rarely, splenorenal fusion present with symptoms of hypersplenism.

12.4 Investigations and Treatment

- Splenorenal fusion is a very rare congenital anomaly in which splenic tissue is present within the renal capsule.
- Splenorenal fusion may arise as a developmental anomaly secondary to the fusion of nephrogenic mesoderm and splenic anlage in the second month of intrauterine life.
- It may also be secondarily acquired as a result of splenosis after trauma or splenectomy.
- Splenorenal fusion can mimic primary or secondary renal tumors on radiological investigations.
- Although very rare, splenorenal fusion should be considered when a patient present with a renal mass.
- Most commonly, splenorenal fusion is found incidentally during radiological evaluation of other unrelated condition.
- This however is not the case always and splenorenal fusion can cause symptoms.
- It is important to keep this in mind to avoid unnecessary surgery.
- Most reported cases of splenorenal fusion underwent total nephrectomy because of suspicion or confusion of malignant tumors.
- Preoperative diagnosis of splenorenal fusion is important but CT scan and MRI although helpful; they cannot distinguish splenorenal fusion from other renal neoplasms.
- The gold standard diagnostic investigation of choice is nuclear scintigraphy (splenic scan).
  - The ^{99m}technetium (^{99m}Tc) sulfur colloid test of the liver and spleen was first used to diagnose accessory spleens and/or splenosis due to its ability to localize the reticuloendothelial system.
  - The diagnosis can be established with ^{99m}Tc-sulfur-colloid scan or ^{99m}Tc-labeled, denatured RBC scans.
  - Scintigraphy using ^{99m}Tc heat-damaged red blood cells (RBCs) or indium^{111}-labeled platelets is more sensitive and specific for splenic uptake, making these tests the current diagnostic tools of choice.
  - RBC scintigraphy was shown to be more sensitive than the sulfur colloid test.

- Fine-needle aspiration (FNA) or percutaneous biopsy under ultrasound or CT-scan guidance can be used to confirm the diagnosis. Although useful, the samples obtained for these tests may not be sufficient enough to confirm the diagnosis and the final diagnosis will be confirmed by histopathological examination of the resected tumor.
- Once the diagnosis is suspected, the patient can be managed accordingly and nephrectomy should be avoided.
- In certain situations, partial nephrectomy is a better alternative to nephrectomy.

Further Reading

1. Gonzalez-Crussi F, Raibley S, Ballantine TV, Grosfeld JL. Splenorenal fusion: heterotopia simulating a primary renal neoplasm. Am J Dis Child. 1977;131:994–6.
2. Rosenthal JT, Bedetti CD, Labayen RF, Christy WC, Yakulis R. Right splenorenal fusion with associated hypersplenism. J Urol. 1981;126:812–4.
3. Dodds WJ, Taylor AJ, Erickson SJ, Stewart ET, Lawson TL. Radiologic imaging of splenic anomalies. AJR Am J Roentgenol. 1990;155:805–10.
4. Sty JR, Conway JJ. The spleen: development and functional evaluation. Semin Nucl Med. 1985;15:276–98.
5. Forino M, Davis GL, Zins JH. Renal splenic heterotopia, a rare mimic of renal neoplasia: case report of imaging and fine-needle aspiration biopsy. Diagn Cytopathol. 1993;9:565–9.
6. Putschar WG, Manion WC. Splenicgonadal fusion. Am J Pathol. 1956;32:15–33.
7. Maillard JC, Menu Y, Scherrer A, Witz MO, Nahum H. Intraperitoneal splenosis: diagnosis by ultrasound and computed tomography. Gastrointest Radiol. 1989;14:179–80.
8. Berman AJ, Zahalsky MP, Okon SA, Wagner JR. Distinguishing splenosis from renal masses using ferumoxide-enhanced magnetic resonance imaging. Urology. 2003;62:748.
9. Kwok CM, Chen YT, Lin HT, Su CH, Liu YS, Chiu YC, et al. Portal vein entrance of splenic erythrocytic progenitor cells and local hypoxia of liver, two events cause intrahepatic splenosis. Med Hypotheses. 2006;67:1330–2.
10. Carr NJ, Turk EP. The histological features of splenosis. Histopathology. 1992;21:549–53.
11. Al Ahmad A, Jourabian M, M. Pipelzadeh Splenorenal fusion in a 26-month-old girl. Pediatr Radiol. 2009;39:735–8.
12. Hiradfar M, Zabolinejad N, Banihashem A, A.M. Kajbafzadeh Renal splenic heterotopia with extramedullary hematopoiesis in a thalassemic patient, simulating renal neoplasm: a case report. J Pediatr Hematol Oncol. 2007;29:195–7.
13. Page JB, Lenz DL, Wong C. Right-sided intrarenal splenosis mimicking a renal carcinoma Sci. World J. 2006;6:2442–4.
14. Bhatt S, MacLennan G, Dogra V. Renal pseudotumors. Am J Roentgenol. 2007;188:1380–7.
15. Yuan S, Vaughan M, Agoff SN. Left-sided splenorenal fusion with marked extramedullary hematopoiesis and concurrent lithium toxicity. A case report and review of the literature. Arch Pathol Lab Med. 2003;127:e1–3.
16. Lamin E, Smith ZL, Ramchandani P. Intrarenal splenosis diagnosed in an incidentally found left renal mass. Urol Case Rep. 2015;3:132–4.
17. Bock DB, King BF, Hezmall HP. Splenosis presenting as a left renal mass indistinguishable from renal cell carcinoma. J Urol. 1991;146:152–4.
18. Kiser JW, Fagien M, Clore FF. Splenosis mimicking a left renal mass. AJR. 1996;167:1508–9.
19. Bhatt S, Mac Lennan G, Dogra V. Renal pseudotumors. AJR. 2007;188:1380–7.
20. Paterson A, Frush DP, Donnelly LF, Foss JN, O'Hara SM, Bisset GS. A pattern-oriented approach to splenic imaging in infants and children. Radiographics. 1999;19:1465–85.

21. Balli O, Karcaaltincaba M, Karaosmanoglu D, Akata D. Multidetector computed tomography diagnosis of fusion of pancreas and spleen confirmed by magnetic resonance imaging. J Comput Assist Tomogr. 2009;33:291–2.
22. Rosenthal JT, Bedetti CD, Labayen RF, et al. Right splenorenal fusion with associated hypersplenism. J Urol. 1981;126:812–4.
23. Obley DL, Slasky BS, Bron KM. Right-sided splenorenal fusion with arteriographic, ultrasonic, and computerized tomographic correlation. Urol Radiol. 1982;4:221–5.
24. Varma DR, Sirineni GR, Rao MV, et al. Sonographic and CT features of splenogonadal fusion. Pediatr Radiol. 2007;37:916–9.
25. Gonzalez-Crussi F, Raibley S, Ballantine TV, et al. Splenorenal fusion: heterotopia simulating a primary renal neoplasm. Am J Dis Child. 1977;131:994–6.
26. Constantine E, Schmeller N, Hofstetter A. Fusion of an ectopic spleen with the left kidney. Preoperative diagnosis: adenocarcinoma of the kidney. Urologe A. 1985;24:227–8.
27. Tallman IM. Accessory spleens in epididymis and spermatic cord. Virchows Arch. 1926;259:237.
28. Von Hochstetter A. Spleen tissue in the left ovary of the left individual part of a human thoracopagus. Virchows Arch. 1953;324:36–54.

13 Wandering Spleen

13.1 History and Introduction

- The spleen is normally located in the left upper quadrant of the abdomen, protected by the rib cage and fixed in place by ligamentous attachments.
- Wandering spleen is an interesting condition that results from congenital absence or weakness of ligamentous attachment of the spleen to the surrounding structures. Normally, these ligamentous attachments help to keep the spleen located in its normal position.
- Panoralus in 1653 was the first to describe the spleen as a "ductless gland".
- In 1667, Van Horne was the first to recognize and describe wandering spleen (Fig. 13.1).
- In 1854, Józef Dietl, a Polish physician was the first to describe wandering spleen in a child (Figs. 13.2 and 13.3).
- In 1856 Józef Dietl (1804–1878), a Polish physician described the second case of wandering spleen and in 1863 Dietl himself described a third case of a wandering spleen.
  - He described wandering spleen as a life-threatening condition because it can lead to an extensive peritonitis and consequently death.
  - He also stated that it was not a patients' temperament but rather relaxation, extension, or the hypoplasia of splenic ligaments that made a spleen wander.
  - He described wandering spleen as a condition seen in women who are emaciated and exhausted by extensive work.
  - He treated wandering spleen using quinine.
  - He believed quinine decreased the size of the spleen and improved his patients' mood.
  - He also used an abdominal compression binder made of plain linen or rubber to be used for patients with wandering spleen.
  - He was against surgical removal of the spleen which was recommended by Friedrich Kűchenmeister (1821–1890) and considered the procedure too risky.

A. H. Al-Salem, *The Spleen*, https://doi.org/10.1007/978-981-99-6191-7_13

Fig. 13.1 Johannes van Horne, Joannis van Horne (surname Latinized as Hornius, 2 September 1621–5 January 1670) was a Dutch anatomist. He published an illustrated atlas of myology. He was a professor of anatomy and surgery at Leiden University. In 1667, Van Horne was the first to recognize and describe wandering spleen (Wikipedia, the free encyclopedia)

Figs. 13.2 and 13.3 Józef Dietl (24 January 1804–18 January 1878) was an Austro-Polish physician. He was born to an Austrian father and Polish mother. He studied medicine in Lviv and Vienna. Dietl described the kidney ailment known as "Dietl's Crisis" as well as its treatment. He was the first to describe a wandering spleen in a child in 1854 (Wikipedia, the free encyclopedia)

   - Instead, he used a knife to pierce the abdominal wall to provoke limited inflammation and local adhesion.
- In 1549, Adrian Zacarelli performed the first splenectomy for an enlarged spleen.
- In 1877, Matin in Berlin performed the first splenectomy for a wandering spleen.
- In 1878, Vincenz Czerny (1840–1916), in Heidelberg, made the next successful splenectomy of a wandering spleen (Fig. 13.4).
- In 1895, Ludwik Rydygier (1850–1920) performed the first splenopexy to attach a wandering spleen to the peritoneum. He fixed the lower end of the spleen in a pocket made in the parietal peritoneum (Fig. 13.5).
- In 1895, Hall carried out splenopexy by making a lumbar incision into the abdomen and fixing the spleen in it by means of tamponade.
- One of the safest and easiest methods of splenopexy is Bardenheuer's method. In this technique, the spleen lies with its inferior pole in a retroperitoneal pouch; its pedicle is fixed to the peritoneal wound, and its body is suspended from the tenth rib.
- At the beginning of the twentieth century, splenopexy became a standard surgical procedure to treat a wandering spleen.

Fig. 13.4 Vincenz Czerny (19 November 1842–3 October 1916) was a German Bohemian surgeon. His main contributions were in the fields of oncological and gynecological surgery. In 1887 Czerny performed the first open partial nephrectomy for renal carcinoma. In 1877, Matin in Berlin performed the first splenectomy for a wandering spleen. In 1878, Vincenz Czerny (1840–1916), in Heidelberg, made the next successful splenectomy of a wandering spleen (Wikipedia, the free encyclopedia)

Fig. 13.5 Ludwik Antoni Rydygier (21 August 1850–25 June 1920) was a Polish surgeon, professor of medicine, rector of the University of Lwów. He was one of the most distinguished Polish and worldwide known surgeons in the late 19th and early twentieth century. In 1881 he was the first to perform a peptic ulcer resection. Wandering spleen was first described by Van Horne, a Dutch physician, in 1667 on an autopsy. In 1877, Martin, a German obstetrician, performed the first splenectomy for a wandering spleen. In 1895, Ludwik Rydygier (1850–1920) performed the first splenopexy for a wandering spleen. He attached a wandering spleen to the peritoneum. He fixed the lower end of the spleen in a pocket made in the parietal peritoneum (Wikipedia, the free encyclopedia)

13.2 Etiology

- Embryologically, the spleen develops in the dorsal mesogastrium, and as a result of rotation of the gut it moves posterolateral to the left side.
- Fusion of the dorsal mesogastrium to the posterior abdominal wall and the left kidney forms the various ligaments including the lienorenal ligament, which contains the tail of the pancreas and the splenic artery. Failure of fusion produces an abnormally long pedicle of the spleen.
- The spleen is typically located in the left upper quadrant of the abdomen where it is held in position by various suspensory ligaments.
- These ligaments are:
  - The gastrosplenic ligament
  - The phrenocolic ligament

- The splenorenal ligament
- Wandering spleen, which is also called floating spleen, ectopic spleen, ptotic spleen, or splenoptosis, can wander anywhere in the abdominal or pelvic cavity attached only by its relatively long vascular pedicle with no ligamentous attachments to the surrounding structures.
- The splenic vessels pass within the pedicle of the spleen and torsion of the pedicle will result in a partial or complete infarction of the spleen.
- Wandering spleen or hypermobile spleen results from the abnormally long or maldeveloped suspensory ligaments of the spleen.
- As a result of this, wandering spleen has excessive mobility and can migrate from its normal position in the left hypochondrium to other parts of the abdomen or pelvis.
- The causes of the wandering spleen are either congenital or acquired.
  - Congenital causes:

    Wandering spleen occurs secondary to developmental anomalies of dorsal mesogastrium.

    The dorsal mesogastrium does not fuse with posterior peritoneum during the fifth and sixth week of fetal development leading to absence or abnormal development of one or more suspensory splenic ligaments.
  - Acquired causes:

    Wandering spleen can develop during adulthood secondary to injuries or other underlying conditions that may weaken the suspensory ligaments of the spleen. These include:

    Connective tissue disease

    Abnormal laxity of the splenic ligaments caused by multiparity

    Accident or injury

    Hormonal changes

    Splenomegaly

    Abdominal wall weakness
- Among adults, wandering spleen is seen most commonly in females of child bearing age.
  - This suggests that pregnancy may contribute to ligamentous lengthening due to laxity of the abdominal wall and hormonal changes during pregnancy.
  - This is especially common in multiparous women.
- Some congenital anomalies, such as hypermobile colon and prune belly syndrome, are known to be associated with wandering spleen.
- Children with prune belly syndrome may have wandering spleen because of the underdevelopment of the ligaments that normally hold the spleen in the left upper part of the abdomen.
- There are other known conditions that have been associated with wandering spleen. These include:
  - Absence or abnormal enlargement of a kidney
  - Infectious mononucleosis
  - Malaria
  - Sickle cell anemia
  - Hodgkin's disease

13.3 Clinical Features

- Wandering spleen is a rare condition in which the spleen is mobile and can be found in the lower abdomen or in the pelvic cavity.
- Wandering spleen develops as a result of laxity of the peritoneal attachments of the spleen.
- The true incidence of wandering spleen is unknown and is difficult to determine because the condition may be underdiagnosed.
- An estimated incidence of wandering spleen was reported of less than 0.2% and accounts for 0.002% of all performed splenectomies.
- Wandering spleen is most commonly diagnosed in young children as well as women between the ages of 20 and 40 years.
- In adults it is more commonly seen in females with a female to male ration of 7:1.
- Most women diagnosed to have wandering spleen are of reproductive age.
- Children make one-third of all reported cases of wandering spleen.
  - It is commonly seen in children under the age of ten years.
  - In children, there is no gender difference in terms of occurrence and the male-female ratio in children is 1:1.
- The clinical presentation of wandering spleen is variable.
  - It can be asymptomatic usually discovered incidentally during physical and radiographic examination for other abdominal complaints.
  - It may present as an asymptomatic, palpable abdominal mass.
  - A classic diagnostic feature of wandering spleen is appearing and disappearing abdominal mass.

    The patient will present with an abdominal mass which is palpable in the lower part of the abdomen but upon reexamination at another time, the mass may not be palpable any more.
  - It may also present with intermittent abdominal pain and discomfort. This due to torsion and spontaneous detorsion of the wandering spleen.
  - Wandering spleen can present with acute or intermittent symptoms due to torsion of the wandering spleen.

    Torsion of a wandering spleen is diagnosed in about 0.2–0.3% of patients who require splenectomy.

    Wandering spleen can present acutely with severe abdominal pain due to acute splenic torsion with subsequent infarction.

    Splenic torsion may be acute or chronic.

    Acute torsion of a wandering spleen may mimic peritonitis, acute appendicitis, twisted ovarian cysts, or ischemic bowel obstruction.

    Chronic torsion of a wandering spleen may present as an abdominal mass, which may be located in any part of the abdominal cavity.
- The most common presentation of a wandering spleen is a palpable abdominal mass with non-specific abdominal symptoms or intermittent abdominal pain and discomfort due to congestion resulting from torsion and spontaneous detorsion of the spleen.
- Other symptoms of a wandering spleen may include:

  - A bulging abdominal mass
  - Chronic constipation
  - Bloating
  - Nausea
  - Vomiting
  - Frequent difficulty in urination
  - Menstrual problems in women
- Many patients with wandering spleen are however asymptomatic.
- Clinical features seen in patients with wandering spleen include:
  - Apparent enlargement in the size of the spleen.
  - A change from the spleen's original position to another location, usually either in other parts of the abdomen or into the pelvis.
  - This feature of the wandering spleen and its ability to move from one location to another in the abdominal cavity is commonly attributed to the spleen's pedicle being abnormally long.
  - Wandering spleen can be asymptomatic and found incidentally as a palpable abdominal mass. This may be confused with an abdominal tumor.
  - Rarely torsion of the wandering spleen can also cause abdominal pain or present as a tender abdominal swelling
  - Wandering spleen may cause chronic constipation.
  - Wandering spleen may cause spleen-related complications such as hypersplenism, thrombocytopenia and lymphoma.

13.4 Investigations

- There are several radiological investigations which can be used to diagnose wandering spleen. These investigations include:
  - Abdominal ultrasonography
  - Magnetic resonance imaging
  - Abdominal CT-scan
  - Isotope scan
- Multiple imaging modalities can be used to confirm the diagnosis of wandering spleen.
- Abdominal ultrasonography will reveal:
  - A solid mobile abdominal mass
  - Absence of the spleen from its normal position
  - Doppler ultrasonography of the splenic vessels can be used to evaluate blood flow of a wandering spleen.
  - Duplex ultrasonography is more specific, but it is operator dependent, and bowel gases can also obscure the findings.
- Abdominal CT scan:
  - This remains the investigation of choice to diagnose wandering spleen
  - It can also demonstrate the spleen's circulation and the viability of splenic parenchyma.

- The typical CT findings for wandering spleen are:
  Absence of the spleen from its normal location anterior to the left kidney and posterior to the stomach.
  A lower abdominal or pelvic mass with homogenous or heterogenous splenic parenchyma, and an attenuation value less than that of normal splenic tissue.
  Multislice spiral CT is helpful in confirming the diagnosis of wandering spleen early, before the spleen progresses to infarction.
  The absence of the spleen from its normal anatomical position and the presence of soft tissue mass in the abdomen or pelvis resembling the spleen on a CT scan are highly suggestive of a wandering spleen.
  In addition, a "whirl sign" with no or partial enhancement of the splenic shape mass on intravenous contrast-enhanced CT scan in a patient presenting with the acute abdomen is strongly suggestive of torsion of the wandering spleen.

- Nuclear scintigraphy (splenic scan):
  - This is the gold standard diagnostic investigation of choice for wandering spleen.
  - The diagnosis can be established with ^{99m}Tc-sulfur-colloid scan or ^{99m}Tc-labeled, denatured RBC scans.
  - Scintigraphy using ^{99m}Tc heat-damaged red blood cells (RBCs) or indium^{111}-labeled platelets is more sensitive and specific for splenic uptake, making these tests the current diagnostic investigation of choice.
  - RBC scintigraphy was shown to be more sensitive than the sulfur colloid test.

13.5 Treatment

- The treatment of wandering spleen depends on:
  - The severity of symptoms.
  - The size, location, and function of the spleen.
- The main aim of treatment is to preserve the spleen and maximizing its function as much as possible.
- Splenic preservation is highly recommended for young patients who are known to be at increased risk for overwhelming post-splenectomy sepsis.
- Today, the only recommended treatment for wandering spleen is operative.
- Conservative management of asymptomatic wandering spleen is known to be associated with a 65% complication rate.
- There are several treatment options of wandering spleen and these include:
  - Splenopexy (fixation of the spleen):
    This is the treatment of choice
    This can be done either through an open technique or more recently laparoscopically.
  - Splenectomy:

This is indicated only if there is torsion with ischemia and infarction of the spleen after unwinding the spleen through detorsion.
Splenectomy is indicated for infracted spleen and sometimes for huge splenomegaly precluding splenopexy.
For all other cases, splenopexy is the treatment of choice.
When a wandering spleen is associated with splenomegaly that causes chronic abdominal pain, and/or deficiencies of one or more of the blood elements such as thrombocytopenia, neutropenia or anemia the treatment of choice is splenectomy.

- Acute abdominal pain associated with wandering spleen is considered a surgical emergency.
- The are several techniques of splenopexy described in the literature:
  - Splenopexy in an extra peritoneal pouch.
  - Disconnecting the gastrocolic ligament, placing the spleen at its anatomical position, and then replacing the stomach and colon; suturing the greater curvature of the stomach to the anterior abdominal wall.
  - Suturing the splenic hilum to the splenic bed.
  - Splenic snood fixation with absorbable mesh wrap.
  - The best and easiest procedure for splenopexy is Bardenheuer's procedure, where the spleen is placed in a retroperitoneal pouch with the body hanging from the tenth rib and the pedicle attached to the peritoneal incision.
- Currently, laparoscopic approach is the preferred technique and is used extensively to treat various splenic conditions because:
  - It is less painful
  - Better cosmesis
  - Early ambulation
  - Less morbidity
  - Faster recovery
- Laparoscopic splenopexy is currently the treatment of choice for wandering spleen unless:
  - The spleen is abnormally enlarged (splenomegaly)
  - The spleen is infarcted
  - There are signs of hypersplenism
- The sandwich technique was described in the literature for laparoscopic splenopexy where two meshes are used to sandwich the spleen.
- Splenectomy is indicated in the following conditions:
  - When the spleen is enlarged
  - When the spleen is infarcted
  - When the spleen is ruptured
  - When there are signs of hypersplenism
- This can be done laparoscopically or through laparotomy.

Further Reading

1. Bhanumathi V, Balkishan B, Masood SV. Torsion of wandering spleen in a woman presenting as emergency. Indian J Surg. 2013;75:59–61.
2. Calik A, Bilgin Y, Kucuktulu U, Cinel A. Intestinal obstruction caused by splenic volvulus: report of a case. Surg Today. 1996;26:704–6.
3. Blouhos K, Boulas KA, Salpigktidis I, Barettas N. Ectopic spleen: an easily identifiable but commonly misdiagnosed entity until manifestation of complications in adulthood. Int J Surg Case Rep. 2014;5(8):451–4.
4. Ayuning N, Niam MS, Jhiew P. Emergency laparoscopic splenectomy for torsion of wandering spleen in a geriatric patient: a case report—open access. Int J Surg Case Rep. 2019;61:91–5.
5. Flores-ríos E, Méndez-díaz C, Rodríguez-garcía E, Pérez-ramos T. Wandering spleen, gastric and pancreatic volvulus and right-sided descending and sigmoid colon. J Radiol Case Rep. 2015;9:18–25.
6. Magowska A. Wandering spleen: a medical enigma, its natural history and rationalization. World J Surg. 2013;37:545–50.
7. Ben Ely A, Zissin R, Copel L, Vasserman M, Hertz M, Gottlieb P, et al. The wandering spleen: CT findings and possible pitfalls in diagnosis. Clin Radiol. 2006;61:954–8.
8. Rafie BA, Abuhamdan OJ, Trengganu NS, Althebyani BH, Almatra BS. Torsion of a wandering spleen as a cause of portal hypertension and mesenteric varices: a rare aetiology. J Surg Case Rep. 2018;2018:1–4.
9. Lane TM, South LM. Management of a wandering spleen. J R Soc Med. 1999;92:84–5.
10. Dietl J. O ruchomej s´ledzionie, uwagi czerpane z ogle˛dzin pos´miertnych (About a wandering spleen from post mortem examination). Pamie˛tnik Tow Lek Warsz. 1856;12:111–23.
11. Abell I. Wandering spleen with torsion of the pedicle. Ann Surg. 1933;98:722–35.
12. Memari M, Nikzad M, Nikzad H, Taherian A. Wandering spleen in an adult man associated with the horseshoe kidney. Arch Trauma Res. 2013;2:129–32.
13. El-helou E, Alimoradi M, Sabra H, Naccour J, Zaarour M, Chebli FA, et al. Splenoptosis in young female, case report. Int J Surg Case Rep. 2020;77:214–8.
14. Awan M, Luis J, Al A, Chander V. Torsion of wandering spleen treated by laparoscopic splenopexy: a case report. Int J Surg Case Rep. 2019;62:58–61.
15. Colombo F, Amore PD, Crespi M, Sampietro G, Foschi D. Torsion of wandering spleen involving the pancreatic tail. Ann Med Surg. 2019;2020(50):10–3.
16. Martin BA. A successful case of splenotomy. Br Med. 1878;1:191–2.
17. Rydygier L. O przyszywaniu s´ledziony. Splenopexy (About sewing on of the spleen Splenopexy). Prz Lek. 1895;34:65–6.
18. Palanivelu C, Rangarajan M, Senthilkumar R, Parthasarathi R, Kavalakat AJ. Laparoscopic mesh splenopexy (Sandwich technique) for wandering spleen. JSLS. 2007;11:246–51.
19. Fiquet-Francois C, Belouadah M, Ludot H, Defauw B, Mcheik JN, Bonnet JP. Wandering spleen in children: multicenter retrospective study. J Pediatr Surg. 2010;45(7):1519–24.
20. Kinori I, Rifkin MD. A truly wandering spleen. J Ultrasound Med. 1988;7(2):101–15.
21. Allen KB, Andrews G. Pediatric wandering spleen—the case for splenopexy: review of 35 reported cases. J Pediatr Surg. 1989;24(5):432–5.
22. Stringel G, Soucy P, Mercer S. Torsion of the wandering spleen: splenectomy or splenopexy. J Pediatr Surg. 1982;17(4):373–5.
23. Taori K, Ghange N, Prakash A. Wandering spleen with torsion of vascular pedicle: early diagnosis with multiplaner reformation technique of multislice spiral CT. Abdom Imaging. 2004;29(4):479–81.
24. Soleimani M, Mehrabi A, Kashfi A, Hamidreza F, Buchler MW, Kraus TW. Surgical treatment of patient with wandering spleen: report of six cases with a review of the literature. Surg Today. 2007;37:261–9.
25. Seashore JH, McIntosh S. Elective splenopexy for wandering spleen. J Pediatr Surg. 1990;25(2):270–2.

26. Caracciolo F, Bonatti PL, Castrucci G, Fusco A, Citterio F. Wandering spleen: treatment with colonic displacement. J R Coll of Surg Edinb. 1986;31:242–4.
27. El Bouhaddouti H, Lamrani J, Louchi A, El Yousfi M. Torsion of a wandering spleen. Saudi J Gastroenterol. 2010;16(4):288–91.
28. Kapan M, Gumus M, Onder A, Gumus H. A wandering spleen presenting as an acute abdomen: case report. J Emerg Med. 2012;43(5):e303–5.
29. Buehner M, Baker MS. The wandering spleen. Collective review. Surg Gyne Obst. 1992;175:373–87.
30. Dawson JHM, Roberts NG. Management of the wandering spleen. Aust N Z J Surg. 1994;64(6):441–4.
31. Nemcek AA, Miller FH, Fitzgerald SW. Acute torsion of a wandering spleen: diagnosis by CT and duplex Doppler and color flow sonography. A J R. 1991;157(2):307–9.
32. Gayer G, Zissin R, Apter S, Atar E, Portnoy O, Itzchak Y. CT findings in congenital anomalies of the spleen. Br J Radiol. 2001;74(884):767–72.
33. Satydas T, Nasir N, Bradpiece HA. Wandering spleen: case report and literature review. J R Coll of Surg Edin. 2002;47:512–4.

Splenopancreatic Fusion 14

14.1 Introduction

- Splenopancreatic fusion is a very rare congenital anomaly that is known to be associated with other congenital disorders, mainly trisomy 13.
- The preoperative diagnosis of splenopancreatic fusion is important. This can be confused with pancreatic tumors. This is specially so in patients scheduled to have distal pancreatectomy or splenectomy. This is to avoid possible intraoperative and postoperative complications.
- Fusion of the spleen and pancreatic tail is very rare.
- Splenopancreatic fusion is usually asymptomatic discovered incidentally.
- Preoperative diagnosis of splenopancreatic fusion can be improved by the use of multidetector computed tomography or magnetic resonance imaging

14.2 Embryology

- The spleen develops from the primitive mesoderm during the fifth week of intrauterine life.
- It arises as multiple mesenchymal buds that condense between the folds of the dorsal mesogastrium during the fifth week of gestation.
- The splenic primordium grows out from the dorsal mesogastrium and rotates from the midline to the left side of the abdomen. This rotation occurs because of intestinal rotation. The spleen becomes localized in the left upper quadrant of the abdomen while acquiring its vascular supply and lymphoid tissues.
- During the tenth week, the stomach and duodenum rotate counterclockwise. As a result of this, the mesogastrium and mesoduodenum swing to the left and this will ultimately position the pancreatic tail and spleen in the left upper quadrant of the abdomen.
- The splenic buds or primordia coalesce to give rise to a single splenic mass by the end of the 12th week.

A. H. Al-Salem, *The Spleen*, https://doi.org/10.1007/978-981-99-6191-7_14

- There are several congenital malformations that can occur during the developmental stages of the spleen. These include:
  - Asplenia:
    This is characterized by total failure of splenic development.
  - Hyposplenia:
    This is a maldevelopment of the spleen and it is characterized by a spleen that is small in size and hypoplastic.
  - Polysplenia:
    This is characterized by the presence of more than one spleen.
  - Lobulated spleen:
    This is characterized by persistence of the embryonic lobulation of the spleen.
  - Spleno-visceral fusion:
    Splenogonadal fusion
    Splenopancreatic fusion
    Splenolienal fusion
    Splenoadrenal fusion
    Splenorenal fusion
  - Accessory spleens:
    These are characterized by the presence of small spleens in addition to the main spleen.
    Accessory spleens are the most commonly encountered developmental abnormality of the spleen.
    Accessory spleens develop when multifocal origins of primitive splenic mesenchymal tissues fail to fuse within the dorsal mesogastrium to form one spleen.
    Accessory spleens are relatively common. The reported incidence of accessory spleens at autopsy was 10%–30%.
    Accessory spleens are morphologically, histologically and functionally identical to normal spleens.
    Typically, accessory spleens are small in size and their size usually does not exceed 2 cm.
    They are mostly single but can be multiple.
    Approximately 75% of accessory spleens are seen at the splenic hilum.
    Accessory spleens can also be found at other sites including the pancreatic border, the splenogastric omentum, the greater omentum and pelvis.
    Accessory spleens are usually asymptomatic but they present with abdominal pain, or present acutely when complicated by torsion, hemorrhage, infarction, or rupture.
- Embryologically:
  - The pancreas is made up of two parts and is formed by the union of the ventral and dorsal buds.
  - These two buds arise in two different regions in the distal foregut.
  - The ventral bud arises from the hepatic diverticulum.
  - It forms the inferior portion of the pancreatic head and uncinate process.

  - The dorsal bud arises from the duodenum.
  - It forms the superior portion of the head, body, and tail of the pancreas.
  - The two pancreatic buds fuse together after a 180° rotation of the duodenum at the end of sixth week of intrauterine life.
  - The head and body of the pancreas grow within the dorsal mesoduodenum and extend to the dorsal mesogastrium.
  - The tail of the pancreas on the other hand lies within the dorsal mesogastrium.
- During these embryological stages of pancreatic development, several malformations can occur including:
  - One or multiple accessory spleens around the pancreatic tail
  - Intrapancreatic accessory spleen
  - Fusion of the pancreatic tail and spleen
  - Ectopic pancreatic tissue in the spleen or accessory spleen
  - Ectopic splenic tissue in the pancreas
- Fusion of the pancreatic tail and spleen can develop as the two organs are located in the dorsal mesoduodenum and dorsal mesogastrium, respectively, and are thus in close proximity.

DEVELOPMENTAL ANOMALIES OF THE SPLEEN

(1) **ASPLENIA**
(2) **HYPOSPLENIA**
(3) **POLYSPLENIA**
(4) **LOBULATED SPLEEN**
(5) **ACCESSORY SPLEENS**
(6) **SPLENOVICERAL FUSION**
- **SPLENOGONADAL FUSION**
- **SPLENOPANCREATIC FUSION**
- **SPLENOLIENAL FUSION**
- **SPLENOADRENAL FUSION**
- **SPLENORENAL FUSION**

(7) **WANDERING SPLEEN**

14.3 Clinical Features and Diagnosis

- Splenopancreatic fusion is a rare congenital abnormality, usually seen as part of the splenopancreatic field abnormalities.
- The spectrum of anomalies includes the followings:
  - Ectopic splenic tissue in the pancreas (Fig. 14.1)
  - Ectopic pancreatic tissue in the spleen or accessory spleen
  - Fusion of the tail of the pancreas and splenic hilum

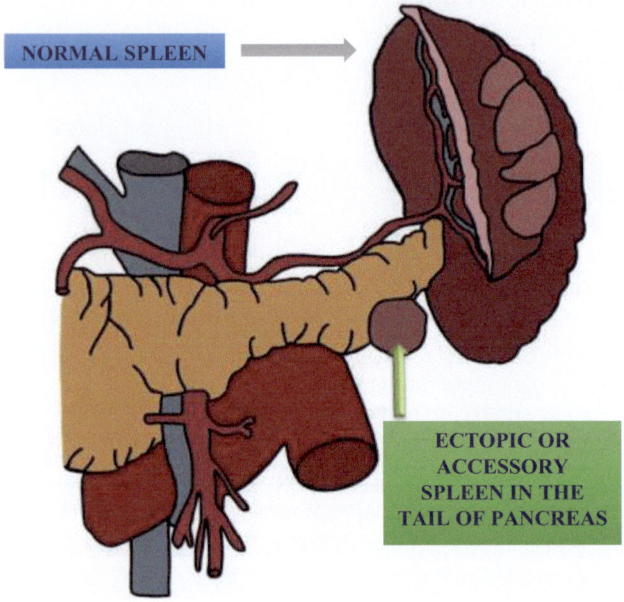

Fig. 14.1 Diagrammatic representation of ectopic splenic tissue in the pancreas

- Splenopancreatic fusion is commonly seen associated with trisomy 13.
- Other associated conditions include:
  - Osteocraniostenosis syndrome
  - Congenital heart defect
  - Oligohydramnios
  - Prune belly syndrome
  - Schinzel-Giedion syndrome
- Schinzel-Giedion syndrome:
  - This is characterized by the followings:
    Large fontanelles
    Ocular hypertelorism
    A wide, broad forehead
    Midface retraction
    A short, upturned nose
    Macroglossia
    A short neck
  - Other associated anomalies include:
    Cardiac anomalies
    Widened and dense long bone cortices
    Cerebral ventriculomegaly
    Hydronephrosis
    Abnormal fundi
  - The associated skeletal anomalies include:

A sclerotic skull base
A wide supraoccipital-exoccipital synchondrosis
Increased cortical density or thickness
Broad ribs
- Other highly supportive features include:
Neuroepithelial tumors (found in 17%)
Hypertrichosis
Brain abnormalities

- Although splenopancreatic fusion may be more common in patients with trisomy 13, splenopancreatic fusion should not be interpreted as pathognomonic to trisomy 13.
- Splenopancreatic fusion is very rare.
- Among the different types of spleno-visceral fusions, splenopancreatic fusion is probably the most common.
- Splenopancreatic fusion was reported in 0.7% of autopsies.
- Splenopancreatic fusions are usually asymptomatic.
- Their main clinical significance lies in the confusion they cause when discovered as they may be misinterpreted as pancreatic tumors. These patients may be subjected to unnecessary surgery.
- The diagnosis of splenopancreatic fusion is also important in patients who will undergo splenectomy or distal pancreatectomy to avoid possible intraoperative complications.
- The diagnosis of splenic heterotopia can be confirmed with 99 m Technetium heat-damaged red blood cell scintigraphy combined with single-photon emission CT.
- This is also valuable in distinguishing an intrapancreatic accessory spleen from pancreatic tumors.
- Triple-phase contrast-enhanced multidetector-row computed tomography and magnetic resonance imaging has been shown to be valuable in diagnosing splenopancreatic fusion.
- The preoperative diagnosis of splenopancreatic fusion is very important for surgeons specially for patients planned to undergo distal pancreatectomy or splenectomy. This is to avoid possible intraoperative or postoperative complications such as bleeding or pancreatic ductal leaks and pancreatic fistula.

Further Reading

1. Yang XY, Heller DS, Baergen RN. Splenopancreatic field abnormality in trisomy 13. Pediatr Dev Pathol. 2002;5(4):414–5.
2. Gomi K, Sato Y, Tanaka M, Ijiri R, Kato K, Aoki I, Tanaka Y. Specificity of splenopancreatic field abnormality in trisomy 13 syndrome: macroscopic and histological analysis in 21 autopsy cases. Pathol Int. 2009;59(3):147–51.
3. Peres LC, Barbosa GH, Careta RS, Nassif CM, de Pina-Neto JM, Giuliani LR, Martinhago CD, Gomy I. Splenopancreatic field abnormality is not unique to trisomy 13. Pediatr Dev Pathol. 2004;7(1):91–4.

4. Lehman AM, McFadden D, Pugash D, Sangha K, Gibson WT, Patel MS. Schinzel-Giedion syndrome: report of splenopancreatic fusion and proposed diagnostic criteria. Am J Med Genet A. 2008;146A(10):1299–306.
5. Balli O, Karcaaltincaba M, Karaosmanoglu D, Akata D. Multidetector computed tomography diagnosis of fusion of pancreas and spleen confirmed by magnetic resonance imaging. J Comput Assist Tomogr. 2009;33(2):291–2.
6. Brookes M, Zietman AL. Chapter 42: Duodenum and pancreas. In: Brookes M, Zietman AL, editors. Clinical embryology. A color atlas and text. Florida: CRC Press; 1998. p. 172–5.
7. Moore KL, Persaud TNV. Chapter 12: The digestive system. In: Moore KL, Persaud TNV, editors. The developing human: clinically oriented embryology, vol. 6. Philadelphia: W. B. Saunders Company; 1998. p. 271–302.
8. Fitzgerald MJT, Fitzgerald M. Chapter 19: Abdomen: the foregut and its mesenteries. In: Fitzgerald MJT, Fitzgerald M, editors. Human embryology. London: Bailliere Tindall; 1994. p. 119–27.
9. Javors BR, Mori H, Meyers MA, Wachberg RH. Chapter 2: Clinical embryology of the abdomen: normal and pathologic anatomy. In: Meyers MA, editor. Dynamic radiology of the abdomen: normal and pathologic anatomy, vol. 5. New York: Springer; 2000. p. 32–44.
10. Dodds WJ, Darweesh RM, Lawson TL, Stewart ET, Foley WD, Kishk SM, Hollwarth M. The retroperitoneal spaces revisited. AJR Am J Roentgenol. 1986;147(6):1155–61.
11. Brookes M, Zietman AL. Chapter 39: Stomach and spleen. In: Brookes M, Zietman AL, editors. Clinical embryology. A colour atlas and text. Florida: CRC Press; 1998. p. 160–3.
12. Kim SH, Lee JM, Han JK, Lee JY, Kim KW, Cho KC, Choi BI. Intrapancreatic accessory spleen: findings on MR imaging, CT, US and scintigraphy, and the pathologic analysis. Korean J Radiol. 2008;9(2):162–74.
13. Spencer LA, Spizarny DL, Williams TR. Imaging features of intrapancreatic accessory spleen. Br J Radiol. 2010;83(992):668–73.
14. Morgenstern L, Skandalakis JE, Hiatt JR, Phillips EH, Morgenstern L, editors. Anatomy and embryology of the spleen. Surgical diseases of the spleen. Berlin, Heidelberg: Springer; 1997. p. 15–24.
15. Yildiz AE, Ariyurek MO, Karcaaltincaba M. 2013; Splenic anomalies of shape, size, and location: pictorial essay. Sci World J. 2013;321810
16. Varga I, Babala J, Kachlik D. Anatomic variations of the spleen: current state of terminology, classification, and embryological background. Surg Radiol Anat. 2018;40:21–9.
17. Mohammadi S, Hedjazi A, Sajjadian M, Ghrobi N, Moghadam MD, Mohammadi M. Accessory spleen in the splenic hilum: a cadaveric study with clinical significance. Med Arch. 2016;70:389–91.
18. Varga I, Galfiova P, Adamkov M, et al. Congenital anomalies of the spleen from an embryological point of view. Med Sci Monit. 2009;15:RA269–76.
19. Rashid SA. Accessory spleen: prevalence and multidetector CT appearance. Malays J Med Sci. 2014;21:18–23.
20. Ko HS, Goo HW, Yoon CH. Unusual infarction of the accessory spleen or polysplenia in two children: case report. J Korean Radiol Soc. 2004;51:555–8.
21. Babcock TL, Coker DD, Haynes JL, Conklin HB. Infarction of an accessory spleen causing an acute abdomen. Am J Surg. 1974;127:336–7.
22. Onuigbo WI, Ojukwu JO, Eze WC. Infarction of accessory spleen. J Pediatr Surg. 1978;13:129–30.
23. Hems TE, Bellringer JF. Torsion of an accessory spleen in an elderly patient. Postgrad Med J. 1990;66:838–9.
24. Sheth H, Chaudhari S, Sinha Y, Prajapati R. Infarcted accessory spleen masquerading as a mesenteric cyst. BMJ Case Rep. 2018, 2018:bcr2018226130.
25. Yousef Y, Cameron BH, Maizlin ZV, Boutross-Tadross O. Laparoscopic excision of infarcted accessory spleen. J Laparoendosc Adv Surg Tech A. 2010;20:301–3.
26. Ishibashi H, Oshio T, Sogami T, Nii A, Mori H, Shimada M. Torsion of an accessory spleen with situs inversus in a child. J Med Investig. 2012;59:220–3.

27. Simon DA, Fleishman NR, Choi P, Fraser JD, Fischer RT. Torsion of an accessory spleen in a child with biliary atresia splenic malformation syndrome. Front Pediatr. 2020;8:220.
28. Ozeki M, Asakuma M, Go N, et al. Torsion of an accessory spleen: a rare case preoperatively diagnosed and cured by single-port surgery. Surg Case Rep. 2015;1:10.
29. Sadro CT, Lehnert BE. Torsion of an accessory spleen: case report and review of the literature. Radiol Case Rep. 2015;8:802.
30. Fritscher-Ravens A, Mylonaki M, Pantes A, Topalidis T, Thonke F, Swain P. Endoscopic ultrasound-guided biopsy for the diagnosis of focal lesions of the spleen. Am J Gastroenterol. 2003;98:1022–7.
31. Renno A, Hill M, Abdel-Aziz Y, Meawad H, Lenhard A, Nawras A. Diagnosis of intrapancreatic accessory spleen by endoscopic ultrasound-guided fine-needle aspiration mimicking a pancreatic neoplasm: a case report and review of literature. Clin J Gastroenterol. 2020;13:287–97.
32. Barawi M, Bekal P, Gress F. Accessory spleen: a potential cause of misdiagnosis at EUS. Gastrointest Endosc. 2000;52:769–72.
33. Ko HJ, Shim JR, Lee TB, et al. Epidermoid cyst in an intrapancreatic accessory spleen in the pancreas head: a case report. BMC Gastroenterol. 2020;20:392.
34. Kim JH, Chung KH, Oh SY, et al. Two cases of epidermoid cysts in the intrapancreatic accessory spleen mimicking pancreatic cystic neoplasm. Korean J Pancreas Biliary Tract. 2014;19:52–8.
35. Im CJ, Kweon JH, Hwang ET, et al. An intrapancreatic accessory spleen mimicking a tumor of the pancreas. Korean J Med. 2009;77:S326–31.
36. Chung YH, Park HJ, Park IS, et al. An epidermoid cyst in an intrapancreatic accessory spleen. Korean J Med. 2009;76:742–5.
37. Nishiguchi S, Habu D, Ishizu H, et al. Accessory spleen in the pelvis diagnosed by Tc-99m phytate scintigraphy. Ann Nucl Med. 2001;15:263–5.
38. Kim SH, Lee JM, Han JK, et al. Intrapancreatic accessory spleen: findings on MR Imaging, CT, US and scintigraphy, and the pathologic analysis. Korean J Radiol. 2008;9:162–74.

Splenic Cysts

15

15.1 Introduction

- Splenic cysts are relatively rare clinical conditions.
- The reported incidence of splenic cysts in the general population is 0.07%.
- Different types of splenic cysts have been described.
- Splenic cysts are classified into:
  - Primary (True splenic cysts)
  - Secondary cysts (pseudocysts)
- Primary true splenic cysts are lined by epithelial cells.
- Primary splenic cysts are subdivided into:
  - Parasitic cyst (60%)
  - Non-parasitic cysts (40%)
- Nonparasitic cysts are commonly congenital cysts.
- These cysts present mostly at young age and are commonly located in the upper pole of the spleen.
- Parasitic cysts are the most common primary cysts and are commonly secondary to Echinococcus granulosus infection.
- The majority of splenic cysts are asymptomatic, but they can present with:
  - Dull and vague upper abdominal pain.
  - This is due to mass effect of the cysts.
  - If the cysts attain a large size, they cause fullness in the left upper quadrant of the abdomen and clinically may be palpable as splenomegaly.
- Non-parasitic splenic cysts account for about 30–40% of all splenic cysts.
- Echinococcal cysts (Hydatid cyst):
  - These are usually seen in the liver and lung and occasionally in the spleen (Figs. 15.1, 15.2 and 15.3).
- Epithelial cysts:
  - These are usually seen in children or young adults.
  - They can be solitary or multiple.
  - They may be seen in association with accessory spleen.

A. H. Al-Salem, *The Spleen*, https://doi.org/10.1007/978-981-99-6191-7_15

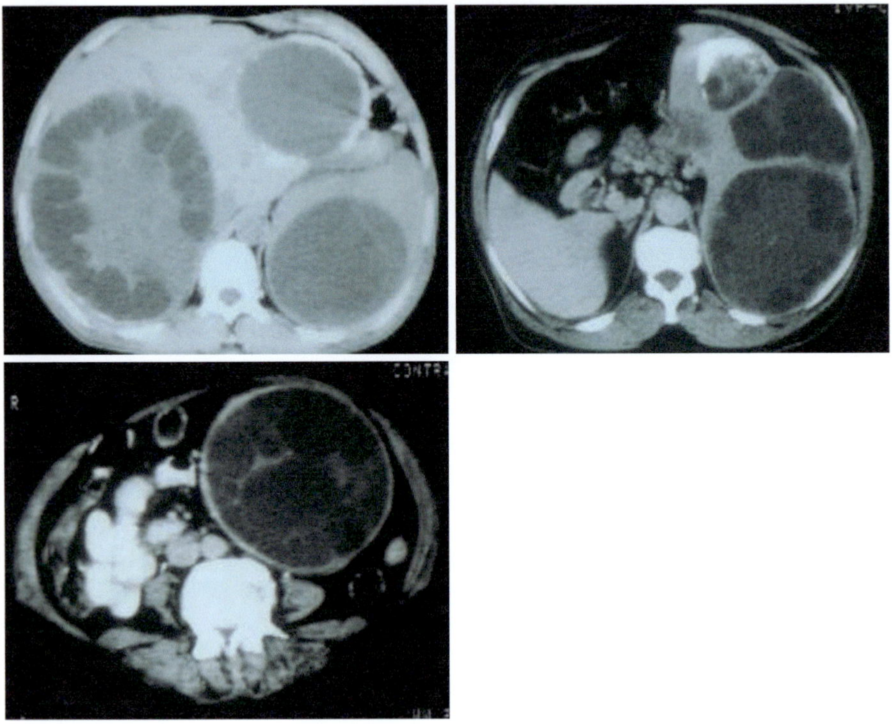

Figs. 15.1–15.3 Abdominal CT-scan showing hydatid cyst involving the liver and spleen in the first one and the spleen in the second one. Note the classic picture of hydatid cysts with daughter cysts

- Epithelial cysts are also called "epithelioid cysts" if they are lined by squamous epithelium.
- Their origin is unknown but they may develop as a result of metaplasia in mesothelial cysts.
- These cysts may reach a large size and will require splenectomy.
- Grossly:

  These cysts have glistening inner surface with marked trabeculation.
- Microscopically:

  These cysts are lined by squamous, columnar, cuboidal or mesothelial-like epithelium but without skin adnexa.

  Rarely, these cysts are mucinous and seen associated with pseudomyxoma peritonei. They stain positively for CEA, CA19–9.

• Mesothelial cysts:
  - These are also called solitary splenic lymphangioma.
  - They are usually primary cysts.
  - They may be secondary following abdominal trauma.
  - Microscopically:

    These cysts are found subcapsular.

    They are usually multicystic.

They may resemble lymphangioma.
  - They stain positive for keratin, HBME-1 and negative for factor VIII-related antigen, CD31, CD34.
- Pseudocyst:
  - These cyst account for 75% of non-parasitic splenic cysts.
  - Pseudocysts are usually secondary to abdominal trauma.
  - Some of these cysts may be epithelial cysts with denuded epithelial lining.
  - They are usually solitary and asymptomatic.
  - Their wall is composed of dense fibrous tissue without an epithelial lining.
  - They are often calcified and contain blood and necrotic debris.
  - They may rupture and cause massive hemoperitoneum.

15.2 Classification and Etiology

- Splenic cysts are rare with a reported incidence of 0.07% in the general population.
- There are four types of splenic cysts:
  - Congenital epidermoid cysts
  - Post-traumatic pseudocysts
  - Hydatid cysts due to Echinococcus granulosus infection
  - Intrasplenic pancreatic pseudocysts
- The classification of splenic cysts was first attempted by Fowler.
- He classified splenic cysts into two types:
  - Nonparasitic cysts as true cysts or primary cysts and false or pseudo cysts based on the presence or absence of an epithelial lining.
- The current classification of splenic cysts was adopted by Mirilas P et al., who classified splenic cysts into:
  - Primary splenic cysts (This includes congenital and neoplastic cysts)
  - Secondary cysts (This includes traumatic and necrotic cysts)
- True cysts are also called congenital cysts and are defined by the presence of an inner endothelial lining.
  - These are developmental in origin
  - Their lining is formed secondary to an in folding of peritoneal mesothelium or collection of mesothelial cells trapped within splenic sulci.
- Palmieri I et al. have established the pathological diagnosis of congenital cysts by immunohistochemistry using cytokeratin, CEA, CA19–9, and Calretinin.
- Primary splenic cysts comprise about 30–40% of splenic cysts and are more common in children than in adults.
- Splenic cysts are generally classified into:
  - Primary (True cysts)
  - Secondary cysts (Pseudocysts)
- Primary cysts are subdivided into:
  - Parasitic cyst (60%)
  - Non-parasitic cysts (40%)
- Post-traumatic splenic cysts are false cysts (pseudocysts) because they lack an endothelial lining.

CLASSIFICATION OF SPLENIC CYSTS

- **PRIMARY (TRUE) CYSTS**
  - **PARASITIC CYSTS (60%)**
  - **NON-PARASITIC CYSTS (40%)**
    CONGENITAL EPIDERMOID CYSTS
    MESOTHELIAL CYSTS
    NEOPLASTIC CYSTS
- **SECONDARY (PSEUDOCYSTS) CYST**

CLASSIFICATION OF SPLENIC CYSTS

- **PRIMARY CYSTS**
  - **CONGENITAL CYSTS**
  - **POST-TRAUMATIC CYSTS (WITH CELLULAR LINING)**
  - **INFLAMMATORYY CYSTS**
  - **NEOPLASTIC BENIGN CYSTS**
    DERMOID CYSTS
    EPIDERMOID CYSTS
    LYMPHANGIOMA
    HEMANGIOMA
- **SECONDARY CYSTS**
  - **TRAUMATIC CYSTS**
  - **DEGENERATIVE CYSTS**
  - **INFLAMMATORY CYSTS**

- In the United States of America, most acquired splenic cysts (approximately 80%) are post-traumatic. These are the sequela of an intrasplenic hematoma with subsequent cyst formation.
- Epidermoid cysts (also known as mesothelial or true cysts) are congenital and constitute approximately 10–20% of nonparasitic splenic cysts.
- It has been suggested that some of the previously classified post-traumatic pseudocysts may represent actually epidermoid cysts where the epithelial lining get denuded as a result of infection and histologically no epithelial lining can be seen. The inner cyst wall of these cysts will show a trabeculated appearance.
- Rarely, a pancreatic pseudocyst may extend along the splenorenal ligament into the splenic hilum and dissect into the splenic parenchyma, resulting in an intrasplenic pseudocyst.
- Hydatid cysts are the most common type of splenic cysts worldwide.

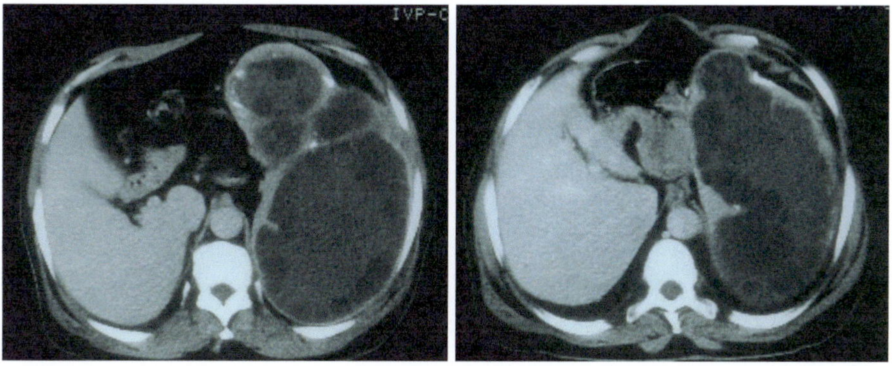

Figs. 15.4 and 15.5 Abdominal CT-scan showing hydatid cyst of the spleen

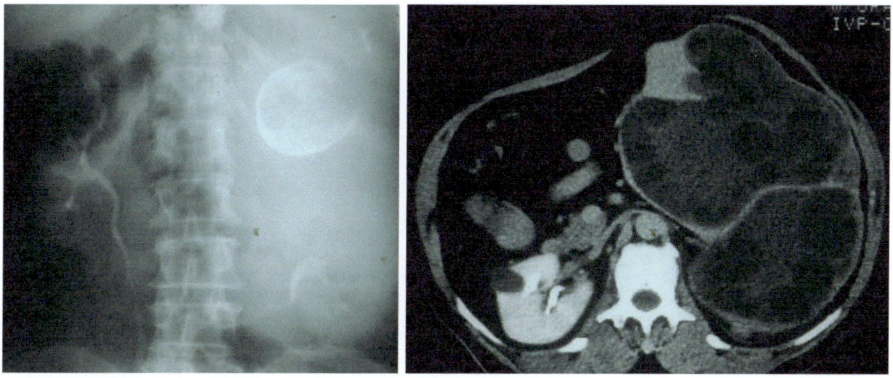

Figs. 15.6 and 15.7 Plain abdominal x-ray showing calcified cyst in the left upper quadrant and CT scan of the abdomen showing large multiple hydatid cysts of the spleen. Note the calcified wall and the characteristic multiple daughter cysts

- The spleen is the third most frequently affected site of hydatid cysts after the liver and the lung.
- The spleen is involved in 2% of cases of hydatid cysts (Figs. 15.4, 15.5, 15.6, 15.7, 15.8 and 15.9).
  - Hydatid cysts are caused by Echinococcus granulosis or mulitlocularis.
  - Their typical appearance is made up of an inner germinal layer, the endocyst, and an outer laminated layer, the ectocyst.
  - Hydatid cysts are seen more in endemic areas like the Mediterranean and Eastern Europe and rarely in the United States of America.
  - Splenic hydatid cyst should be suspected if:

    The cyst has calcified walls

    There are other associated intra-abdominal cysts especially in the liver or lungs

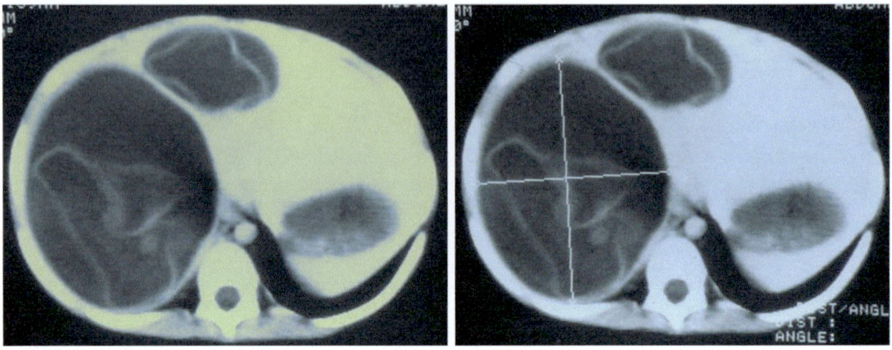

Figs. 15.8 and 15.9 Abdominal CT-scan showing hydatid cysts of the liver and spleen. Note the inner germinal layer

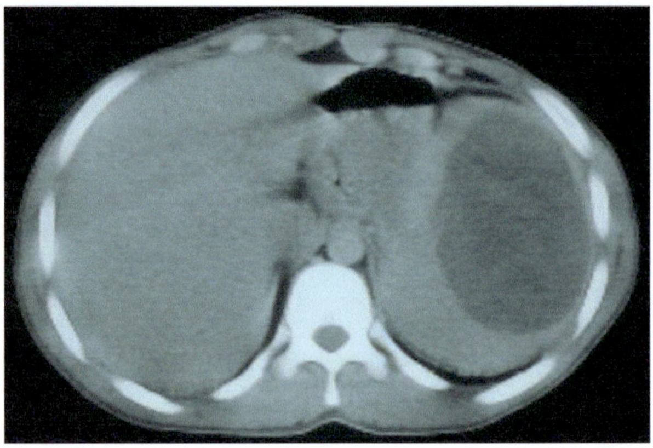

Fig. 15.10 Abdominal CT-scan showing a solitary congenital cyst of the spleen

The cyst has intraluminal daughter cysts. The presence of daughter cysts is highly suggestive of hydatid cysts.

- Primary cysts are congenital in origin and appear as simple cysts (Fig. 15.10).
  - They have a trabeculated lining with epidermoid, transitional, or mesothelial epithelium. They can contain more than one type of epithelial lining.
  - They are caused by a defect in mesothelial migration leading to unfolding of the peritoneal mesothelium or from entrapment within the splenic sulci.
  - The epithelial cells are often positive for CA 19–9 and CEA on immunohistochemistry.
  - They also stain positive for keratin which is characteristic of epithelium, and negative for factor 8 which is characteristic of endothelium.
- Another exceedingly rare form of non-parasitic primary splenic cysts is dermoid cysts.
  - These cysts contain dermal appendages like hair follicles and sweat glands.
- Secondary splenic cysts are characterized by their lack of epithelial lining.
  - They are also labelled as pseudocysts.

- Secondary splenic cysts:
  Typically have smooth borders
  Are unilocular
  Can be thick-walled with focal calcifications
- Most of these cysts are post-traumatic and form after an intrasplenic hematoma liquefies and fails to resorb.
- Traumatic splenic cysts are often characterized by a hemorrhagic interior and if they rupture, they will cause hemoperitoneum.
- Secondary splenic cysts can also develop after a splenic infarct when this area of the spleen liquefies.

• Splenic peliosis:
  - This is a rare clinical entity characterized by multiple cystic and blood-filled cavities in the spleen.
  - They develop as a result of splenic sinusoidal dilation.
  - They are found incidentally and usually are asymptomatic but they can rupture and lead to intra-abdominal hemorrhage.

15.3 Clinical Features

- Most splenic cysts are asymptomatic.
- They are usually discovered during radiological evaluation of other intraabdominal conditions.
- Splenic cysts are frequently found incidentally on abdominal ultrasound, computed tomography (CT) scan, or magnetic resonance imaging (MRI).
- Splenic cysts can be symptomatic and their usual clinical presentation includes:
  - Left upper abdominal pain or discomfort
  - Feeling of fullness
  - Early satiety
  - Nausea and vomiting
  - Left shoulder pain
  - Upper abdominal distension
  - Left pleuritic chest pain
  - Shortness of breath
  - Infected splenic cysts can present with fever, vomiting or a raised white blood cells count.
  - Splenomegaly is usually present when the cyst size is more than 6 cm.
  - Splenic cysts larger than 5 cm are more prone to complications including:
  - Hemorrhage, rupture or infection and surgical treatment is therefore recommended.
  - Hypertension may develop as a result of compression of the cyst on the left kidney

15.4 Investigations

- There are several radiological investigations which can be used to diagnose splenic cysts.
  - On abdominal ultrasound, splenic cysts are usually anechoic, with increased through-transmission.
  - They tend not to have septations unless complicated.
  - Splenic cysts are of low intensity on T1-WI and very high intensity on T2-WI relative to the normal splenic parenchyma, and they do not enhance. Variable SI may be identified within false splenic cysts on T1-WI owing to proteinaceous or hemorrhagic content.
  - On abdominal CT scan, splenic cysts typically are well-defined, fluid-attenuation, unilocular with imperceptible walls. There may be enhancement of the internal trabeculae.
  - On MRI, they are well-defined, rounded masses that have low signal intensity on T1-weighted images and high signal intensity on T2-weighted imaging and they do not enhance.
- Abdominal ultrasound:
  - On abdominal ultrasound splenic cysts are anechoic or hypoechoic intrasplenic lesions.
  - They can have internal echoes from debris.
  - There may be shadowing from calcifications in the cyst wall, internal septation, and wall trabeculation.
- Abdominal CT scan (Figs. 15.11, 15.12 and 15.13):
  - On CT scan the wall of the cyst is thin, with clear and sharp definition from the surrounding parenchyma of the spleen.
  - The rim of the cyst cavity does not enhance with intravenous contrast.
- Abdominal MRI:

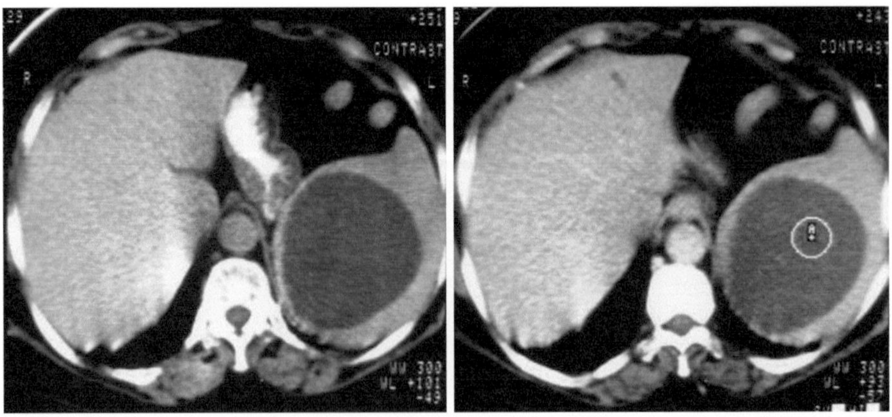

Figs. 15.11 and 15.12 Abdominal CT-scan showing a large solitary congenital cyst of the spleen

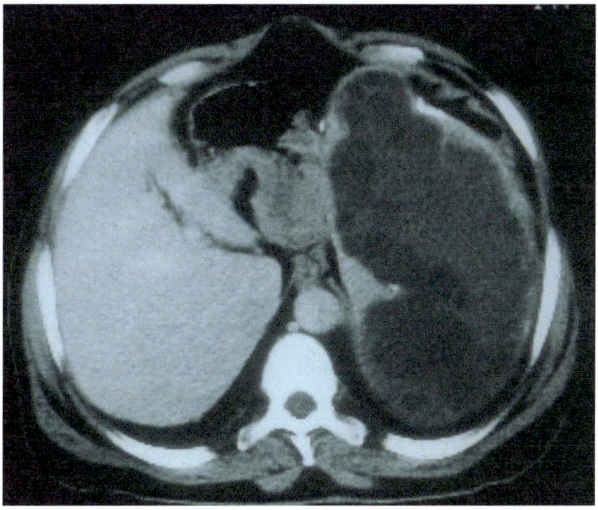

Fig. 15.13 Abdominal CT-scan showing the spleen occupied by a large hydatid cyst. Note also the multiple small daughter cysts

  - On MRI splenic cysts appear as well-defined cysts with a homogeneous fluid cavity.
- The spleen is the third common site of hydatid disease after the liver and lung.
- The diagnosis of splenic hydatid cyst often is made incidental because of the asymptomatic nature of splenic hydatid cysts.
- Parasitic cysts are usually calcified and have daughter cysts.
- Serologic tests can be used to confirm a parasitic splenic cyst.

15.5 Management

- The management of splenic cyst depends on several factors.
- These include:
  - The type of cyst, parasitic or non-parasitic cysts
  - The size of the cyst
  - The presence or absence of symptoms and complications
- Small asymptomatic splenic cysts (less than 5 cm in diameter) especially non-parasitic cysts are best treated conservatively and followed up with serial ultrasounds.
- These cysts are likely to resolve spontaneously.
- Larger cysts (more than 5 cm in diameter) and symptomatic cysts must be treated surgically.
- These cysts are known to be associated with complications including:
  - Hemorrhage
  - Rupture
  - Infection with subsequent peritonitis and abscess formation
- Surgical management options of splenic cyst include:

- Percutaneous drainage
- Total splenectomy
- Partial splenectomy
- Marsupialization
- Fenestration
- Laparoscopic deroofing of the cyst

* Some of these procedures are associated with a high recurrence rate and intraoperative bleeding.
* Partial and total splenectomy is the appropriate methods of treatment for splenic cyst which is associated with low rate of recurrences and postoperative complications.
* Spleen-preserving surgery is currently the treatment of choice of splenic cysts.
* It is estimated that preserving 25% of the spleen's parenchyma size through partial splenectomy provides sufficient immunologic function and protection for the patient.
* Total splenectomy is indicated in the following conditions:
  - Polycystic cases
  - Large cysts
  - Hilar location of the cyst
  - Cyst with fully parenchymal coverage
  - Uncontrollable massive bleeding
  - Inaccessible cysts
* Patients with splenic hydatid cysts should be covered with albendazole 10 to 15 mg/kg per day (with maximum dose of 800 mg) as post-operative medical treatment for 6 months.
* Occasionally patients will present with cyst rupture or infection, which requires surgical treatment.
* Small, asymptomatic non-parasitic splenic cysts or pseudocysts can be monitored and followed up regularly by ultrasound.
  - Small splenic cysts less than 5 cm can be observed with ultrasounds to monitor their growth or involution over time.
* Parasitic splenic cysts should be treated surgically.
* Parasitic cysts should not be drained percutaneously due to the risk of seeding and recurrence or anaphylaxis.
* "PAIR" (Puncture of cyst, aspiration of cyst contents, injection with 3% hypertonic saline, alcohol, or 0.5% silver nitrate to sterilize the cyst, and re-aspiration, in conjunction with anthelminthics).
  - This method of treatment has been described if the patient is at increased risk for surgery, and the cyst is less than 5 cm in diameter.
* Spleen preserving surgery should be adopted if the splenic cyst is small in size, solitary and located on the periphery of the spleen.
* If spleen preservation is not possible, splenectomy should be performed.
* Simple cysts can be managed in a variety of ways:
  - Total splenectomy
  - Partial splenectomy

  - Decapsulation
  - Cyst fenestration
  - Simple deroofing of the cyst
  - Percutaneous drainage is known to be associated with a high recurrence rate
- These procedures can all be performed laparoscopically.
- The trend now is moving towards spleen preservation whenever possible.
- Decapsulation:
  - This involves suctioning out the cyst, opening the anterior cyst wall and then circumferentially excising the cyst wall using the Liga Sure or Harmonic Scalpel.
  - This technique leaves the portion of the cyst wall that is stuck to the spleen, which leads to less blood loss.
  - This is technique is preferred over simple unroofing, which is known to be associated with a higher recurrence rate.
- Partial splenectomy leads to less recurrence and preserve functional splenic tissue.
- Patients undergoing partial or total splenectomy should receive vaccines (pneumococcal, meningococcal, and Hemophilus influenzae) 2 weeks prior to surgery, or 2 weeks after surgery if splenectomy was performed as an emergency.
- Symptomatic cysts should always be treated surgically.
- Laparotomy and splenectomy have been the treatment of choice for congenital splenic cysts in the past. Currently, this can be done laparoscopically.
- Splenic conservation is currently the procedure of choice to treat splenic cyst to avoid OPSI (overwhelming post splenectomy infection) in children.
- Splenic preservation techniques include:
  - Decapsulation
  - Deroofing
  - Marsupialization
  - Cyst excision
  - Partial splenectomy
- Currently, all splenic preservation procedures can be performed laparoscopically.
- Splenectomy is performed treatment only if:
  - The cyst is located in the hilum of the spleen making dissection and cyst excision difficult.
  - The cyst is large involving almost whole of the spleen and replacing nearly all of the splenic parenchyma.
  - There are dense peri splenic adhesions
  - The splenic cyst is grossly infected
- Splenic cysts involving only a part of the spleen projecting to the surface can be easily treated with:
  - Marsupialization
  - Deroofing
  - Cyst excision

- Deep seated splenic cysts which are not accessible are best treated with partial splenectomy and splenorrhaphy or in certain situations splenectomy.
- Splenic auto transplantation should always be considered in those undergoing splenectomies especially for children. This could contribute to overall protection against OPSI.

Further Reading

1. Fowler RH. Nonparasitic benign cystic tumors of the spleen. Int Abstr Surg. 1953;96:209–27.
2. Eckert J, Deplazes P. Biological, epidemiological, and clinical aspects of echinococcosis, a zoonosis of increasing concern. Clin Microbiol Rev. 2004;17:107–35.
3. Gianom D, Wildisen A, Hotz T, Goti F, Decurtins M. Open and laparoscopic treatment of nonparasitic splenic cysts. Dig Surg. 2003;20:74–8.
4. Pachter HL, Hofstetter SR, Elkowitz A, Harris L, Liang HG. Traumatic cysts of the spleen—the role of cystectomy and splenic preservation: experience with seven consecutive patients. J Trauma. 1993;35:430–6.
5. Mirilas P, Mentessidou A, Skandalakis JE. Splenic cysts: are there so many types? J Am Coll Surg. 2007;204:459–65.
6. Chin EH, Shapiro R, Hazzan D, Katz LB, Salky B. A ten-year experience with laparoscopic treatment of splenic cysts. JSLS. 2007;11:20–3.
7. Macheras A, Misiakos EP, Liakakos T, et al. Non-parasitic splenic cysts: a report of three cases. World J Gastroenterol. 2005;11:6884–7.
8. Comitalo JB. Laparoscopic treatment of splenic cysts. JSLS. 2001;5:313–6.
9. Tagaya N, Oda N, Furihata M, et al. Experience with laparoscopic management of solitary symptomatic splenic cysts. Surg Laparosc Endosc Percutan Tech. 2002;12:279–82.
10. Karlo CA, Stolzmann P, Do RK, Alkadhi H. Computed tomography of the spleen: how to interpret the hypodense lesion. Insights Imaging. 2013;4:65–76.
11. Balzan SM, Riedner CE, Santos LM, Pazzinatto MC, Fontes PR. Posttraumatic splenic cysts and partial splenectomy: report of a case. Surg Today. 2001;31:262–5.
12. Hansen MB, Moller AC. Splenic cysts. Surg Laparosc Endosc Percutan Tech. 2004;14:316–22.
13. Franquet T, Montes M, Lecumberri FJ, Esparza J, Bescos JM. Hydatid disease of the spleen: imaging findings in nine patients. AJR Am J Roentgenol. 1990;154(3):525–8.
14. Vezakis A, Dellaportas D, Polymeneas G, et al. Two cases of primary splenic hydatid cyst in Greece. Korean J Parasitol. 2012;50:147–50.
15. Kawashima A, Fishman E. Benign splenic lesions. In: Gore RM, Levine MS, Laufer I, editors. Textbook of gastrointestinal radiology. Philadelphia: Saunders; 1994. p. 2251–99.
16. Freeman JL, Jafri SZ, Roberts JL, Mezwa DG, Shirkhoda A. CT of congenital and acquired abnormalities of the spleen. Radiographics. 1993;13:597–610.
17. Calligaris L, Bortul M. Laparoscopic treatment or a nonparasitic splenic cyst: case report. J Laparoendosc Surg. 1996;6:431–4.
18. Ough YD, Nash HR, Wood DA. Mesothelial cysts of the spleen with squamous metaplasia. Am J Clin Pathol. 1981;76:666–9.
19. Palmieri I, Natale E, Crafa F, Cavallaro A, Mingazzini PL. Epithelial splenic cysts. Anticancer Res. 2005;25:515–21.
20. Morgenstern L. Nonparasitic splenic cysts: pathogenesis, classification, and treatment. J Am Coll Surg. 2002;194:306–14.
21. Smith ST, Scott DJ, Burdick JS, Rege RV, Jones DB. Laparoscopic marsupialization and hemisplenectomy for splenic cysts. J Laparoendosc Adv Surg Tech A. 2001;11:243–9.
22. Redmond HP, Redmond JM, Rooney BP, Duignan JP, Bouchier-Hayes DJ. Surgical anatomy of the human spleen. Br J Surg. 1989;76:198–201.

23. Karfis EA, Roustanis E, Tsimoyiannis EC. Surgical management of nonparasitic splenic cysts. JSLS. 2009;13:207–12.
24. Morgenstern L, Shapiro SJ. Partial splenectomy for nonparasitic splenic cysts. Am J Surg. 1980;139:278–81.
25. Williams RJ, Glazer G. Splenic cysts: changes in diagnosis, treatment and aetiological concepts. Ann R Coll Surg Engl. 1993;75:87–9.
26. Mackenzie RK, Youngson GG, Mahomed AA. Laparoscopic decapsulation of congenital splenic cysts: a step forward in splenic preservation. J Pediatr Surg. 2004;39:88–90.
27. Schier F, Waag KL, Ure B. Laparoscopic unroofing of splenic cysts results in a high rate of recurrences. J Pediatr Surg. 2007;42:1860–3.
28. Sinha CK, Agrawal M. Nonparasitic splenic cysts in children: current status. Surgeon. 2011;9:49–53.
29. Gezer HÖ, Oğuzkurt P, Temiz A, İnce E, Ezer SS, et al. Spleen salvaging treatment approaches in non-parasitic splenic cysts in childhood. Indian J Surg. 2016;78:293–8.
30. Garza-Serna U, Ovalle-Chao C, Martinez D, Flores-Villalba E, Diaz-Elizondo JA, et al. Laparoscopic partial splenectomy for congenital splenic cyst in a pediatric patient: case report and review of literature. Int J Surg Case Rep. 2017;33:44–7.
31. Pisters PW, Pachter HL. Autologous splenic transplantation for splenic trauma. Ann Surg. 1994;219:225–35.
32. Leemans R, Harms G, Rijkers GT, Timens W. Spleen autotransplantation provides restoration of functional splenic lymphoid compartments and improves the humoral immune response to pneumococcal polysaccharide vaccine. Clin Exp Immunol. 1999;117:596–604.
33. Kenney CD, Hoeger YE, Yetasook AK, et al. Management of non-parasitic splenic cysts: does size really matter? J Gastrointest Surg. 2014;18:1658–63.
34. Arikanoglu Z, Taskesen F, Gumus H, et al. Selecting a surgical modality to treat a splenic hydatid cyst: total splenectomy or spleen-saving surgery? J Gastrointest Surg. 2012;16:1189–93.
35. Matsubayashi H, Kuraoka K, Kobayashi Y, et al. Ruptured epidermoid cyst and haematoma of spleen: a diagnostic clue of high levels of serum carcinoembryonic antigen, carbohydrate antigen 19-9 and Sialyl Lewis x. Dig Liver Dis. 2001;33:595–9.
36. Sardi A, Ojeda HF, King D Jr. Laparoscopic resection of a benign true cyst of the spleen with the harmonic scalpel producing high levels of CA 19-9 and carcinoembryonic antigen. Am Surg. 1998;64:1149–54.
37. Davidson J, Tung K. Splenic peliosis: an unusual entity. Br J Radiol. 2010;83:e126–8.
38. Ingle SB, Hinge Ingle CR, Patrike S. Epithelial cysts of the spleen: a minireview. World J Gastroenterol. 2014;20:13899–903.
39. Dachman AH, Ros PR, Murari PJ, et al. Nonparasitic splenic cysts: a report of 52 cases with radiologic-pathologic correlation. AJR Am J Roentgenol. 1986;147:537–42.
40. Musy PA, Roche B, Belli D, et al. Splenic cysts in pediatric patients—a report on 8 cases and review of the literature. Eur J Pediatr Surg. 1992;2:137–40.
41. Rasheed K, Zargar SA, Telwani AA. Hydatid cyst of spleen: a diagnostic challenge. N Am J Med Sci. 2013;5:10–20.
42. Keckler SJ, Peter SD, Tsao K, et al. Laparoscopic excision of splenic cysts: a comparison to the open approach. Eur J Pediatr Surg. 2010;20:287–9.
43. Mertens J, Penninckx F, DeWever I, et al. Long-term outcome after surgical treatment of nonparasitic splenic cysts. Surg Endosc. 2007;21:206–8.
44. Wu HM, Kortbeek JB. Management of splenic pseudocysts following trauma: a retrospective case series. Am J Surg. 2006;191:631–4.

Splenic Abscess 16

16.1 Introduction

- Splenic abscess is a relatively rare clinical condition.
- The reported incidence is 0.05–0.7% in autopsy studies.
- The incidence of splenic abscess is however increasing due to the followings:
  - Increasing number of immunocompromised patients who are particularly more susceptible to develop splenic abscesses.
  - Increasing number of trauma patients.
  - Increasing number of patients with neoplastic conditions.
  - The widespread use of diagnostic imaging investigations such as computerized tomography (CT) and ultrasonography (US) for the evaluation of abdominal complaints.
  - The signs and symptoms of splenic abscesses are nonspecific and often misdiagnosed but the widespread use of modern imaging techniques has improved the diagnosis.
- Approximately two third of splenic abscesses in adults are solitary and one third are multiple.
- This is different in children where the majority of splenic abscesses are multiple and minority is solitary.
- Several factors have been described to predispose to splenic abscess.
- Sepsis continues to be the commonest predisposing factor.
- Among these, infective endocarditis is considered the commonest predisposing factor for splenic abscess.
- Other predisposing factors include:
  - Beta Thalassemia
  - Sickle cell anemia
- Tere are several causes of splenic abscesses and these include:
  - The most common cause of splenic abscess is hematogenous spread originating from an infectious focus elsewhere in the body.
  - Subacute infective endocarditis:

A. H. Al-Salem, *The Spleen*, https://doi.org/10.1007/978-981-99-6191-7_16

  - This is known to be associated with systematic embolization in 22–50% of cases and this is associated splenic abscess in a 10–20% of cases.
  - Other infective causes include:
    - Typhoid fever
    - Paratyphoid fever
    - Malaria
    - Urinary tract infection
    - Pneumonias
    - Osteomyelitis
    - Otitis media
    - Mastoiditis
    - Pelvic infections
  - Other infection that may contiguously involve the spleen leading to splenic abscess include:
    - Pancreatic abscess
    - A retroperitoneal abscess
    - A subphrenic abscesses
    - Diverticulitis
  - Splenic trauma is another well-recognized etiologic factor of splenic abscess.
  - Splenic trauma will result in splenic hematoma that may get infected.
  - Splenic infarction can also become infected and lead to splenic abscess.
  - Splenic infarctions are seen as a complication of the following conditions
    - Sickle cell disease
    - Leukemia
    - Polycythemia
    - Vasculitis
- Splenic abscesses are commonly seen in the following conditions:
  - Patients with neoplasia and immunosuppressive drugs
  - Patients with acquired immunodeficiency syndrome (AIDS) and HIV infection
  - Following abdominal trauma
  - Patients with widespread infection
  - Patients with splenic infarction
  - Patients with diabetes
  - Patients with subacute bacterial endocarditis
- The common clinical features of splenic abscess include:
  - Fever is the most common symptom
  - Left upper quadrant abdominal pain
  - Left upper quadrant abdominal tenderness
- There are several causative organisms which can cause splenic abscess including:
  - Staphylococci
  - Streptococci
  - Escherichia coli
  - M. Tuberculosis
  - Salmonella typhi

  - K. pneumonia
  - Fungi
- Splenic abscess can develop as a result of the followings:
  - Septic emboli:

    Septic emboli can reach the spleen through the blood stream and settle in the parenchyma of the spleen.

    These septic emboli can result from the following conditions:

    Vegetations from subacute bacterial endocarditis is the commonest cause of splenic abscess.

    Central line sepsis

    Bacteremia and fungemia in immunocompromised patients on chemotherapy and AIDS
  - Altered splenic architecture:

    This is seen patients following splenic trauma and who are treated conservatively.

    The splenic trauma can lead to splenic hematoma which when become infected can lead to abscess formation.

    Splenic infarcts due to leukemic deposits, sickle cells disease or vasculitis can also and in the presence of bacteremia become infected leading to abscess formation.
  - Contagious spread from surrounding pathology:

    Necrotizing pancreatitis with abscess of the lesser sac can spread to the spleen causing splenic abscess.

    Gastric or colonic perforations

    Subphrenic abscess

    A retroperitoneal abscess
- Abdominal ultrasound is diagnostic of splenic abscess (Figs. 16.1 and 16.2).
- Abdominal CT-scan is a more valuable investigation to diagnose splenic abscess:
  - It allows more accurate anatomical localization of site and size of the abscess (Figs. 16.3 and 16.4).

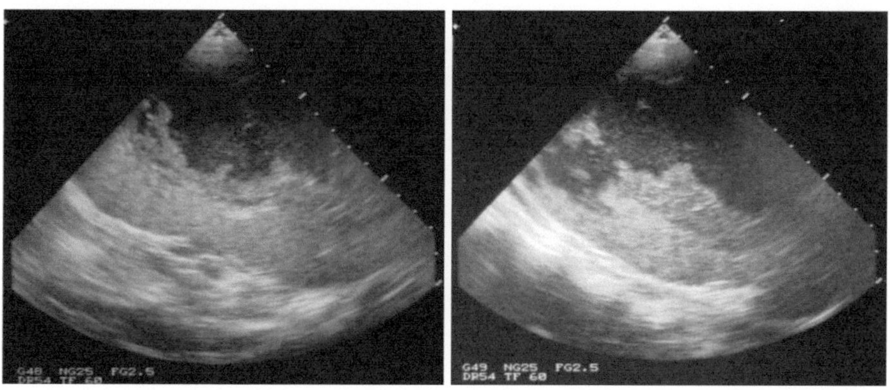

Figs. 16.1 and 16.2 Abdominal ultrasounds showing a large splenic abscess

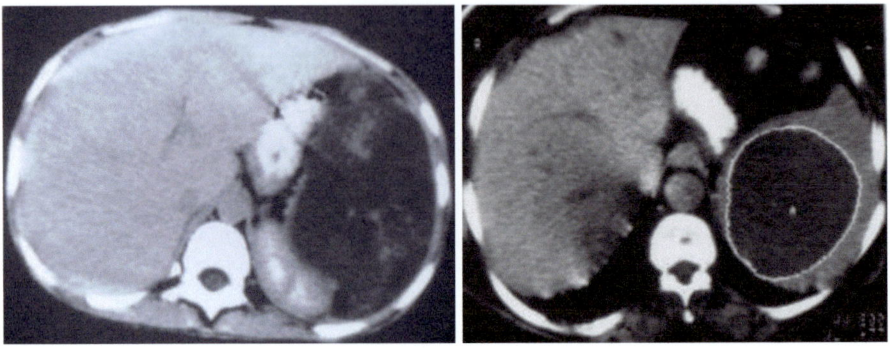

Figs. 16.3 and 16.4 Abdominal CT-scan showing splenic abscess. Note the large size of the abscess in the first photograph which is almost replacing all of the splenic tissue. In the second photograph there is a well localized splenic abscess

- Treatment of splenic abscess depends on several factors including:
  - The patient's overall condition
  - The presence or absence of comorbidities
  - The primary disorder
  - The size and location of the abscess
- Empiric broad-spectrum antibiotics should be the initial management of splenic abscesses.
- The choice of antibiotics is subsequently altered based on the culture results.
- Many splenic abscesses can be treated adequately without splenectomy.
- Percutaneous drainage of splenic abscess is an effective and less invasive treatment method than surgical intervention in selected patients with splenic abscess.
- Recently, percutaneous drainage of splenic abscess has gained acceptance among treating physicians.
- The reported success rate of percutaneous drainage ranges from 67% to 100%.
- Percutaneous drainage of splenic abscess also preserves the spleen and avoids the risk of overwhelming post splenectomy sepsis.
- Percutaneous drainage can also be used as a temporary measure to bridge the gap to elective surgery in patients who are clinically unstable or in patients who have multiple comorbidities and cannot undergo splenectomy.
- Percutaneous drainage is useful in patients who have unilocular or bilocular abscesses and if the character of the abscess content allows percutaneous drainage.
- The following conditions will not usually respond to percutaneous drainage.
- These include the followings:
  - Multilocular abscesses
  - A splenic abscess with ill-defined cavities
  - A splenic abscess with septations
  - A splenic abscess with necrotic debris
  - A splenic abscess with very thick pus

- Surgery as a treatment for splenic abscess is reserved for patients:
  - Who are stable
  - Who are not candidate to percutaneous drainage
- Splenectomy can be performed laparoscopically or via the open technique.
- In the past, total splenectomy was the treatment of choice for splenic abscess (Figs. 16.5, 16.6, 16.7 and 16.8).
- Currently and to obviate the risk of subsequent overwhelming post splenectomy sepsis, splenic preservation is being advocated. This is even in those with splenic abscess.
  - A variety of procedures have been advocated including:
    - Partial splenectomy
    - CT-guided percutaneous catheter drainage
    - Non-interventional treatment of splenic abscess with antibiotics.
- The management of splenic abscess include:

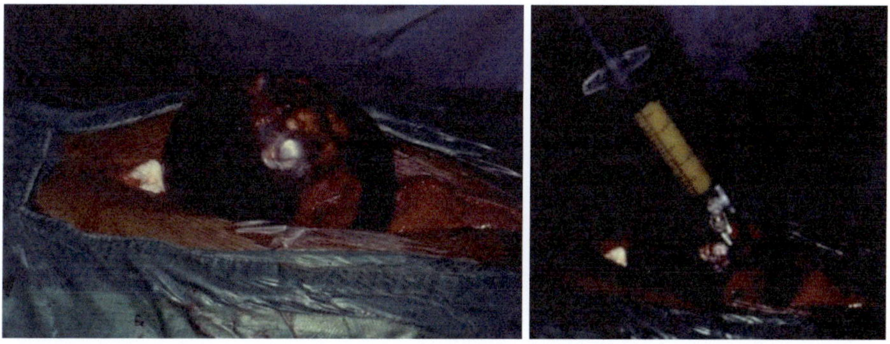

Figs. 16.5 and 16.6 Clinical intraoperative photographs showing splenic abscess. Note the abscess being aspirated

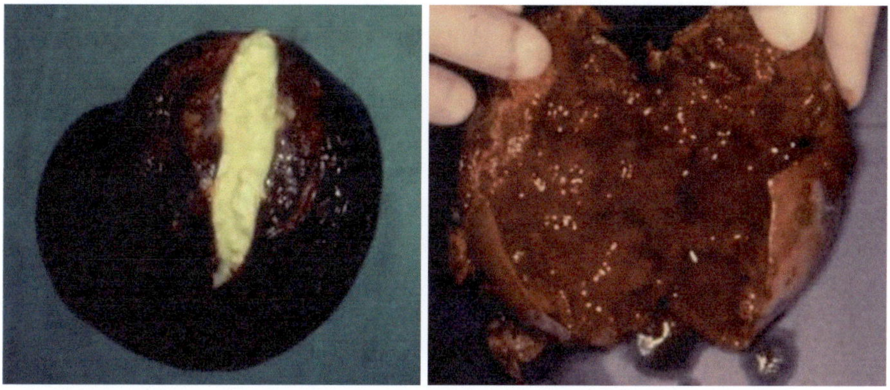

Figs. 16.7 and 16.8 Clinical photographs showing excised spleen with a large splenic abscess. In the first photograph, the splenic abscess was secondary to tuberculosis. Note also the large abscess cavity in the second photograph which is almost replacing the splenic tissue totally

- Medical therapy with antibiotics
- Percutaneous drainage (PCD) and antibiotics

  Percutaneous drainage of splenic abscess is an effective alternative to splenectomy.

  It preserves the spleen and the immunologic function by avoiding splenectomy especially in young patients.

  The success rate of percutaneous drainage for splenic abscess has been reported between 67% and 100%.

  Percutaneous drainage of splenic abscess is most likely to be successful when the abscess is localized and unilocular and when its content is not thick and liquefied enough to be drained.

  Percutaneous aspiration or drainage of splenic abscess may be used as a bridge to surgery for patients who are critically ill and at high risk for surgery.
- Partial splenectomy
- Splenectomy

16.2 Etiology and Pathophysiology

- The spleen is an uncommon site of infection and abscess formation.
- This is attributed to the intrinsic immunological functions of the spleen which provide protection for the spleen against suppurative infections.
- Splenic abscesses however can develop as a result of the followings:
  - Septic embolization

    Septic emboli can travel via the blood stream and reach the spleen to settle in the splenic parenchyma and lead to abscess formation.

    This is seen commonly in patients with subacute bacterial endocarditis where septic vegetations or emboli travel from the valves of the heart through the blood stream and settle in the spleen parenchyma to develop a splenic abscess.

    This is also seen in patients with:

    Central venous line sepsis

    Bacteremia from other source of infection

    Fungemia in immunocompromised patients

    Patients on chemotherapy

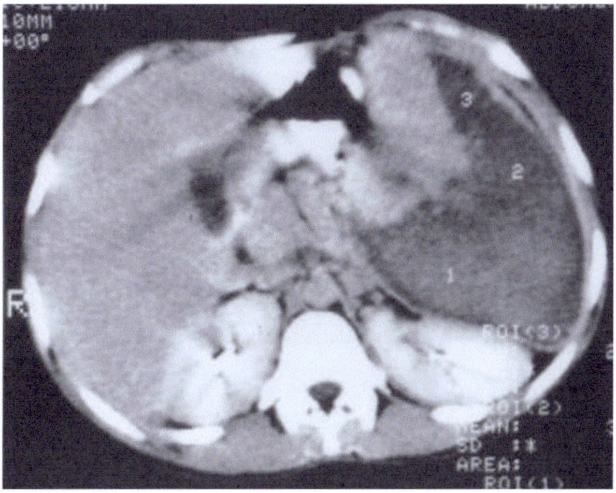

Fig. 16.9 Abdominal CT-scan showing a large splenic infarction in a patient with sickle cell disease. In the presence of bacteremia this can become infected and form a splenic abscess

Patients with acquired immunodeficiency syndrome (AIDS)

Single or multiple septic foci can in turn lead to the development of single or multiple splenic abscesses.

- Altered splenic architecture

  Splenic trauma can alter the architecture of the spleen and conservatively treated splenic trauma patients can develop a splenic abscess as a result of infection of a splenic hematoma.

  Splenic infarcts due to sickle cells disease, leukemic deposits, or vasculitis can also develop splenic abscess as a result of infection of the splenic infarcts (Fig. 16.9).

- Contagious spread of infection from surrounding pathology

  Necrotizing pancreatitis with abscess formation of lesser sac can spread to spleen causing splenic abscess.

  Other cause of contagious spread to the spleen include:

  Spread from gastric perforation

  Spread from colonic perforation

  Spread from liver abscess

  Spread from subphrenic abscess

  Spread from a retroperitoneal abscess

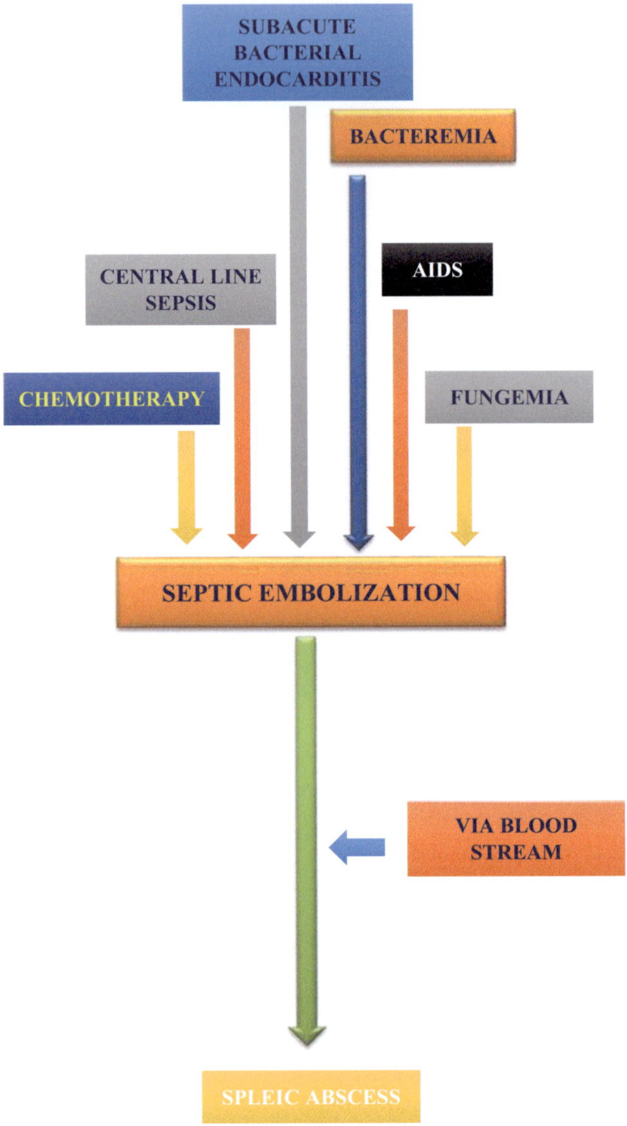

– Micro abscesses of the spleen in HIV-Infected patients

  Multiple micro abscesses of spleen can develop in patients with AIDS. These ae usually diagnosed with a high-resolution ultrasound (7.5 MHz). These micro abscesses can go unrecognized using conventional ultrasound by 3.5 MHz probe.

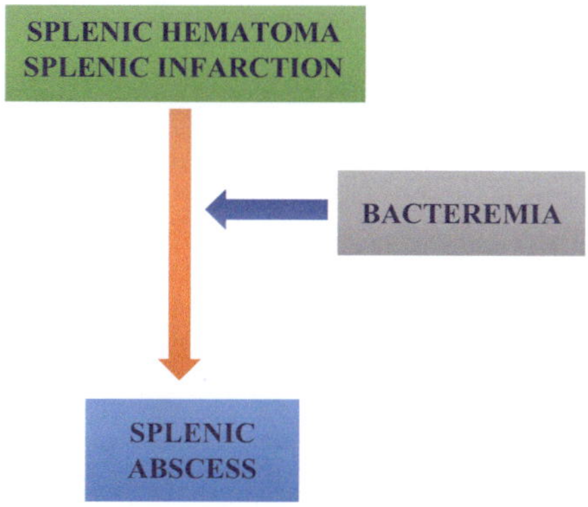

- Splenic abscesses are caused by more than one organism (Polymicrobial) in 50% of cases.
- There are several causative organisms for splenic abscess.
- A Wide spectrum of strains of organism were isolated on culture of pus from a splenic abscess namely:
  - Streptococci
  - Staphylococci
  - Enterococci
  - Escherichia Coli
  - Klebsiella Pneumonia
  - Pseudomonas Aeruginosa
  - Proteus
  - Salmonella strains
  - Mycobacterium tuberculosis
  - Anaerobes like Bacteroides
  - Fusobacterium
  - Actinomycetes
  - Brucellosis
  - Pepto streptococcus
  - Propionibacterium acnes
  - Clostridium
  - Mucromycosis
  - Burkholderia pseudo mallei
  - Fungi like Candida and Aspergillus
  - Melioidosis is reported as the commonest cause in countries like Taiwan and Singapore

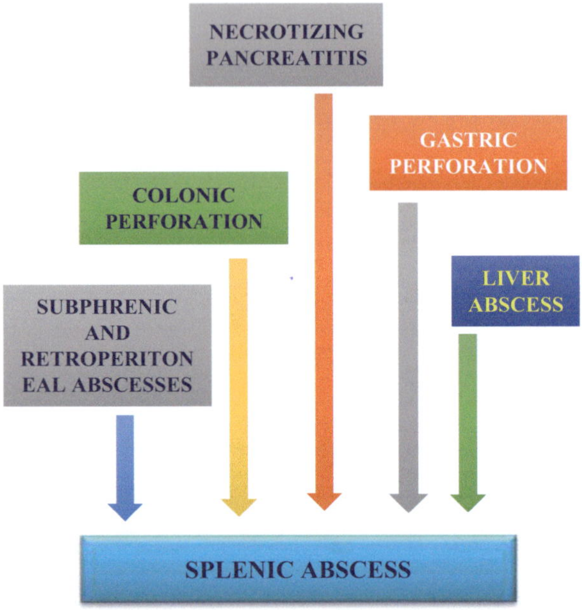

- In HIV-infected patients, Salmonella spp., and Mycobacterium tuberculosis are among the common causative organisms of splenic abscess.
- Sickle cell anemia is also known to be associated with Salmonella as a causative organism for splenic abscess, but more recent studies noted a predominance of staphylococcus infection associated with this condition.

ETIOLOGY OF SPLENIC ABSCESS

- **SEPTIC EMBOLIZATION**
  - **SUBACUTE BACTERIAL ENDOCARDITIS**
  - **CENTRAL VENOUS LINE SEPSIS**
  - **BACTEREMIA**
  - **FUNGEMIA IN IMMUNOCOMROMIZED PATIENTS**
  - **PATIENTS ON CHEMOTHERAPY**
  - **PATIENTS WITH AQUIRED IMMUNODEFICIENCY SYNDROME**
- **ALTERED SPLENIC ARCHITECTURE**
  - **SPLENIC TRAUMA AND SPLENIC HEMATOMA**
  - **SPLENIC INFARCTION**
    SICKLE CELL DISEASE
    SICKLE CELL TRAIT
    LEUKEMIC DEPOSITS
    VASCULITIS

- **CONTAGIOUS SPREAD FROM ADJACENT PATHOLOGY**
  - **NECROTIZING PANCREATITIS**
  - **SPREAD FROM GASTRIC PERFORATION**
  - **SPREAD FROM COLONIC PERFORATION**
  - **SPREAD FROM LIVER ABSCESS**
  - **SPREAD FROM A SUBPHRENIC ABSCESS**
  - **SPREAD FROM A RETROPERITONEAL ABSCESS**

16.3 Clinical Features

- Splenic abscess is a relatively rare clinical condition and remains a diagnostic challenge.
- Abscesses of the spleen have been reported since the time of Hippocrates.
- Hippocrates postulated that a patient with splenic abscess will follow one of three courses:
  - The patient may die
  - The abscess may heal
  - The abscess might become chronic and the patient lives with the disease
- The frequency of splenic abscess is a reported as 0.05–0.7% of the general population.
- The signs and symptoms of splenic abscess are variable and not specific.
- The classic triad of splenic abscess include:
  - Left upper quadrant abdominal pain
  - Fever
  - Splenomegaly
- This triad of symptoms is present in only one third of the patients.
- Fever when present as a symptom of splenic abscess can be moderate and continuous or intermittent.
- Splenomegaly is found in patients with a large splenic abscess. This is usually associated with tenderness.
- Splenomegaly can also be found in patients with previous splenomegaly that was complicated by splenic abscess.
- Some of the patients may present with Left pleuritic pain and reactionary left pleural effusion can be seen radiologically.
- As a result of irritation of the under surface of diaphragm some patients may present with left shoulder tip pain or hiccoughs.
- Many patients do not show any of these features and present only with fever and leukocytosis. In these patients the diagnosis of splenic abscess will be confirmed by abdominal ultrasound or abdominal CT scan.
- The reported mortality of splenic abscess is still high but with early diagnosis and appropriate management the mortality can be decreased to less than 10%.

- Although the signs and symptoms of splenic abscess have been well described, they are not very specific.
- The history and physical examination are not sufficiently reliable to make the diagnosis of splenic abscess.
- Splenic abscess remains a substantial diagnostic challenge.
- The symptoms of splenic abscess can be variable and depend on the site, size, and clinical progression of the abscess.
- Patients with splenic abscess may present with variable symptoms and signs including:
  - Fever may be the only manifestation of splenic abscess.
  - Fever (>90%) can be moderate, continuous, intermittent, or even absent.
  - Abdominal pain, which is either generalized or more commonly localized to the left upper quadrant.
  - Abdominal pain typically develops occurs suddenly and usually localized to the left hypochondrium (>39%). The presence of abdominal pain usually signifies peri splenitis.
  - The pain may radiate to left chest or left shoulder.
  - Involvement and irritation of the diaphragmatic pleura can cause left shoulder pain. This is called Kehr's sign.
  - Pleuritic chest pain from involvement of the left lung base (>15%) is aggravated by coughing or forced expiration.
  - Nausea, vomiting, anorexia, general malaise and weakness.
  - Abdominal tenderness is present in only half of the cases and most often in the left upper abdominal quadrant.
  - Costovertebral tenderness may also be noted.
  - Splenomegaly can be detected in about 30% of patients.
  - Chest findings are nonspecific and include:
    Dullness at the left lung base (>30%)
    Left basilar crepitations (>21%)
    Elevation of the left hemidiaphragm (>15%)
  - Other less frequent findings include:
    Splenic friction rubs
    Hepatomegaly
    Tenderness at costovertebral angle
    Ascites

16.4 Investigations

- The diagnosis of splenic abscess remains a diagnostic challenge. This is because the symptoms and signs of splenic abscess are most frequently nonspecific.
- A complete blood count may show leukocytosis in 60–80% of cases.
- Blood urea and creatinine
- Serum electrolytes
- Liver function tests

- The hemoglobin may be normal but more commonly there is an associated anemia.
- C reactive proteins are usually elevated
- It is important to do blood culture and sensitivity
- Aspirated pus from the abscess should be sent for culture and sensitivity including anaerobes, fungi and tuberculosis (Fig. 16.10).
- Abdominal and chest radiographs (Figs. 16.11, 16.12, 16.13, 16.14, and 16.15):
  - Abnormalities are detected in 50–80% of chest films and 25% of abdominal films. These include:
    - Left basilar infiltrates
    - An associated Left Pleural effusion
    - Elevated left hemidiaphragm
    - Shift of abdominal viscera
    - Plain abdominal x-ray may show an abnormal soft tissue shadow in the left upper quadrant of the abdomen
    - Plain abdominal x-ray may show gas collection with a fluid level in the left hypochondrial region.
- Abdominal ultrasound (Figs. 16.16, 16.17, 16.18 and 16.19)
  - This is the initial valuable investigation and most splenic abscesses are diagnosed by abdominal ultrasound as it is highly sensitive.
  - It can be repeated easily and can be done at bedside but it is operator dependent.
  - It is also valuable to have aspiration biopsy for culture and sensitivity under ultrasound guidance.
  - It is cost effective, noninvasive and readily available.
  - It is valuable in documenting the site, size and number of splenic abscess.
  - It is also useful in documenting whether the abscess is unilocular or multilocular.
  - Abdominal ultrasound is the initial investigation of most suspected splenic abscesses.
  - Ultrasonography typically demonstrates an area of decreased or absent echogenicity, sometimes with regular areas of echo density (debris) or gas pattern within the lesion.

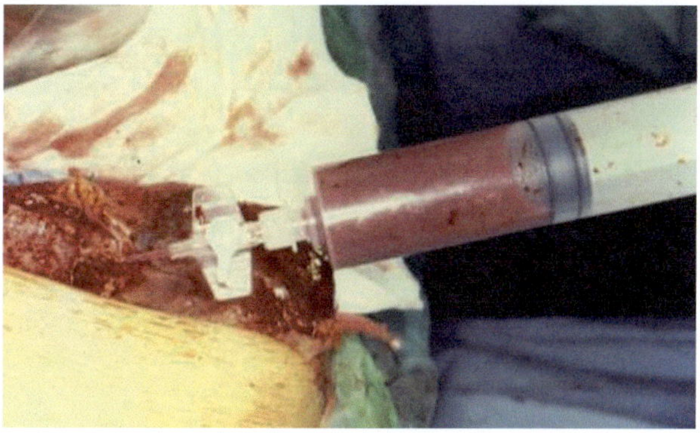

Fig. 16.10 A clinical intraoperative photograph showing aspiration of a large splenic abscess. The aspirate should be sent for culture and sensitivity including anaerobes, fungi and tuberculosis

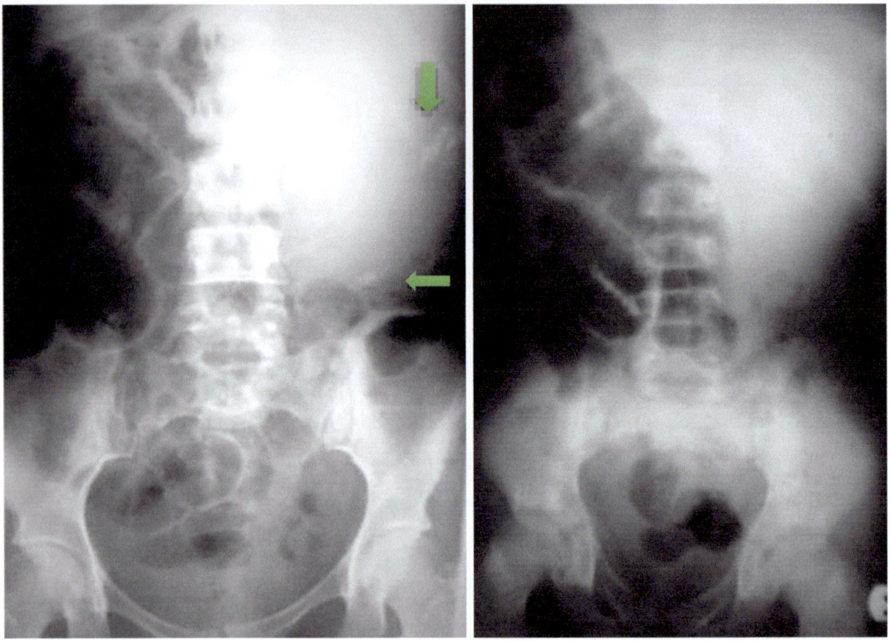

Figs. 16.11 and 16.12 Plain abdominal radiographs of patients with splenic abscesses. Note the large soft tissue density in the left upper quadrant. Note also the calcification in the first radiograph. Note also the shift of the abdominal viscera to the right side of the abdomen. Note also the slightly dilated small intestines suggestive of a paralytic ileus secondar to the adjacent infection in the spleen

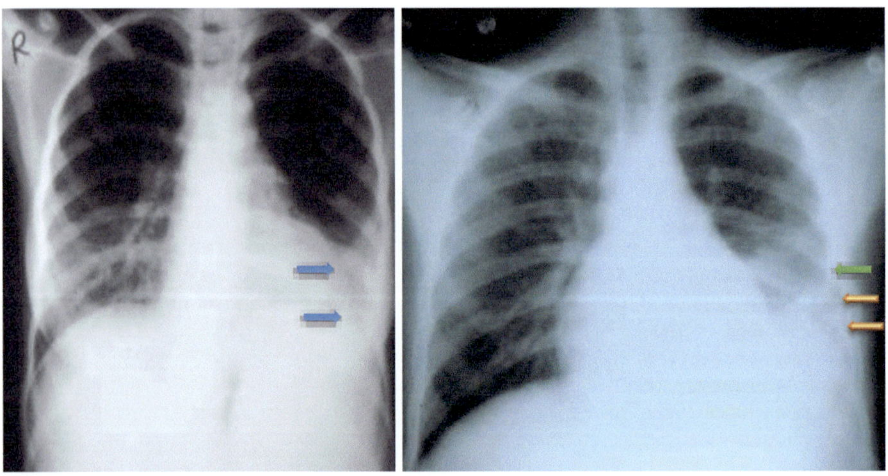

Figs. 16.13 and 16.14 Chest radiographs for two patients with splenic abscess. Note the left basal pneumonia in the first radiograph and the associated left pleural effusion in the second photograph

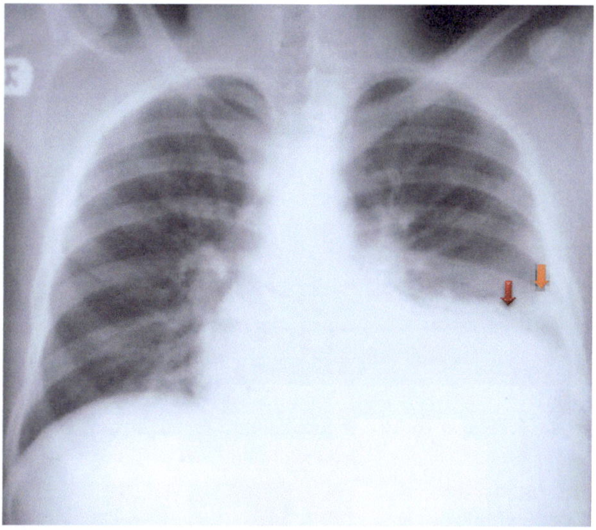

Fig. 16.15 Chest radiograph showing left pleural effusion in a patient with splenic abscess

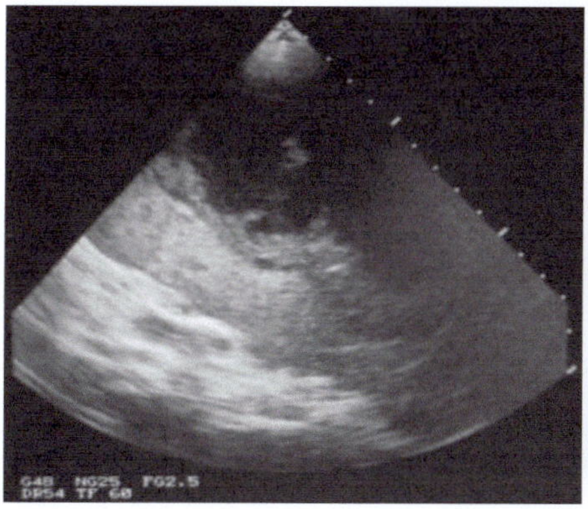

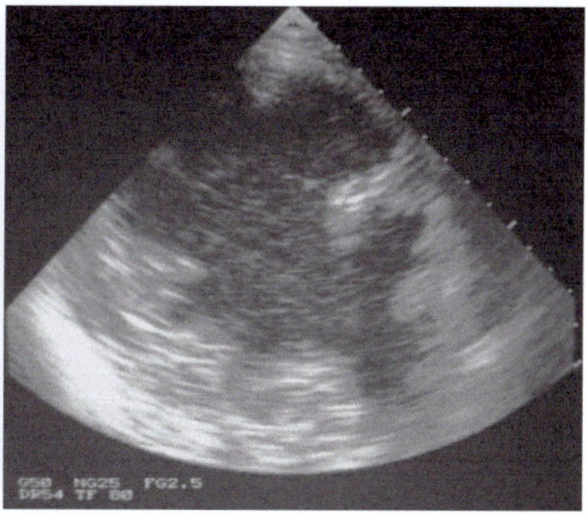

Figs. 16.16 and 16.17 Abdominal ultrasound showing splenic abscess

  - Splenomegaly can frequently be demonstrated by ultrasound.
  - High resolution (7.5 MHz) ultrasonography can detect micro-abscesses in patients infected with human immunodeficiency virus that are missed with conventional ultrasonography
- Abdominal CT scan (Figs. 16.20, 16.21, 16.22, 16.23, 16.24, 16.25 and 16.26)
  - Abdominal CT scan is very valuable and accurate investigation for the diagnosis of splenic abscess.
  - CT scan appears to be the single most sensitive investigation for the detection of splenic abscess. This is specially so if enhancement by intravenous contrast is used.
  - The reported sensitivity of CT-Scan for the diagnosis of splenic abscess is about 100%.

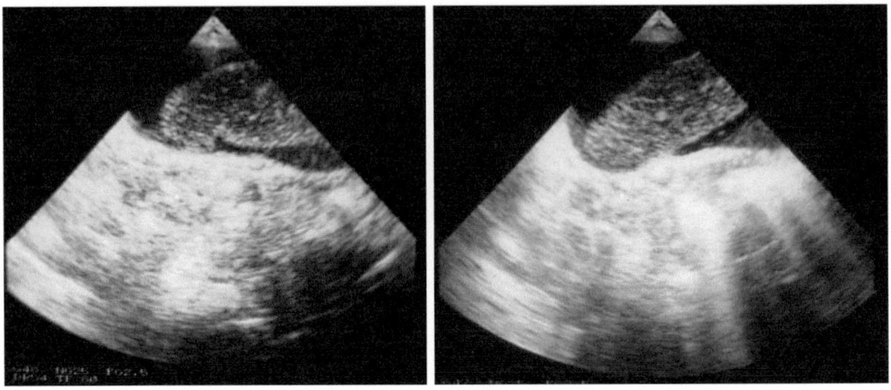

Figs. 16.18 and 16.19 Abdominal ultrasound showing ruptured large splenic abscess

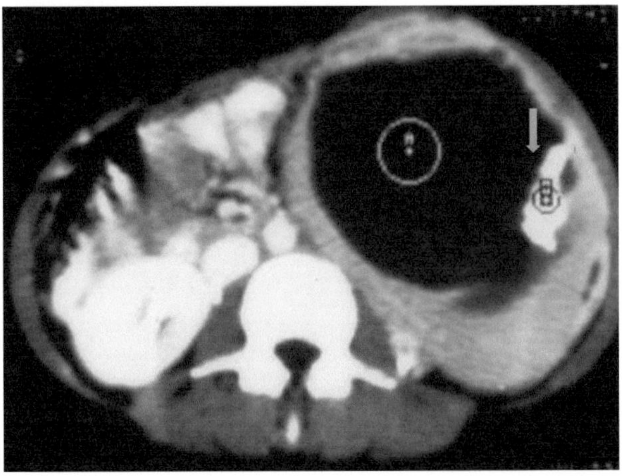

Fig. 16.20 Abdominal CT-scan showing a large splenic abscess. Note the area of calcification

- Splenic abscess on CT-scan will appear as an area of low-density fluid or necrotic tissue lesion within the relatively homogenous spleen that fails to enhance after intravenous contrast.
- Abdominal CT scan clearly outlines:
  - The site of splenic abscess
  - The size of splenic abscess
  - Whether the abscess is unilocular or multilocular
  - The number of abscesses
- CT guided drainage or aspiration of splenic abscess can also be performed. This is important to obtain a sample from the abscess for culture and sensitivity.
- Enhancement of the rim of the abscess cavity is seen in a minority of cases. This makes it difficult to differentiate splenic infarctions from splenic abscesses on contrast enhanced CT-scan.

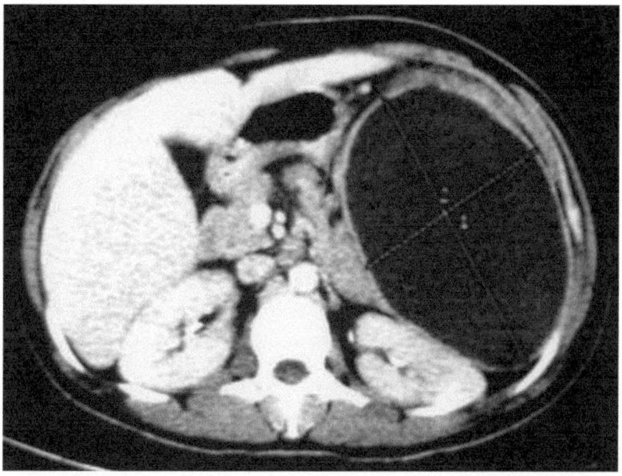

Fig. 16.21 Abdominal CT scan showing a large splenic abscess occupying the whole spleen

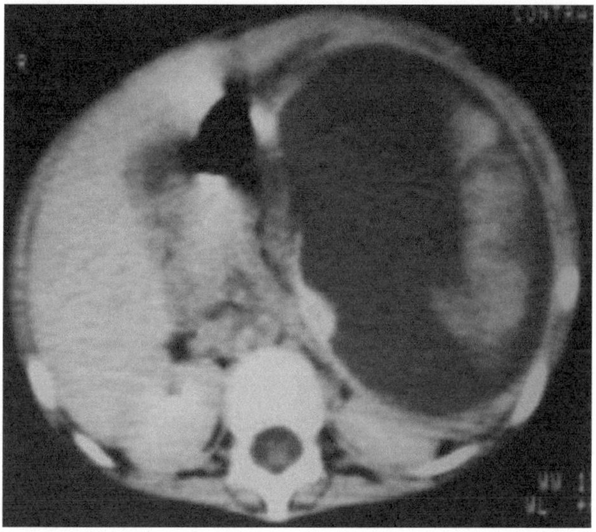

Fig. 16.22 Abdominal CT scan showing a large splenic abscess occupying the whole spleen. Note also the sequestrum of splenic tissue in the middle of abscess cavity

- The use of radionuclide scans primarily technetium 99 m sulfur-colloid liver spleen scans, has largely been replaced by ultrasonography, computed tomography and magnetic resonance imaging.
- Magnetic Resonance Imaging (MRI):
  - Magnetic resonance imaging is sensitive for detection of other abdominal abscesses.
  - It is however costly and offer fewer advantages than ultrasonography or computed tomography.

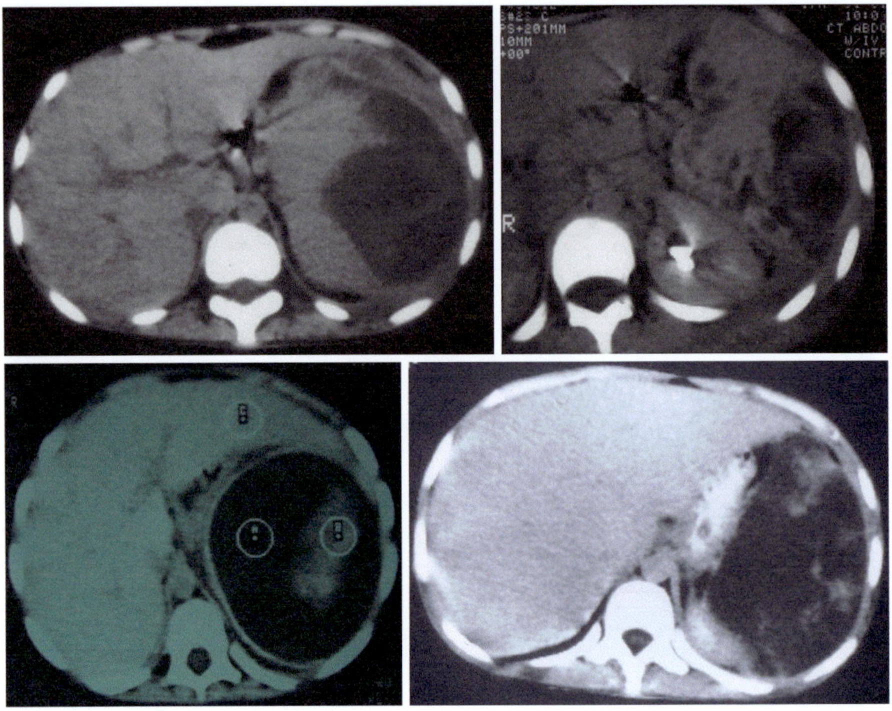

Figs. 16.23–16.26 Abdominal CT-scans showing different sizes and shapes of splenic abscess. The abscess can be localized or involve the whole spleen. Note the small multiple abscesses in the left lower picture. The splenic abscess in the right lower photograph shows a large splenic abscess involving the whole spleen

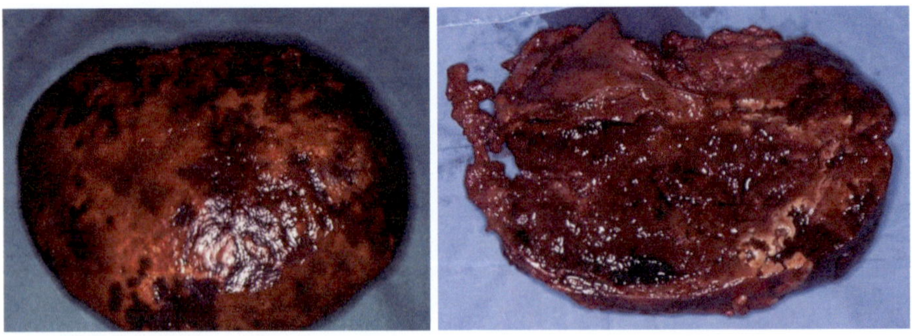

Figs. 16.27 and 16.28 Clinical photographs showing a very large splenic abscess involving the whole spleen. Note the abscess cavity in the second photograph which is occupying almost the whole spleen

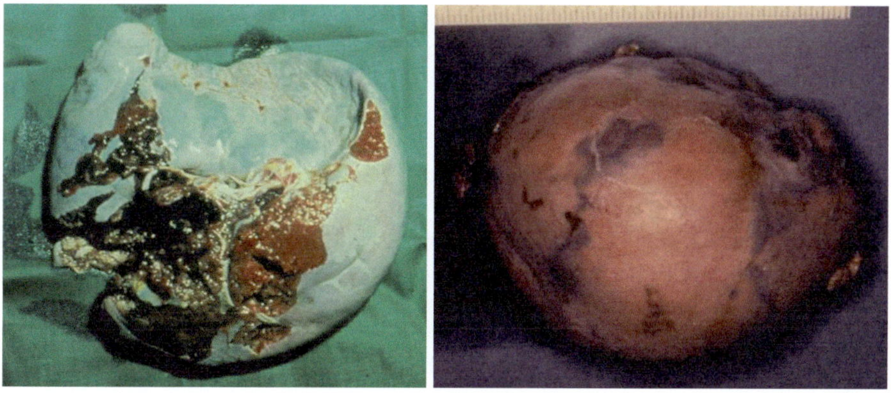

Figs. 16.29 and 16.30 Clinical photographs showing excised spleens for large splenic abscess. These were done through open laparotomy. Currently, splenectomy is done laparoscopically in selected cases of splenic abscess

16.5 Management

- Splenic abscesses should be evaluated, diagnosed and treated early.
- Untreated splenic abscess has a high mortality.
- Splenectomy has been the traditional treatment of choice and remains the gold standard for the treatment of splenic abscess (Figs. 16.27, 16.28, 16.29 and 16.30).
- There are several options to treat splenic abscess.
- These depend on:
  - The site of the abscess
  - The size of the abscess
  - Whether the abscess is unilocular or multilocular
  - The general condition of the patient
- In the past, antibiotic therapy and splenectomy were considered the treatments of choice for splenic abscess.

FACTORS AFFECTING THE MANAGEMENT OPTIONS OF SPLENIC ABSCESS

- **THE SITE OF THE ABSCESS**
- **THE SIZE OF THE ABSCESS**
- **THE GENERALCONDITION OF THE PATIENT**
- **WHETHER THE ABSCESS IS UNILOCULAR OR MULTILOCULAR**

- It is important to remember that the spleen has several important immunological functions, and splenectomy carries an increased morbidity rate with the danger of post splenectomy sepsis.
- Thus, percutaneous drainage of splenic abscesses is currently used instead of surgical treatment with good results.
- Broad spectrum antibiotics should be started early and changed according to the culture and sensitivity results.
  - Antibiotics are the first line of therapy for splenic abscess.
  - Antibiotics play an important role in the treatment of splenic abscess.
  - Broad spectrum empirical therapy should be initiated as soon as splenic abscess is suspected, pending surgical or percutaneous drainage.
  - Antibiotics should include agents against streptococci, staphylococci, and anaerobic gram-negative bacilli.
  - After blood or abscess culture results are obtained antibiotic coverage can be modified accordingly.
  - Antibiotics are also used for patients who are not fit for anesthesia and surgery and for those who underwent aspiration that was not successful.
  - Patients with multiloculated or multiseptated abscesses who are not amenable to percutaneous drainage should be treated with broad spectrum antibiotics.
  - The response of patients on antibiotics should be assessed both clinically and by abdominal ultrasound.
  - Absence of fever, improvement in the general condition of the patient are two indicators of a good clinical response.
  - A decrease in the size of the abscess by ultrasound is also an indication of a good response to antibiotics.
  - Failure of antibiotic treatment is indicated by:
    - Deterioration in the general condition of patient
    - Increasing spikes of fever
    - Marked tachycardia
    - Localized guarding in the left hypochondrium
    - Signs of overwhelming peritonitis
    - Ultrasonography may show an increase in the size of the abscess or multiple abscesses with or without involvement of adjacent organs.
- The decision to intervene either surgically by doing splenectomy or percutaneous drainage of pus is critical and has to be reached early as delay can led to septicemia and multiple organ failure.
  - If splenectomy is the treatment, it is advisable to vaccinate these patients against encapsulated bacterial pathogens as early as possible.
  - Besides the more common organisms isolated from splenic abscesses, mycobacteria, *Candida,* and *Aspergillus* should also be considered.
  - These organisms account for a small but significant number of splenic abscesses specially patients who are immunocompromised.
  - Fungal abscesses are known to respond more favorably to antifungal treatment, because they result more often from a disseminated infection.

TREATMENT OPTIONS FOR SPLENIC ABSCESS

- **ANTIBIOTICS**
- **ULTRASOUND AND CT-SCAN GUIDED PERCUTANEOUS DRAINAGE**
- **ULTRASOUND AND CT-SCAN GUIDED PERCUTANEOUS ASPIRATION**
- **OPEN OR LAPAROSCOPIC TOTAL SPLENECTOMY**
- **OPEN DRAINAGE OF THE ABSCESS**
- **OPEN OR LAPAROSCOPIC PARTIAL SPLENECTOMY**

- Ultrasound and CT-scan guided percutaneous aspiration and drainage of splenic abscesses:
- This is known to be associated with a lower initial morbidity and mortality than splenectomy and it allow preservation of the spleen.
  - Percutaneous drainage can be done under ultrasound or CT-scan guidance.
  - With increasing experience US guided percutaneous drainage is becoming safe and effective procedure.
  - Percutaneous drainage is more successful:

    When the abscess is more superficial under the capsule of the spleen and near the parietal wall.

    When the abscess is unilocular.

    Multilocular or centrally located abscess is not suitable for percutaneous drainage.
  - For a multilocular abscess, percutaneous drainage can be used to obtain a pus sample for culture and sensitivity to select the appropriate antibiotics.
  - Percutaneous drainage is a better choice in moribund patients.
  - Preservation of the spleen is the biggest advantage of percutaneous drainage.
  - Percutaneous drainage can also be used as a temporary procedure to patients who are critically ill or those who have several comorbidities that make them not fit for surgery. This procedure can buy them sometime to make them ready once surgery becomes a necessity.
  - The success rates of percutaneous drainage of splenic abscess have ranged from 50% to 90%.
  - In general, smaller (<5 cm) solitary or unilocular abscesses have a higher success rate with percutaneous drainage.
  - Small micro abscesses, as well as complex, phlegmonous or multilocular abscesses and those with thick fluid have a lower success rate with percutaneous drainage.
  - Percutaneous drainage is also useful initially, in those patients who are unstable or who present an unacceptably high surgical risk.
  - Failure to achieve effective drainage or failure of the patient's condition to improve is an indication for definitive surgery.

- Aspirated abscess fluid should be sent for Gram staining, bacterial, fungal and mycobacterial culture.
- Bleeding continues to be a concern when considering percutaneous drainage and it is important to check the coagulation profile of the patient prior to percutaneous drainage.
- Ultrasound guided percutaneous drainage is currently the most acceptable method of splenic abscess treatment. It is more successful when:
  - The abscess is unilocular or bilocular
  - The abscess has a discrete wall
  - The abscess has no internal septations
  - The abscess content is liquid enough to be drained
  - The abscess is located at the periphery and at the middle or lower pole of the spleen.
  - There are one or two abscess cavities
- Ultrasound guided percutaneous drainage is indicated when the risks of general anesthesia, surgical drainage, or splenectomy are high.
- The advantages of percutaneous drainage include:
  - External drainage without significant risks of intra-abdominal spillage
  - Avoidance of perioperative complications of surgery
  - Less intervention times
  - Less cost
  - Better cosmetic appearance and better acceptance by the patient
  - Less pain and easier nursing care
  - Preservation of immunologic function of the spleen especially in young patients
- There are contraindications to percutaneous drainage. These include:
  - Multiloculated splenic abscess
  - Debris-filled and thick pus-filled abscess
  - Multiple small abscesses
  - Uncontrollable coagulopathy
  - Poorly defined abscess on CT scan or ultrasound
  - Ascites
  - Difficult and unsafe route for drainage

INDICATIONS FOR ULTRASOUND GUIDED PERCUTANEOUS DRAINAGE OF SPLENIC ABSCESS

- **THE ABSCESS IS UNILOCULAR OR BILOCULAR**
- **THE ABSCESS HA A DISCRETE WALL**
- **THE ABSCESS HAS NO INTERNAL SEPTATIONS**
- **THE ABSCESS CONTENT IS LIQUID ENOUGH TO BE DRAINED**
- **THE ABSCESS IS LOCATED AT THE PERIPHERY AND AT THE MIDDLE OR OWER POLE OF THE SPLEEN**
- **THERE ARE ONE OR TWO ABSCESS CAVITIES**

- Relative contraindications to percutaneous drainage of splenic abscess:
  - Splenic abscess secondary to contagious sepsis in colon, pancreas, or stomach or a subdiaphragmatic abscess.
  - Ruptured splenic abscess
  - Phlegmon or a poorly localized abscess
- Complications of percutaneous drainage:
  - Bleeding
  - Iatrogenic injuries to left kidney, colon, stomach, diaphragm, and intestines.
  - Iatrogenic injury to the pleura leading to pneumothorax, hemothorax or empyema.
  - Enteric fistula

- Needle aspiration is much simpler and less aggressive than percutaneous catheter drainage, allowing shorter hospital stay and lower costs.
- Needle aspiration is used primarily in the treatment of splenic abscess with the maximal diameter less than 5 cm.
- If needle aspiration was insufficient, percutaneous drainage is the alternative choice.
- In patients with splenic abscesses longer or equal to 5 cm in maximal diameter, percutaneous drainage is used.

CONTRAINDICATIONS TO ULTRASOUND GUIDED PERCUTANEOUS DRAINAGE OF SPLENIC ABSCESS

- **MULTILOCULAR SPLENIC ABSCESS**
- **DEBRIS-FILLED AND THICK PUS FILLED SLENIC ABSCESS**
- **MULTIPLE SMALL ABSCESSES**
- **ASCITES**
- **DIFFICULT AND UNSAFE ROUTE FOR DRAINAGE**

RELATIVE CONTRAINDICATIONS TO PERCUTANEOUS DRAINAGE OF SPLENIC ABSCESS

- **SPLENIC ABSCESS SECONDARY TO CONTAGIOUS SEPSIS**
- **RUPTURED SPLENIC ABSCESS**
- **PHLEGMON OR A POORLY LOCALIZED ABSCESS**

COMPLICATIONS OF PERCUTANEOUS ULTRASOUND QUIDED DRAINAGE OF SPLENIC ABSCESS

- **BLEEDING**
- **IATROGENIC INJURIES TO THE LEFT KIDNEY, COLON, STOMACH, DIAOHRAGM AND INTESTINES**
- **IATROGENIC INJURY TO PLEURA LEADING TO PNEUMOTHORAX, HEMOTHORAX OR EMPYEMA**
- **ENTERIC FISTULA**

- Open or laparoscopic splenectomy:
  - Splenectomy whether done openly or through laparoscopy is the definite treatment of splenic abscess.
  - Laparoscopic splenectomy is currently safe and effective in selected patients.
  - In some cases of splenic abscess, the spleen can be conserved by partial splenectomy.
  - Septectomy remains the only alternative choice where:

    The abscess is multifocal or multiseptated

    Other measures including antibiotic therapy or percutaneous drainage are unsuccessful.

    There is concomitant pathology like an associated bowel perforation.

    There are known complications of percutaneous drainage of splenic abscess like bleeding or injuries to other intrabdominal organs.
  - Splenectomy is known to be associated with intraoperative and postoperative complications.
  - Another disadvantage of splenectomy is the loss of the immunological functions of the spleen.
- Open drainage of splenic abscess by laparotomy
  - Open drainage is used when the abscess cannot be drained percutaneously.
  - Open drainage of splenic abscesses can be done by a smaller incision or via an extraperitoneal approach through the bed of 11th rib.
  - Postoperatively the splenic bed can be continuously washed with saline by wide bore drainage tubes so as to remove the thick debris.
- Overwhelming Post splenectomy Sepsis (OPSS) carries a significant mortality risk, especially for young patients who have undergone splenectomy.
- Whenever splenectomy is considered, these patients should be immunized against Streptococcus pneumoniae, Meningococcus, and Hemophilus influenzae type b and should receive oral antibiotics prophylaxis.

Further Reading

1. Fotiadis C, Lavranos G, Patapis P, Karatzas G. Abscess of spleen: report of three cases. World J Gastroenterol. 2008;14(19):3088–91.
2. Sreekar H, Saraf V, Pangi AC, Sreeharsha H, Reddy R, Kamat G. Indian J Surg. 2011;73(6):398–402.
3. Bernabeu-Wittel M, Villanueva JL, Pachon J, Alarcon A, Lopez-Cortes LF, Viciana P, et al. Etiology, clinical features and outcome of splenic microabscesses in HIV-infected patients with prolonged fever. Eur J Clin Microbiol Infect Dis. 1999;18:324–9.
4. Chun C, Raff M, Contreas L. Changing clinical spectrum of splenic abscess: a multicenter study and review of literature. Am J Surg. 1987;154:27–34.
5. Jayssand R, Brasme L, Vernet-Garnier V, Deville JF, Atkas B, et al. Splenic abscess complicating Salmonella Typhi infection. Eur J Clin Microbiol Infect Dis. 2000;19:399–400.
6. Zerem E, Bergsland J. Ultrasound guided percutaneous treatment for splenic abscesses: the significance in treatment of critically ill patients. World J Gastroenterol. 2006;12(45):7341–5.
7. Chang KC, Chuah SK, Changchien CS, Tsai TL, Lu SN, Chiu YC, et al. Clinical characteristics and prognostic factors of splenic abscess: a review of 67 cases in a single medical center of Taiwan. World J Gastroenterol. 2006;12(3):460–4.
8. Westh H, Reines E, Skibsted L. Splenic abscesses: a review of 20 cases. Scand J Infect Dis. 1990;22(5):569–73.
9. Lee MC, Lee CM. Splenic abscess: an uncommon entity with potentially life-threatening evolution. Can J Infect Dis Med Microbiol. 2018;2018:8610657.
10. Al-Salem AH, Qaisaruddin S, Al Jam'a A, Al-Kalaf J, El-Bashier AM. Splenic abscess and sickle cell disease. Am J Hematol. 1998;58(2):100–4.
11. Smyrniotis V, Kehagias D, Voros D, Fotopoulos A, Lambrou A, Kostopanagiotou G, et al. Splenic abscess. An old disease with new interest. Dig Surg. 2000;17(4):354–7.
12. Nieciecki M, Kozuch M, Czarniecki M, Mlosek RK, Michno A, Olszewski W, et al. How to diagnose splenic abscesses? Acta Gastroenterol Belg. 2019;82(3):421–6.
13. Schäberle W, Eisele R. Percutaneous ultrasound-controlled drainage of large splenic abscesses. Chirurg. 1997;68(7):744–8.
14. Völk M, Strotzer M. Diagnostic imaging of splenic disease. Radiologe. 2006;46(3):229–43.
15. Iñiguez A, Butte JM, Zuñiga JM, Torres J, Llanos O. Splenic abscesses. Report of seven cases. Rev Med Chil. 2008;136(1):38–43.
16. Fotiadis C, Lavranos G, Patapis P, Karatzas G. Abscesses of the spleen: report of three cases. World J Gastroenterol. 2008;14(19):3088–91.
17. Bayer AS, Bolger AF, Taubert KA, Wilson W, Steckelberg J, Karchmer AW, et al. Diagnosis and management of infective endocarditis and its complications. Circulation. 1998;98(25):2936–48.
18. Zubair CM, Haider MM, Musa MA, Khosa ZA. Challenges in the management of multiple splenic abscesses caused by extensively drug-resistant (XDR) Salmonella Typhi. J Coll Physicians Surg Pak. 2021;30(7):855–7.
19. Pangonis S, Patamasucon P, Fitzpatrick E. Pneumococcal sepsis complicated by splenic abscesses and purpura fulminans in a 15-month-old child: case report and review of the literature. J Investig Med High Impact Case Rep. 2016;4(1):2324709616636398.
20. Albanopoulos K, Archontovassilis F, Alexakis N, Pantelidaki A, Bramis C, Leandros E. Splenic abscess in a patient with Wegener's granulomatosis treated with laparoscopic splenectomy. Mt Sinai J Med. 2006;73(7):1045–8.
21. Lee CH, Leu HS, Hu TH, Liu JW. Splenic abscess in southern Taiwan. J Microbiol Immunol Infect. 2004;37(1):39–44.
22. Tung CC, Chen FC, Lo CJ. Splenic abscess: an easily overlooked disease? Am Surg. 2006;72(4):322–5.
23. Ulhaci N, Meteoğlu I, Kacar F, Ozbaş S. Abscess of the spleen. Pathol Oncol Res. 2004;10(4):234–6.

24. Villamil-Cajoto I, Lado FL, Van den Eynde-Collado A, Díaz-Peromingo JA. Splenic abscess: presentation of nine cases. Rev Chil Infectol. 2006;23(2):150–4.
25. Al-Salem AH. Splenic complications of sickle cell anemia and the role of splenectomy. ISRN Hematol. 2011;2011:864257.
26. Yilmaz M, Arslan F, Başkan O, Mert A. Splenic abscess due to brucellosis: a case report and a review of the literature. Int J Infect Dis. 2014;20:68–70.
27. Aslam A, Ahmed Shatla ES, Imanullah S, Bedair EM. Splenic abscess: a rare complication of the UVC in newborn. Case Rep Pediatr. 2014;2014:903421.
28. Chong VH, Abdullah MS, Sangkarakumar S, Mathew VV, Telisinghe PU. Chronic splenic abscess as the first manifestation of pancreatic tail tumor. Indian J Surg. 2015;77(Suppl. 3):1504–5.
29. Schiavo L, Scalera G, De Sena G, Ciorra FR, Pagliano P, Barbarisi A. Nonsurgical management of multiple splenic abscesses in an obese patient that underwent laparoscopic sleeve gastrectomy: case report and review of literature. Clin Case Rep. 2015;3(10):870–4.
30. Nassour F, Schoucair NM, Tranchart H, Maitre S, Dagher I. Delayed intra splenic abscess: a specific complication following laparoscopic sleeve gastrectomy. Obes Surg. 2018;28(2):589–93.
31. Li G, Gao L, Zhou J, Ye B, Zhang J, Qu C, et al. Management of splenic abscess after splenic arterial embolization in severe acute pancreatitis: a 5-year single-center experience. Gastroenterol Res Pract. 2019;2019:6069179.
32. Bundy JJ, Hage AN, Srinivasa RN, Gemmete JJ, Srinivasa RN, Jairath N, et al. Intra-arterial ampicillin and gentamicin and the incidence of splenic abscesses following splenic artery embolization: a 20-year case control study. Clin Imaging. 2019;54:6–11.
33. AlZarooni N, AlBaroudi A, AlOzaibi L, AlZoabi O. Splenic abscess as a possible sequela of COVID-19: a case series. Ann Saudi Med. 2021;41(5):307–11.
34. Davido B, Dinh A, Rouveix E, Crenn P, Hanslik T, Salomon J. Splenic abscesses: from diagnosis to therapy. Rev Med Intern. 2017;38(9):614–8.
35. Moll R, Sailer M, Reith HB, Schindler G. CT-gesteuerte Drainagenbehandlung der Milz bei Abscessen und Hamatomen. Klinikarzt. 2004;33:183–8.
36. Carbonell AM, Kercher KW, Matthews BD, Joels CS, Sing RF, Heniford BT. Laparoscopic splenectomy for splenic abscess. Surg Laparosc Endosc Percutan Tech. 2004;14(5):289–91.
37. Hung SK, Ng CJ, Kuo CF, Goh ZNL, Huang LH, Li CH, et al. Comparison of the mortality in emergency department sepsis score, modified early warning score, rapid emergency medicine score and rapid acute physiology score for predicting the outcomes of adult splenic abscess patients in the emergency department. PLoS One. 2017;12(11):e0187495.
38. Liang JT, Lee PH, Wang SM, Chang KJ. Splenic abscess: a diagnostic pitfall in the ED. Am J Emerg Med. 1995;13(3):337–43.
39. Alvi AR, Kulsoom S, Shamsi G. Splenic abscess: outcome and prognostic factors. J Coll Physicians Surg Pak. 2008;18(12):740–3.
40. Klimpel V. Does Kehr's sign derive from Hans Kehr? A critical commentary on its documentation? Chirurg. 2004;75(1):80–3.
41. Handa U, Tiwari A, Singhal N, Mohan H, Kaur R. Utility of ultrasound-guided fine-needle aspiration in splenic lesions. Diagn Cytopathol. 2013;41(12):1038–42.
42. Divyashree S, Gupta N. Splenic abscess in immunocompetent patients managed primarily without splenectomy: a series of 7 cases. Perm J. 2017;21:16–139.
43. Faruque AV, Qazi SH, Arshad M, Anwar N. Isolated splenic abscess in children, role of splenic preservation. Pediatr Surg Int. 2013;29(8):787–90.
44. Okamoto T, Nakamura K, Fukuda K. Endoscopic ultrasound-guided transluminal drainage for aseptic splenic abscess. Endosc Int Open. 2021;9(10):E1581–2.
45. Legrand F, Lecuit M, Dupont B, Bellaton E, Huerre M, Rohrlich PS, et al. Adjuvant corticosteroid therapy for chronic disseminated candidiasis. Clin Infect Dis. 2008;46(5):696–702.
46. Ferraioli G, Brunetti E, Gulizia R, Mariani G, Marone P, Filice C. Management of splenic abscess: report on 16 cases from a single center. Int J Infect Dis. 2009;13(4):524–30.
47. Choudhury SR, Rajiv C, Pitamber S, Akshay S, Dharmendra S. Management of splenic abscess in children by percutaneous drainage. J Pediatr Surg. 2006;41(1):e53–6.

48. Ooi LL, Leong SS. Splenic abscesses from 1987 to 1995. Am J Surg. 1997;174(1):87–93.
49. Bona R. Evaluation of splenomegaly and other splenic disorders in adults. UpToDate. Post TW (ed): UpToDate, Waltham; 2021.
50. de Porto AP, Lammers AJ, Bennink RJ, ten Berge IJ, Speelman P, Hoekstra JB. Assessment of splenic function. Eur J Clin Microbiol Infect Dis. 2010;29:1465–73. https://doi.org/10.1007/s10096-010-1049-1.
51. Nelken N, Ignatius J, Skinner M, Christensen N. Changing clinical spectrum of splenic abscess. A multicenter study and review of the literature. Am J Surg. 1987;154:27–34.
52. Ng KK, Lee TY, Wan YL, Tan CF, Lui KW, Cheung YC, et al. Splenic abscess: diagnosis and management. Hepato-Gastroenterology. 2002;49:567–71.
53. de Bree E, Tsiftsis D, Christodoulakis M, Harocopos G, Schoretsanitis G, Melissas J. Splenic abscess: a diagnostic and therapeutic challenge. Acta Chir Belg. 1998;98:199–202.
54. Ooi LL, Nambiar R, Rauff A, Mack PO, Yap TL. Splenic abscess. Aust N Z J Surg. 1992;62:780–4.
55. Smyrniotis V, Kehagias D, Voros D, Fotopoulos A, Lambrou A, Kostopanagiotou G, et al. Splenic abscess. An old disease with new interest. Dig Surg. 2000;17:354–7.
56. Chou YH, Hsu CC, Tiu CM, Chang T. Splenic abscess: sonographic diagnosis and percutaneous drainage or aspiration. Gastrointest Radiol. 1992;17:262–6.
57. Green SL, Scott LK. Cryptogenic splenic abscess. Va Med. 1986;113:164–6.
58. Ralls PW, Quinn MF, Colletti P, Lapin SA, Halls J. Sonography of pyogenic splenic abscess. AJR Am J Roentgenol. 1982;138:523–5.
59. Carbonell AM, Kercher KW, Matthews BD, Joels CS, Sing RF, Heniford BT. Laparoscopic splenectomy for splenic abscess. Surg Laparosc Endosc Percutan Tech. 2004;14:289–91.
60. Kang M, Saxena AK, Gulati M, Suri S. Ultrasound-guided percutaneous catheter drainage of splenic abscess. Pediatr Radiol. 2004;34:271–3.
61. Thanos L, Dailiana T, Papaioannou G, Nikita A, Koutrouvelis H, Kelekis DA. Percutaneous CT-guided drainage of splenic abscess. AJR Am J Roentgenol. 2002;179:629–32.
62. Tung CC, Chen FC, Lo CJ. Splenic abscess: an easily overlooked disease? Am Surg. 2006;72:322–5.
63. Westh H, Reines E, Skibsted L. Splenic abscesses: a review of 20 cases. Scand J Infect Dis. 1990;22:569–73.
64. Paris S, Weiss SM, Ayers WH Jr, Clarke LE. Splenic abscess. Am Surg. 1994;60:358–61.
65. Ferraioli G, Brunetti E, Gulizia R, Mariani G, Marone P, Filice C. Management of splenic abscess: report on 16 cases from a single center. Int J Infect Dis. 2009;13:524–30.
66. Teich S, Oliver GC, Canter JW. The early diagnosis of splenic abscess. Am Surg. 1986;52:303–7.
67. Llenas-García J, Fernández-Ruiz M, Caurcel L, Enguita-Valls A, Vila-Santos J, Guerra-Vales JM. Splenic abscess: a review of 22 cases in a single institution. Eur J Intern Med. 2009;20:537–9.
68. Allal R, Kastler B, Gangi A, Bensaid AH, Bouali O, Cherrak C, et al. Splenic abscesses in typhoid fever: US and CT studies. J Comput Assist Tomogr. 1993;17:90–3.
69. Torres JR, Gotuzzo E, Istúriz R, Elster C, Wolff M, Northland R, et al. Salmonellal splenic abscess in the antibiotic era: a Latin American perspective. Clin Infect Dis. 1994;19:871–5.
70. Sarr MG, Zuidema GD. Splenic abscess—presentation, diagnosis, and treatment. Surgery. 1982;92:480–5.
71. Gleich S, Wolin DA, Herbsman H. A review of percutaneous drainage in splenic abscess. Surg Gynecol Obstet. 1988;167:211–6.

Splenic Infarction 17

17.1 Introduction

- Splenic infarction is a relatively rare condition and it develops when the blood flow to the spleen is compromised causing tissue ischemia and eventual necrosis.
- Splenic infarction may be the result of arterial or venous occlusion.
- Splenic infarctions are considered to be rare without an underlying predisposing factor.
- The two most common causes of splenic infarction are:
  - Thromboembolic disease
  - Hematologic diseases
- Splenic infarcts are considered a rare cause of abdominal pain but the exact prevalence is unclear.
- The discovery and diagnosis of splenic infarction is thought to be increasing due to the following factors:
  - Increased utilization of abdominal imaging studies
  - Increased use of splenic embolization as a therapy for some of the causes of splenomegaly. Splenic embolization can be complicated by splenic infarction.
  - Increased conservative nonoperative management of traumatic splenic injuries.
- Splenic infarctions are common in patients with sickle cell anemia but these are usually small and repetitive leading to auto splenectomy.
- In sickle cell anemia splenic infarctions are secondary to vasoocclusive crisis affecting the spleen.
- Massive splenic infarction is defined arbitrary as splenic infarction involving more than 50% of the spleen size (Figs. 17.1, 17.2, 17.3 and 17.4).
- Splenic infarcts are initially deep red in color, due to congestion. These subsequently become pale as necrosis and organization of the infarction develop.
- Grossly, splenic infarcts appear as wedge-shaped with the base adjacent to the splenic capsule.

A. H. Al-Salem, *The Spleen*, https://doi.org/10.1007/978-981-99-6191-7_17

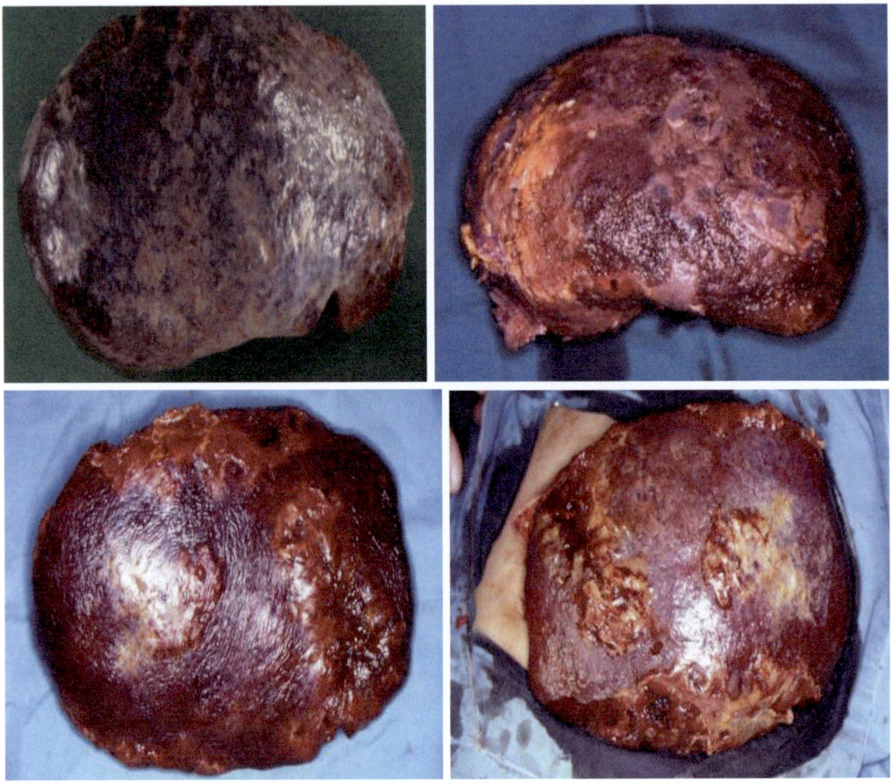

Figs. 17.1–17.4 Clinical photographs showing spleens with variable degrees of infarctions

- They may be single or multiple, and heal forming retracted scars.
- Splenic infarction can be caused by a heterogeneous group of diseases.
- Splenic infarction is very rare and has been reported both in hematological and non-hematological diseases (Figs. 17.5, 17.6 and 17.7).
- These include:
  - Evans syndrome
  - Subacute bacterial endocarditis
  - Polycythemia Vera
  - Paroxysmal nocturnal hemoglobinuria
  - Hb SC disease
  - Hb S-beta-thalassemia
- Splenic infarction is a well-documented complication of hemoglobinopathies.
- Among these hemoglobinopathies at greatest risk of developing splenic infarction are:
  - Sickle-cell hemoglobin C disease
  - Sickle cell-beta-thalassemia
  - Sickle cell trait
  - Sickle cell anemia

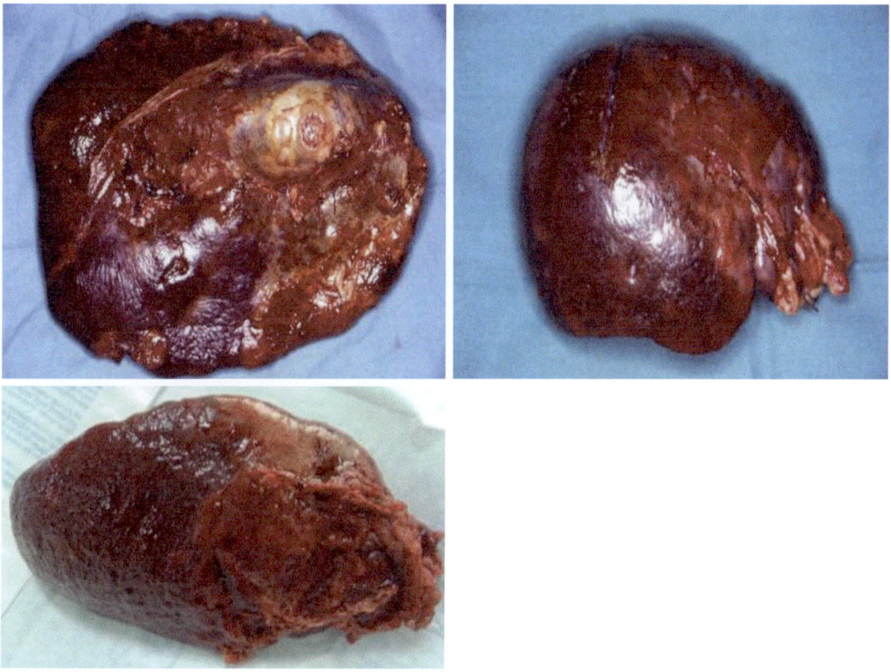

Figs. 17.5–17.7 Clinical photographs showing excised spleens with large infarcts. Note the adherent omentum to the site of infarction

- The precipitating factor for development of splenic infarction in these hemoglobinopathies is hypoxia.
- This can be induced by:
  - High altitude typically in unpressurized airplanes
  - Mountains climbing
  - Massive splenic infarction has been reported to occur at mountain altitudes of 5000 to 7000 feet above sea level.
  - This promotes sickling by metabolic effects such as:
    Acidosis
    Hypoxia
    Dehydration

CAUSES OF SPLENIC INFARCTION

- **EVANS SYNDROME**
- **SUBACUTE INFECTIVE ENDOCARDITIS**
- **SPLENIC ARTERY EMBOLIZATION**
- **POLYCYTHEMIA VERA**
- **PAROXYSMAL NOCTURNAL HEMOGLOBINUREA**
- **HEMOGLOBINOPATHIES**
  - **SICKLE-CELL HEMOGLOBIN C DISEASE**
  - **SICKLE-CELL BETA-THALASSEMIS**
  - **SICKLE CELLTRAIT**
  - **SICKLE CELL ANEMIA**

- Although splenic infarction has been reported in patients with sickle cell trait at high altitude there have also been reported cases of splenic infarction not related to high altitude exposure.
- These infarcts can be massive involving more than 50% of the spleen size (Figs. 17.8 and 17.9).
- The presentation of massive splenic infarction is variable but the majority of patients present with:
  - Sudden onset of severe left upper quadrant abdominal pain.
  - This may be associated with:
    Nausea
    Vomiting
    Fever
    Chills
  - Clinically, there is tender splenomegaly.

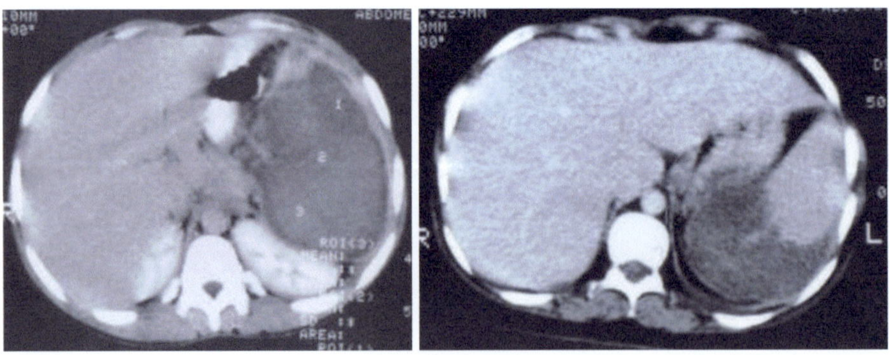

Figs. 17.8 and 17.9 Abdominal CT-scan showing massive splenic infarcts. Note the infarction involving the whole spleen in the first photograph and more than half of the spleen size in the second one

CLINICAL CONDITIONS KNOWN TO BE COMPLICATED BY SPLENIC INFARCTION

- EVANS SYNDROME
- SUBACUTE BACTERIAL ENDOCARDITIS
- POLYCYTHEMIA VARA
- PAROXYSMAL NOCTURNAL HEMOGLOBINUREA
- HBSC DISEASE
- HB S-BETA-THALASSEMIA

- Splenic infarction develops as a result of occlusion of the main splenic vascular supply or its branches, leading to parenchymal ischemia and subsequent tissue necrosis.
- The extent of splenic infarction depends on the affected vessels.

HEMOGLOBINOPATHIES ASSOCIATED WITH SPLENIC INFARCTION

- SICKLE CELL HEMOGLOBIN C DISEASE
- SICKLE CELL-BETA-THALASSEMIA
- SICKLE CELL TRAIT
- SICKLE CELL ANEMIA

- The infarction may be segmental, or it may be global affecting a larger area of the spleen size.
- Splenic infarction can involve part or affect the entire spleen.
- Nonoperative management is the treatment of choice for splenic infarction but this requires monitoring and close follow-up.
- Surgery as the treatment for splenic infarction is indicated in the following conditions:
  - Persistent symptoms
  - The development of splenic infarction related complications such as:
    - Hemorrhage
    - Splenic rupture
    - Abscess formation
    - Persistent pseudocyst
- In these cases, splenectomy or partial splenectomy for segmental infarcts is the treatment of choice.
- Currently, both procedures can be done laparoscopically.

17.2 Anatomy

- The arterial supply to the spleen consists of (Fig. 17.10):
  - The splenic artery which is a branch of the celiac axis
  - The short gastric arteries. These are branches of the left gastroepiploic artery, which supply the upper pole of the spleen.
- Even with occlusion of the main splenic artery, the collateral blood supply from the short gastric arteries often may preserve some or all of the splenic parenchyma.
- Within the spleen, the arterial supply is known to be segmental.
- Occlusion of these secondary branches results in the classic wedge-shaped infarcts.
- Most commonly, these infarcts contract and fibrose over time, as seen in patients with sickle hemoglobinopathies in which repeated attacks of infarction ultimately result in infarctions of the spleen leading to fibrosis and auto splenectomy (Figs. 17.11 and 17.12).

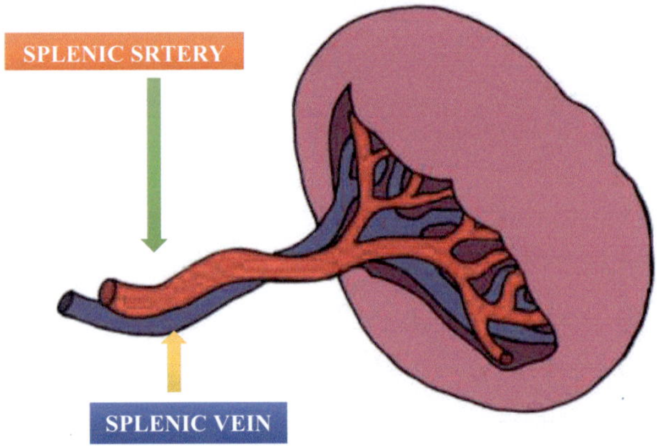

Fig. 17.10 Diagrammatic representation of the main arterial and venous drainage of the spleen. The spleen is also supplied the short gastric arteries (branches of the left gastroepiploic artery), which supply the upper pole of the spleen

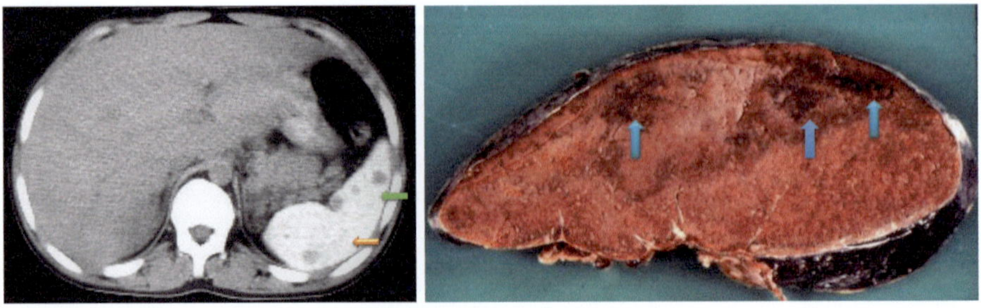

Figs. 17.11 and 17.12 Abdominal CT-scan showing a calcified spleen in a patient with sickle cell disease (Autosplenectomy) and a cut surface of the spleen showing multiple infarcts

17.3 Pathophysiology

- The splenic artery (branching off the celiac artery) supplies blood flow to the spleen in combination with the short gastric arteries (branches off of the left gastroepiploic artery).
- One of the most common causes of splenic infarction is sickle hemoglobinopathies
- Splenic infarction in sickle cell disease is the result of:
  - Hypoxia, dehydration and acidosis which lead to crystallization of the abnormal hemoglobin.
  - This will lead to rigid erythrocyte and rouleaux formation.
  - These abnormal sickled red blood cells will lead to occlusion of the splenic blood vessels and development of splenic infarction.
  - These infarcts are commonly small and repetitive.
  - These infarcts can also be large in size affecting most of the spleen.
  - In homozygous sickle cell disease, multiple infarcts during childhood commonly result in a scarred, contracted, auto infarcted spleen by early adulthood (Auto splenectomy).
  - In heterozygous sickle cell (Patients with sickle trait) splenic infarction can develop as a result of exposure to low oxygen tension from:
    Travel in unpressurized airplanes
    Vigorous activity
    Skiing in high-altitude locations
    Climbing high-altitude mountains
- There are several diseases which cause splenomegaly which put these patients at a higher risk for developing splenic infarction. These diseases include:
  - Chronic myelogenous leukemia
  - Myelofibrosis
  - Gaucher's disease
  - Malarial splenomegaly syndrome
  - AIDS with mycobacterium avium complex
  - Lymphoma
- Rarely, abnormal splenic vascular anatomy may cause infarction.
- Wandering spleen
  - In this condition, the spleen lacks the usual ligaments that fix the spleen to the surrounding structures and as a result the spleen will have a long vascular pedicle.
  - This allows the spleen to move freely which can lead to torsion of the pedicle of the spleen leading to its infarction.
- Infiltrative hematologic diseases will cause congestion of the splenic circulation by abnormal cells.
- In myelofibrosis, the splenic parenchyma is infiltrated by extramedullary hematopoiesis, causing congestion of the splenic circulation. This may be complicated by splenic infarction.

- In malignant hematologic diseases like chronic myeloid leukemia, there is an increase in splenic tissue mass. This will lead to increased splenic oxygen requirements Add to this a decreased oxygen-carrying capacity secondary to the anemia of hypersplenism. These in turn can lead to splenic infarction.
- Thromboembolism is another common cause of splenic infarcts.
  - Splenic embolization may result from various cardiovascular conditions, including:
    - A left atrial or ventricular mural thrombus that formed as a result of acute myocardial infarction or atrial fibrillation.
    - Complications of cardiac catheterization.
    - Subacute bacterial endocarditis.
- Injection of gastric varices secondary to portal hypertension and gastric variceal bleeding can be complicated by splenic infarction.
- Hypercoagulable states can also cause splenic infarction. These include:
  - Hereditary protein C deficiency
  - Acquired hypercoagulable states including:
    - Myeloproliferative disorders
    - Lupus anticoagulant
    - Erythropoietin therapy
- Splenic infarction has also been reported in association with postpartum toxic shock syndrome.
- Splenic vein thrombosis secondary to pancreatitis or surgery can result in venous splenic infarction.
- Unusual causes of splenic infarction include:
  - Malaria
  - Pancreatitis
  - Cocaine abuse
  - Complication of liver transplantation
  - Increased intra-abdominal pressure and the resultant impedance of venous return from the visceral circulation.

17.4 Etiology

- There are several causes of splenic infarction.
- The vast majority of splenic infarction (88%), however, are due to:
  - Infiltrative hematologic diseases that cause congestion of the splenic circulation by abnormal cells.
  - Thromboembolic conditions that lead to obstruction of splenic blood vessels.
- The causes of splenic infarcts include:
  - Malignant hematologic disorders:
    - Leukemia
    - Lymphoma (i.e., Hodgkin, non-Hodgkin)
    - Myelofibrosis

- Benign hematologic disorders:
  Hypercoagulable states (protein C or protein S deficiency)
  Oral contraceptives
  Lupus erythromatosis
  Erythropoietin therapy
  Idiopathic venous thrombosis
  Polycythemia vera
  Sickle hemoglobinopathies (Sickle-cell hemoglobin C disease, Sickle cell-beta-thalassemia, Sickle cell trait, and Sickle cell anemia)
- Embolic disorders:
  Subacute Bacterial Endocarditis
  Atrial fibrillation
  Prosthetic mitral valve
  Paradoxical emboli from the right heart
  Left ventricular mural thrombus following myocardial infarct
  Infected thoracic aortic graft
  HIV-associated mycobacterial infections
- Vascular disorders
- Autoimmune/collagen vascular disease
- Trauma:
  Blunt abdominal trauma
- Torsion of the wandering spleen
- Left-heart catheterization via femoral artery approach
- Sclerotherapy of bleeding gastric varices
- Vasopressin infusion
- Embolization for splenic bleeding
- Operative causes:
  Pancreatectomy
  Liver transplant
- Miscellaneous:
  Splenic vein thrombosis
  Pancreatitis
  Amyloidosis
  Sarcoidosis
  Pancreatic cancer
  Acute respiratory distress syndrome (ARDS)
  Postpartum toxic shock syndrome
  Splenic infarct associated with COVID-19

17.5 Clinical Features

- Splenic infarction is seen associated most commonly with hematologic disorders.
- The frequency of splenic infarcts may be rising because of the following factors:
  - Increased use of radiologic investigations of patients which can lead to subsequent increase in the incidental detection of asymptomatic splenic infarcts.

- Current standard of conservative nonoperative management of blunt splenic injuries.
- Increased use of angiographic embolization for vascular splenic injuries. This can be complicated with the development of splenic infarcts.

- The clinical spectrum of splenic infarction is variable and ranges from asymptomatic infarction to hemorrhagic shock.
- Splenic infarction is a rare cause of abdominal pain.
- Asymptomatic splenic infarcts are usually discovered incidentally on radiologic evaluation or postmortem studies or at laparoscopy or laparotomy for other unrelated indication.
- Symptomatic splenic infarcts can present with massive subcapsular hemorrhage with free rupture into the peritoneal cavity.
- The most common presenting symptom of splenic infarction is left-upper-quadrant abdominal pain (up to 70%).
- Left-upper-quadrant tenderness (35%)
- Other presenting symptoms include:
  - Fever and chills
  - Nausea and vomiting
  - Pleuritic chest pain
  - Left shoulder pain (Kehr's sign).
- Septic thromboemboli may result in splenic abscesses, which present with sepsis and left upper abdominal pain.
- In 16.6% of patients, splenic infarction was the presenting symptom of an underlying disease.
- The diagnosis of splenic infarction should be suspected based on the presence of an underlying disease that is most likely to be complicated by splenic infarction.
- 50% of the patients present with localized left sided upper abdominal pain.
- 36% of the patients have left sided abdominal tenderness
- 32% of the patients have splenomegaly
- 31% of the patients have no signs or symptoms localized to the spleen area
- 36% of the patients have fever >38° C
- 56% of the patients have a WBC count >12,000
- 71% of the patients have elevated LDH level
- 32% of the patients have nausea and or vomiting
- 48% of the patients have acute left upper quadrant abdominal pain
- 36% of the patients have left upper quadrant abdominal tenderness

17.6 Investigations

- A complete blood count may show elevated WBC count.
- Abdominal ultrasound is the main non-invasive investigation.
- Abdominal ultrasound will show a hypoechoic wedge-shaped area of spleen indicative of splenic infarction. Evolution of infarction may appear as hyperechoic with retraction of the splenic capsule.

- Abdominal CT-scan is a more valuable investigation.
- It accurately outlines the site and size of splenic infarction and help in differentiating this from other conditions namely splenic abscess (Figs. 17.13, 17.14, 17.15, 17.16, 17.17 and 17.18).
- Abdominal Computerized tomography (CT) scan performed with intravenous contrast is the current diagnostic investigation of choice in suspected cases of splenic infarction.
- Splenic infarct appears as a wedge-shaped area of splenic tissue with the apex pointed toward the helium of the spleen and the base pointed toward the splenic capsule.
- As the infarction matures, the affected tissue may normalize, liquefy or become contracted or scarred.
- Magnetic resonance imaging (MRI) is another useful diagnostic modality.
- The postcontrast scans clearly show the classic segmental, wedge-shaped, low-attenuation splenic infarct.

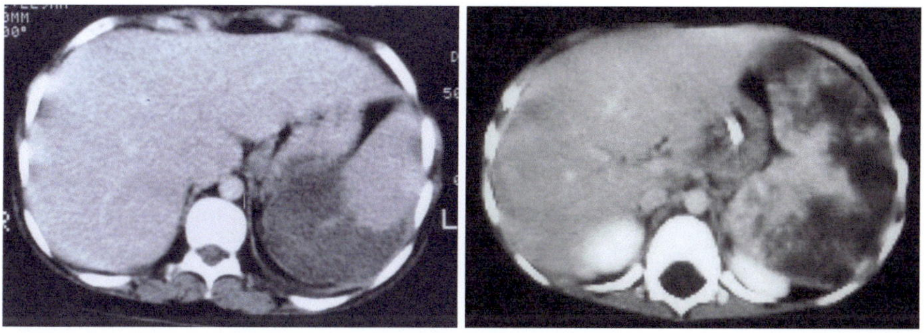

Figs. 17.13 and 17.14 Abdominal Ct-scan showing splenic infarction. Note the large sie of the infarction involving more than 50% of the spleen

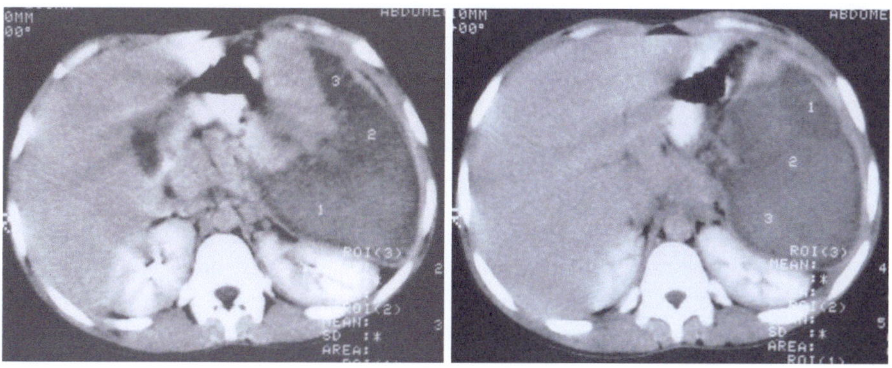

Figs. 17.15 and 17.16 Abdominal CT-scan showing massive splenic infarction of the spleen. Note the size of the infarction which is involving more than 50% of the size of the spleen

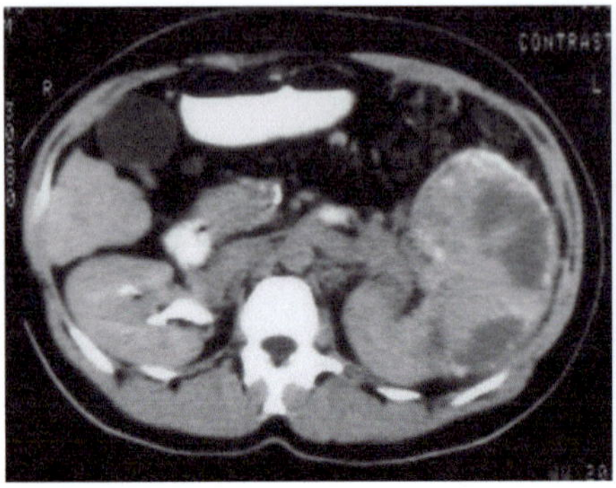

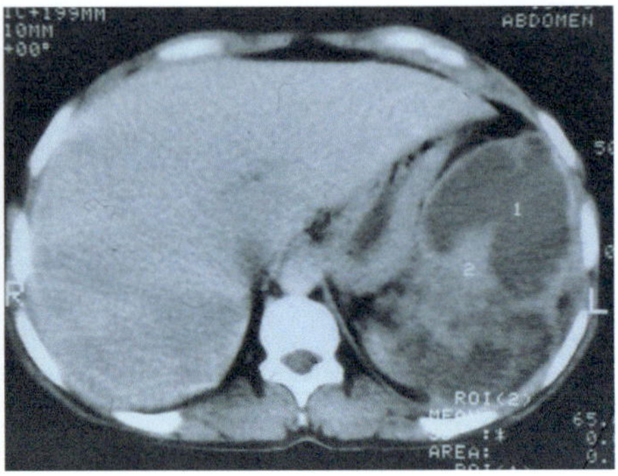

Figs. 17.17 and 17.18 Abdominal CT-scan showing massive splenic infarction of the spleen. Note the variable size of the infarction. There are multiple small infarcts in the first one and the abscess involving the lower part of the spleen

- Less frequently, the entire spleen or most of the spleen may be infarcted, leaving only a rim of contrast-enhancing capsule.
- Angiography is indicated
  - When a vascular lesion is suspected as the etiologic cause, as in cases of arterial embolization.
  - When it is necessary to manage segmental bleeding by embolization.
- Leukocytosis and elevated lactate dehydrogenase (LDH) may be found in splenic infarction. However, these results lack specificity to splenic infarct.

17.7 Management

- The management of splenic infarct is based primarily on the underlying causative disease (Figs. 17.19, 17.20, 17.21, 17.22, 17.23, 17.24 and 17.25).
- An asymptomatic splenic infarct without complication does not require surgical intervention.
- The nonoperative treatment consists of analgesia with either narcotics or nonsteroidal anti-inflammatory drugs (NSAIDs) and close follow-up.
- Splenic infarct in the non-infectious setting may be treated conservatively with analgesics, hydration, anti-emetics and other means of supportive care.
- Abdominal pain due to uncomplicated cases of splenic infarction usually resolves without surgical intervention in 7–14 days.
- There are known complications associated splenic infarction. These include:
  - Development of pseudocyst
  - Development of splenic abscess
  - Development of hemorrhage
  - Splenic rupture
  - Development of splenic artery aneurysm

- In some instances, the infarcted splenic tissue may become infected and lead to abscess formation.
- Infarcted splenic tissue may also undergo a hemorrhagic transformation.
- The indications for surgical intervention include:
  - Persistent symptoms
  - Development of complications such as:
    - Sepsis
    - Abscess
    - Hemorrhage
    - Persistent pseudocyst formation

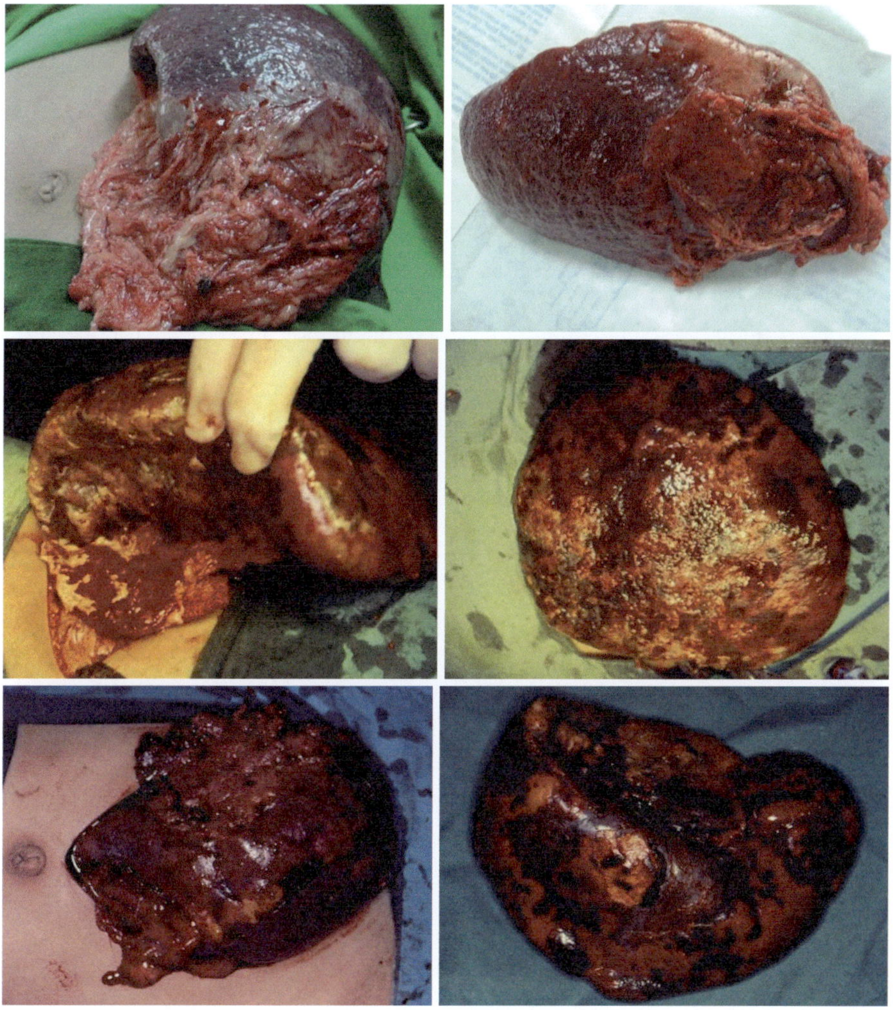

Figs. 17.19–17.25 Clinical photograph showing splenic infacrts

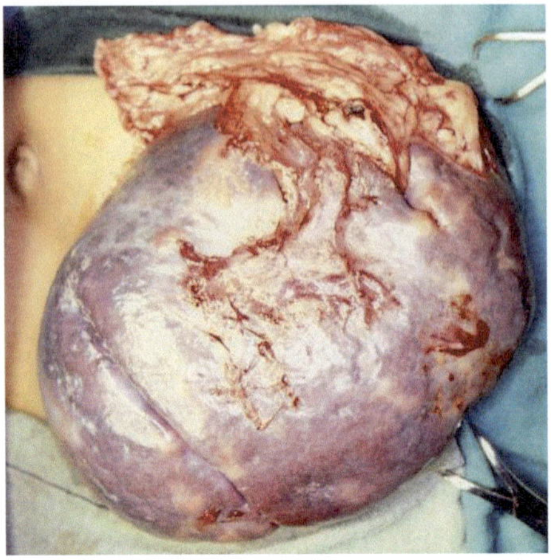

Figs. 17.19–17.25 (continued)

- Patients with hematologic conditions are more likely to require laparotomy for complications of those conditions (69%), especially patients with nonmalignant conditions (90%).
- Patients with hematologic malignancies often (47%) had surgery for acute symptoms of splenic infarction.
- Complications of splenic infarction are a frequent indication (60%) for operation in patients with septic emboli.
- Pseudocysts usually develop as a result of prior trauma, with subsequent organization and liquefaction of the hematoma which become surrounded by a fibrous pseudo capsule.
  - Small (<4 cm) asymptomatic pseudocysts are treated conservatively and usually regress with time.
  - Larger (>5 cm) pseudocysts require treatment because the risk of rupture can be as high as 25%.

INDICATION FOR SURGICAL INTERVENTIONS FOR SPLENIC INFARCTIONS

- **PERSISTENT SYMPTOMS**
- **DEVELOPMENT OF COMPLICATIONS:**
  - **SEPSIS**
  - **ABSCESS FORMATION**
  - **HEMORRHAGE**
  - **PERSISTENT PSEUDOCYST FORMATION**

- Splenic abscess can result from septic emboli or superinfection of a prior splenic infarction.
  - Traditionally, such abscesses have been managed by splenectomy.
  - Currently, percutaneous catheter drainage and appropriate antibiotic therapy can be used to treat selected cases of well-formed, unilocular splenic abscesses.
- Persistent symptoms and development of complications of splenic infarction will require surgery.
- Splenectomy is the standard procedure but splenic preservation is preferable whenever possible.
- Complications, such as bleeding or pseudocyst formation may be managed with partial splenectomy.
- Currently, splenectomy and partial splenectomy can be done using the traditional open technique or laparoscopically.
- Peri splenic inflammation and dense adhesions can make splenectomy difficult.
- Some authors advocate splenectomy for all those with splenic infarction and splenic abscess as there is no point in preserving the residual, partially functioning or non-functioning spleen.
- If splenectomy (rather than splenic preservation) is planned, the patient should receive the pneumococcal, meningococcal and Haemophiles influenzae vaccine at least 2 weeks prior to the operation. In cases of emergency, these immunizations can be given prior to discharge from the hospital.

Further Reading

1. Chapman J, Helm TA, Kahwaji CI. Splenic infarcts. Treasure Island, FL: Stat Pearls; 2021.
2. Ebert EC, Nagar M, Hagspiel KD. Gastrointestinal and hepatic complications of sickle cell disease. Clin Gastroenterol Hepatol. 2010;8(6):483–9.
3. Franklin QJ, Compeggie M. Splenic syndrome in sickle cell trait: four case presentations and a review of the literature. Mil Med. 1999;164(3):230–3.
4. Sheikha A. Splenic syndrome in patients at high altitude with unrecognized sickle cell trait: splenectomy is often unnecessary. Can J Surg. 2005;48(5):377–81.
5. O'Keefe JH Jr, Holmes DR Jr, Schaff HV, Sheedy PF 2nd, Edwards WD. Thromboembolic splenic infarction. Mayo Clin Proc. 1986;61(12):967–72.
6. Ting W, Silverman NA, Arzouman DA, Levitsky S. Splenic septic emboli in endocarditis. Circulation. 1990;82(5 Suppl):IV105–9.
7. Yu LK, Hsu CW, Tseng JH, Liu NJ, Sheen IS. Splenic infarction complicated by splenic artery occlusion after N-butyl-2-cyanoacrylate injection for gastric varices: case report. Gastrointest Endosc. 2005;61(2):343–5.
8. Olson JF, Steuber CP, Hawkins E, Mahoney DH Jr. Functional deficiency of protein C associated with mesenteric venous thrombosis and splenic infarction. Am J Pediatr Hematol Oncol. 1991;13(2):168–71.
9. Torda A. Postpartum toxic shock syndrome associated with multiple splenic infarcts. Med J Aust. 2005;182(2):93.
10. Gupta BK, Sharma K, Nayak KC, Agrawal TD, Binani A, Purohit VP, et al. A case series of splenic infarction during acute malaria in Northwest Rajasthan, India. Trans R Soc Trop Med Hyg. 2010;104(1):81–3.

11. Desai DC, Hebra A, Davidoff AM, Schnaufer L. Wandering spleen: a challenging diagnosis. South Med J. 1997;90(4):439–43.
12. Nores M, Phillips EH, Morgenstern L, Hiatt JR. The clinical spectrum of splenic infarction. Am Surg. 1998;64(2):182–8.
13. Hayashi H, Beppu T, Okabe K, Masuda T, Okabe H, Baba H. Risk factors for complications after partial splenic embolization for liver cirrhosis. Br J Surg. 2008;95(6):744–50.
14. Wu SC, Chen RJ, Yang AD, Tung CC, Lee KH. Complications associated with embolization in the treatment of blunt splenic injury. World J Surg. 2008;32(3):476–82.
15. Jaroch MT, Broughan TA, Hermann RE. The natural history of splenic infarction. Surgery. 1986;100(4):743–50.
16. Joshi SC, Pant I, Shukla AN, Anshari MA. Splenic infarct as a diagnostic pitfall in radiology. J Cancer Res Ther. 2008;4(2):99–101.
17. Pachter HL, Guth AA, Hofstetter SR, Spencer FC. Changing patterns in the management of splenic trauma: the impact of nonoperative management. Ann Surg. 1998;227(5):708–17; discussion 717–9.
18. Antopolsky M, Hiller N, Salameh S, Goldshtein B, Stalnikowicz R. Splenic infarction: 10 years of experience. Am J Emerg Med. 2009;27(3):262–5.
19. Urban BA, Fishman EK. Helical CT of the spleen. AJR Am J Roentgenol. 1998;170(4):997–1003.
20. Balcar I, Seltzer SE, Davis S, Geller S. CT patterns of splenic infarction: a clinical and experimental study. Radiology. 1984;151(3):723–9.
21. Kluger Y, Paul DB, Townsend RN, Diamond DL. Enhanced rim around infarcted, traumatized spleen on computed tomographic scans: case report. J Trauma. 1994;36(3):436–7.
22. Choi G, Kim KA, Lee J, Park YS, Lee J, Choi JW, et al. Ultrasonographic atlas of splenic lesions. Ultrasonography. 2022;41(2):416–29.
23. Goerg C, Schwerk WB. Splenic infarction: sonographic patterns, diagnosis, follow-up, and complications. Radiology. 1990;174(3 Pt 1):803–7.
24. Pachter HL, Hofstetter SR, Elkowitz A, Harris L, Liang HG. Traumatic cysts of the spleen—the role of cystectomy and splenic preservation: experience with seven consecutive patients. J Trauma. 1993;35(3):430–6.
25. Cavazos S, Ratzer ER, Fenoglio ME. Laparoscopic management of the wandering spleen. J Laparoendosc Adv Surg Tech A. 2004;14(4):227–9.
26. Matsushima H, Kuroki T, Adachi T, Kitasato A, Hirabaru M, Hidaka M, et al. Laparoscopic spleen-preserving distal pancreatectomy with and without splenic vessel preservation: the role of the Warshaw procedure. Pancreatology. 2014;14(6):530–5.
27. De Greef E, Hoffman I, Topal B, Broers C, Miserez M. Partial laparoscopic splenectomy for splenic abscess because of salmonella infection: a case report. J Pediatr Surg. 2008;43(5):E35–8.
28. Héry G, Becmeur F, Méfat L, Kalfa D, Lutz P, Lutz L, et al. Laparoscopic partial splenectomy: indications and results of a multicenter retrospective study. Surg Endosc. 2008;22(1):45–9.
29. Javaid U, Young P, Gill G, Bhargava P. Acute complete splenic infarction secondary to COVID-19 infection. Radiol Case Rep. 2022;17(5):1402–6.
30. Kato K, Gleeson TA. Splenic necrosis requiring ultrasound-guided drainage following meningococcal septicaemia. Oxf Med Case Reports. 2019;2019(3):omz020.
31. Mamoun C, Houda F. Splenic infarction revealing infectious endocarditis in a pregnant woman: about a case and brief literature review. Pan Afr Med J. 2018;30:184.
32. Niccoli Asabella A, Altini C, Nappi AG, Lavelli V, Ferrari C, Marzullo A, Loiodice A, Rubini G. Sickle cell diseases: what can nuclear medicine offer? Hell J Nucl Med. 2019;22(1):2–3.
33. Wand O, Tayer-Shifman OE, Khoury S, Hershko AY. A practical approach to infarction of the spleen as a rare manifestation of multiple common diseases. Ann Med. 2018;50(6):494–500.
34. Smalls N, Obirieze A, Ehanire I. The impact of coagulopathy on traumatic splenic injuries. Am J Surg. 2015;210(4):724–9.
35. Bender MA. Sickle cell disease. In: Adam MP, Mirzaa GM, Pagon RA, Wallace SE, LJH B, Gripp KW, Amemiya A, editors. GeneReviews® [Internet]. Seattle, WA: University of Washington, Seattle; 2003.

36. el Barzouhi A, van Buren M, van Nieuwkoop C. Renal and splenic infarction in a patient with familial hypercholesterolemia and previous cerebral infarction. Am J Case Rep. 2018;10(19):1463–6.
37. Nofal R, Zeinali L, Sawaf H. Splenic infarction induced by Epstein-Barr virus infection in a patient with sickle cell trait. J Paediatr Child Health. 2019;55(2):249–51.
38. Pereda MA, Isaac J, Zhang Y, Jayakumar R, Gupta R, Miller ST. Massive splenic infarction in a child with sickle cell disease on chronic transfusion therapy. J Pediatr Hematol Oncol. 2019;41(2):e79–82.
39. Blackwood B, Binder W. Unusual complications from Babesia infection: splenic infarction and splenic rupture in two separate patients. J Emerg Med. 2018;55(5):e113–7.
40. Fernando C, Mendis S, Upasena AP, Costa YJ, Williams HS, Moratuwagama D. Splenic syndrome in a young man at high altitude with undetected sickle cell trait. J Patient Exp. 2018;5(2):153–5.

18 Congestive Splenomegaly

18.1 Introduction

- The spleen plays an important role immunologically and also acts as a blood filter.
- It lies within the left upper quadrant of the peritoneal cavity under the ribs 9–12.
- A normal spleen weighs 150 g and is approximately 11 cm in craniocaudal length.
- The spleen has several important functions including:
  - The spleen act as blood Filter removing both abnormal and senescent red blood cells (RBCs), as well as particulates and microorganisms
  - The spleen has important immunological functions:
    - The spleen provides an interface between adaptive and innate immunity
    - The spleen is important for the synthesis of immunoglobulin M (IgM)
    - The spleen is important for the synthesis of properdin (an essential component of the alternate pathway of complement activation)
    - The spleen is important for the synthesis of tuftsin (an immunostimulatory tetrapeptide)
  - Erythropoiesis, particularly early in fetal life and sometimes as an adaptive response to bone marrow failure.
  - A reservoir of blood cells, including RBCs and platelets that can be utilized under stress.
- Iron metabolism
- The normal spleen is usually not palpable, although it can sometimes be palpable in adolescents and patients who are thin built.
- Splenomegaly is defined as enlargement of the spleen, measured by size or weight.

A. H. Al-Salem, *The Spleen*, https://doi.org/10.1007/978-981-99-6191-7_18

18.2 Causes of Splenomegaly

- Congestive:
  - Liver cirrhosis
  - Heart failure
  - Thrombosis of portal, hepatic or splenic veins
- Malignant causes:
  - Lymphoma
  - Acute and chronic leukemias
  - Polycythemia vera
  - Multiple myeloma
  - Essential thrombocythemia
  - Primary myelofibrosis
  - Primary splenic tumors
  - Metastatic solid tumors
- Infections
  - Viral Hepatitis
  - Infectious mononucleosis
  - Cytomegalovirus
  - Salmonella
  - Brucella
  - Tuberculosis
  - Malaria
  - Schistosomiasis
  - Toxoplasmosis
  - Leishmaniasis
  - Infective endocarditis
  - Fungal infections
- Inflammation
  - Sarcoidosis
  - Serum sickness
  - Systemic lupus erythematosus
  - Rheumatoid arthritis (Felty syndrome)
- Infiltrative, non-malignant causes
  - Gaucher's disease
  - Niemann-Pick disease
  - Amyloidosis
  - Other lysosomal storage diseases (e.g., mucopolysaccharidoses)
  - Langerhans cell histiocytosis
  - Hemophagocytic lymph histiocytosis
  - Rosai-Dorfman disease
- Hematological causes
  - Acute and chronic hemolytic anemias
  - Sickle cell disease
  - Thalassemia

  - Sickle-beta thalassemia
  - Following use of recombinant human granulocyte colony-stimulating factor
- A spleen weight of 400–500 g indicates splenomegaly, while a weight of more than 1000 g is labelled as massive splenomegaly.
- Poulin et al. defined splenomegaly as:
  - Moderate splenomegaly if the largest dimension of the spleen is 11–20 cm
  - Severe splenomegaly if the largest dimension of the spleen is greater than 20 cm.
- In the past, splenomegaly was a clinical diagnosis based on an enlarged palpable spleen.
- Currently, mild splenomegaly can be diagnosed using radiological imaging studies.
- Congestive splenomegaly is caused by portal hypertension, which may be:
  - Prehepatic: This can be caused by:
    Thrombosis of splenic veins
    Portal vein thrombosis or stenosis
    Congestive heart failure
  - Hepatic: Caused by Liver cirrhosis.
  - Post hepatic: Budd-Chiari syndrome (thrombosis of hepatic veins).
- Congestive splenomegaly may be complicated by hypersplenism.

CAUSES Of CONGESTIVE SPLENOMEGALY

- **PREHEPATIC**
  - **THROMBOSIS OF SPLENIC VEIN**
  - **THROMBOSIS OF PORTAL VEIN**
  - **CONGESTIVE HEART FAILURE**
  - **PORTAL VEIN STENOSIS**
- **HEPATIC**
  - **LIVER CIRRHOSIS**
- **POST HEPATIC**
  - **THROMBOSIS OF HEPATIC VEINS**
  - **(BUDD-CHIARI SYNDROME)**

- Congestive splenomegaly is caused by portal hypertension, which may be due to:
  - Budd-Chiari syndrome (thrombosis of hepatic veins)
  - Liver cirrhosis
  - Congestive heart failure
  - Portal vein stenosis
  - Thrombosis of splenic veins or portal vein
- Portal vein thrombosis may be due to inflammation, extrinsic pressure, trauma, tumor or idiopathic
- Portal vein stenosis may be due to extension of obliterative process at birth in umbilical vein and ductus venosus into portal vein.

- Banti syndrome:
  - This is idiopathic portal hypertension associated with fibroelastosis in portal tracts, dilated capillaries, and phlebosclerosis; associated with hypersplenism (Anemia, leukopenia and thrombocytopenia).
- Histologically, congestive splenomegaly is characterized by:
  - Grossly the spleen is large, firm in consistency, dark with fibrosis of capsule.
  - Microscopically, the spleen is characterized by:

    Dilated veins and sinuses
    Fibrosis of red pulp
    Hemosiderin laden macrophages
    Iron and calcium containing fibrotic nodules (Gamna-Gandy bodies) secondary to hemorrhage
    No prominent lymphoid follicles

18.3 Pathophysiology and Etiology

- Splenomegaly of Portal Hypertension (PHT) is a passive congestive splenomegaly.
- Congestive splenomegaly not only involves splenic congestion, but is also accompanied by congestion of splenic red pulp, vascular proliferation, fibrosis, lymphoid hyperplasia and activation.
- In PHT, spleen angiogenesis can regulate the portosystemic collateral circulation and increase the spleen blood flow, which is an important pathologic and physiologic sign of the spleen.
- If a spontaneous portosystemic shunt develops, it will not show as splenomegaly.
- Hypersplenism will develop after splenomegaly decompensation.
- In addition, the collateral vessels of splenic peri splenic ligaments (lienorenal ligament, splenophrenic ligament, and splenocolic ligament) become abundant in splenomegaly. A wide collateral circulation can be established retroperitoneally and in the pericardial area, and the spleen may be called a "shunt bridge". This was used in Warren shunt surgery to divert blood flow from gastro-splenic region to achieve selective shunt depressurization.
- In congestive splenomegaly, the spleen becomes enlarged and engorged with blood because of impaired blood flow through the splenic vein, which empties into the portal vein.
- Splenic enlargement develops as a result of passive congestion and engorgement of the spleen.
- Such impairment of blood flow in the splenic vein may be caused by:
  - Liver disease
  - Portal vein or splenic vein thrombosis
  - Constrictive pericarditis
  - Congestive heart failure

- Banti syndrome is characterized by abnormal enlargement of the spleen (splenomegaly) due to obstruction of blood flow in splenic, hepatic or portal veins and abnormally increased blood pressure in these veins.
- Obstruction of the portal, hepatic, or splenic veins can be secondary to several factors including:
  - Congenital abnormalities of the veins
  - Thrombosis of the veins
  - Inflammation and obstruction of veins such as liver cirrhosis
  - Increased arsenic intake
  - Patients taking long-term azathioprine, particularly after kidney transplantation
- Common Causes of Chronic congestive splenomegaly include:
  - Infectious mononucleosis
  - Splenic infiltration
  - Hematological malignancy
  - Portal hypertension
  - Syphilis
  - Endocarditis
- Other Causes of Chronic congestive splenomegaly include:
  - Visceral leishmaniasis (kala-azar)
  - Chronic myelogenous leukemia
  - Myelofibrosis
  - Malaria
  - Splenic marginal zone lymphoma

18.4 Clinic Features

- Splenomegaly is an enlargement of the spleen that often results from blood-borne infections. The red pulp expands as neutrophils accumulate in the spleen and encroaches on the lymphoid follicles.
- In congestive splenomegaly there is an obstruction to the venous outflow from the spleen which will lead to an increase in the amount of red pulp. Initially, the red pulp is congested, but over time it becomes fibrotic.
- The main cause of congestive splenomegaly is liver cirrhosis.
- Splenomegaly may arise as a symptom of a number of diseases, including certain systemic infections, inflammatory diseases, hematologic diseases, inherited spleen disorders, splenic cysts, and neoplastic diseases.
- In congestive splenomegaly, the spleen becomes engorged with blood because of impaired flow through the splenic vein, which empties into the portal vein. This may be caused by liver disease, portal vein or splenic vein pathology, constrictive pericarditis, or congestive cardiac failure.
- Massive splenomegaly can place a heavy burden on the heart and circulatory system. The enlarged spleen can absorb more than half the cardiac output, whereas the normal spleen absorbs only 5 percent.

- In addition, splenomegaly often causes an increase in the volume of blood plasma and this together with increased retention of red blood cells in the spleen, may result in anemia.
- The following features are indicative of chronic congestive splenomegaly:
  - Abdominal pain
  - Chest pain
  - Back pain
  - Splenic encroachment
  - Anemia
  - Cytopenia
  - Fatigue
  - Frequent infections
  - Easy bruising and bleeding
- Complications of Chronic congestive splenomegaly
  - Anemia
  - Increased bleeding tendency
  - Ruptured spleen
- Banti's syndrome
  - Guido Banti (Professor of pathological anatomy, Florence, Italy) described Banti's Syndrome in 1894 as congestive splenomegaly without any intrahepatic or extrahepatic blockage (Fig. 18.1).

Fig. 18.1 Guido Banti (8 June 1852–8 January 1925) was an Italian physician and pathologist. He also performed innovative studies on the heart, infectious diseases and bacteriology, splenomegaly, nephrology, lung disease, leukemia and motor aphasia. He gave his name to Banti's disease Banti's syndrome, named for Guido Banti, is a chronic congestive enlargement of the spleen resulting in premature destruction of the red blood cells by the spleen (Wikipedia, the free encyclopedia).

- This syndrome is characterized by abnormal spleen enlargement (splenomegaly) caused by blood flow obstruction in some veins and abnormally high blood pressure (hypertension) in the hepatic, portal veins of the liver and the splenic veins of the spleen.
- Banti's syndrome can be caused by a variety of conditions that cause obstruction and excessively high blood pressure within the splenic veins or the liver (hepatic or portal veins).
- These include:
  Congenital anomalies of the hepatic or portal veins
  Blood clots in the hepatic or portal veins
  Inflammation and obstruction of veins in the liver (vascular obstruction), such as cirrhosis.
  In certain cases, increased arsenic intake is a cause of Banti's syndrome.
  Patients on long-term azathioprine, particularly following kidney transplantation can develop Banti's syndrome.
- The symptoms of Banti's syndrome include:
  Weakness
  Anemia
  Fatigue
  Splenomegaly
  Bleeding esophageal varices
  Ascites
  Hypersplenism (Anemia, Leukopenia, Thrombocytopenia)

18.5 Diagnosis, Management and Prognosis

- Splenic venography and magnetic resonance imaging are useful investigations to confirm the diagnosis of Banti's syndrome.
- Active bleeding can be treated with vasoconstrictor medications or other portal hypertension treatments.
- Recurrent bleeding can be treated by a surgical shunt to redirect blood flow and decrease the pressure in the portal vein.
- In patients with portal hypertension, beta-blockers are given as a preventative measure.
- Ethamolin was approved by the US Food and Drug Administration (FDA) in 1988 as an orphan drug for bleeding esophageal varices.
- The management of Banti's syndrome is to control gastrointestinal bleeding caused by portal hypertension and treat associated hypersplenism.
- Variceal ligation and endoscopic sclerotherapy were shown to be equally effective in 95 percent of patients with acute gastrointestinal bleeding, with a recurrence rate of roughly 20% and 3% recurrent bleeding, respectively.
- Some patients may also require a shunt procedure as a last resort to decompress the pressure in the portal vein.

- Patients with symptomatic hypersplenism including severe anemia requiring repeated blood transfusions, or recurrent splenic infarction should be treated surgically.
- Successful medical treatment of the primary disorder in such cases can lead to regression of the hypersplenism without the need for surgery.
- Splenectomy can still be used to help control or stage the underlying disease in cases of splenomegaly.
- Both open and laparoscopic splenectomy procedures are feasible.
- Banti's syndrome has a favorable prognosis, with a 5-year survival rate of 100% after successful varices ligation.

Further Reading

1. Jeker R. Hypersplenism. Ther Umsch. 2013;70(3):152–6.
2. McKenzie CV, Colonne CK, Yeo JH, Fraser ST. Splenomegaly: pathophysiological bases and therapeutic options. Int J Biochem Cell Biol. 2018;94:40–3.
3. Chu HB, Liu X, Zhao JH, Xu Y, Wang L, Wang T, et al. Subtotal splenectomy for splenomegaly in cirrhotic patients. Int J Clin Exp Pathol. 2014;7(8):4981–90.
4. Chu HB, Zhang TG, Zhao JH, Jian FG, Xu YB, Wang T, et al. Assessment of immune cells and function of the residual spleen after subtotal splenectomy due to splenomegaly in cirrhotic patients. BMC Immunol. 2014;15:42.
5. Zhu X, Han W, Wang L, Chu H, Zhao J, Xu Y, et al. Penicillar arterioles of red pulp in residual spleen after subtotal splenectomy due to splenomegaly in cirrhotic patients: a comparative study. Int J Clin Exp Pathol. 2015;8(1):711–8.
6. Chen SR, Zhao JH, Wang L, Xu Y, Zhu S, Wang T, et al. Changes in nerve fibers and microvessel density in residual spleen after subtotal splenectomy due to portal hypertension. Int J Clin Exp Pathol. 2016;9(2):1557–67.
7. Yu D, Li Y, Xu YB, Tang J, Li K, Chu H. Dysregulated matrix metalloproteinases and tissue inhibitors of metalloproteinase in residual splenic tissue after subtotal splenectomy due to portal hypertension. Int J Clin Exp Pathol. 2017;10(2):1273–82.
8. Chu H, Han W, Wang L, Xu Y, Jian F, Zhang W, et al. Long-term efficacy of subtotal splenectomy due to portal hypertension in cirrhotic patients. BMC Surg. 2015;15:89.
9. Achiwa S, Hirota S, Kako Y, Takaki H, Kobayashi K, Yamakado K. Radiological anatomy of spontaneous splenorenal shunts in patients with chronic liver disease. Jpn J Radiol. 2017;35(4):206–14.
10. Lv Y, Yee Lau W, Wu H, Han XY, Gong X, Liu N, et al. Causes of peripheral cytopenia in hepatitic cirrhosis and portal hypertensive splenomegaly. Exp Biol Med (Maywood). 2017;242(7):744–9.
11. Han X, Lv Y, Li Y, Deng J, Qiu Q, Liu N, et al. Distribution characteristics of cells in splenomegaly due to hepatitis B-related cirrhotic portal hypertension and their clinical importance. J Int Med Res. 2018;46(7):2633–40.
12. Sharma N, Bajpai M, Kumar A, Paul S, Jana M. Portal hypertension: a critical appraisal of shunt procedures with emphasis on distal splenorenal shunt in children. J Indian Assoc Pediatr Surg. 2014;19(2):80–4.

19 Splenectomy and Immunizations

19.1 Introduction

- The spleen is an important organ and has several valuable functions. These functions include:
  - Immunological functions:
    The spleen has the ability to produce antibodies and efficiently destroy invading microorganisms specially encapsulated bacteria.
    Splenectomy is known to be associated with an increased risk of overwhelming post-splenectomy infection.
  - The spleen is composed of red pulp and white pulp
    The red pulp filters the blood, removing foreign material, damaged and worn out red blood cells.
    It also functions as a storage site for iron, red blood cells and platelets.
    The white pulp is considered the largest secondary lymphoid organ in the body.
    The white pulp contains splenic macrophages which are important and help destroy and get rid of encapsulated bacteria such as Streptococcus pneumoniae.
  - Erythropoiesis
    The spleen is a site for the development of new red blood cells, particularly in situations where the bone marrow has been compromised by various disorders such as leukemia.
- The spleen was originally called the *organum plenum mysterii* by Galen.
- The first splenectomy was performed by Andirano Zaccarello in 1549.
- Andriano Zaccarello performed the first splenectomy on a Neapolitan woman with massive splenomegaly in 1549.

THE FIRST SPLENECTOMY WAS PERFORMED BY ANDIRANO ZACCARELLO IN 1549

A. H. Al-Salem, *The Spleen*, https://doi.org/10.1007/978-981-99-6191-7_19

Traditionally, surgical removal of the spleen was done via an open approach using either an upper midline or a left subcostal or transverse incision.

With the recent advances of minimally invasive surgery techniques, laparoscopic splenectomy became the standard procedure for elective splenectomy both in children and adults.

The first laparoscopic splenectomy was performed by Delaitre and Maignien in 1991.

THE FIRST LAPAROSCOPIC SPLENECTOMY WAS PERFORMED BY DELAITRE AND MAIGNIEN IN 1991

- It is advisable to delay elective splenectomy until the patient is at least 6 years old to minimize the risk of overwhelming post splenectomy sepsis.
- Splenectomy can be done laparoscopically or through an open surgery.
- Laparoscopy is the preferred procedure now. This is specially so in cases where the spleen is not massively enlarged and when the procedure is performed electively.
- Today, most elective splenectomies are done laparoscopically, except in those with massive splenomegaly. Recently, laparoscopic splenectomy was shown to be safe and feasible even in those with massive splenomegaly.
- Open splenectomy is performed in emergency cases and when the spleen is massively enlarged.
- Vaccination against *S. pneumoniae*, *H. influenza* and *N. meningitidis* should be given at least 2 weeks pre-operatively if possible to minimize the chance of overwhelming post-splenectomy infection (OPSI) or 2 weeks postoperatively.

19.2 Indications and Contraindications for Splenectomy

- Absolute indications for splenectomy:
  - Splenic trauma
  - Splenic rupture
  - Hereditary spherocytosis
  - Splenic abscess
  - Splenic cyst
  - As part of radical surgical removal of locally advanced gastric carcinoma and pancreatic carcinoma
  - Angioma of the spleen
  - Primary splenic malignancy
  - Uncontrolled or recurrent bleeding esophageal varices secondary to splenic vein thrombosis
- Relative indications:

- Congenital or acquired hemolytic anemia
- Thalassemia
- Acute, chronic myeloid or chronic lymphatic leukemia
- Lymphoma (Hodgkin's)
- Polycythemia vera
- Myelofibrosis
- Acute or chronic ITP
- Parasitic infestation
- Malarial splenomegaly
- Felty's syndrome
- Angioma of the spleen
- Splenic cysts
- Malignant metastases to the spleen
- Tropical or non-tropical splenomegaly
- Intrahepatic or extrahepatic portal hypertension
- Amyloidosis
- Gaucher's disease

- Contraindications for splenectomy:
  - Uncorrectable coagulopathy
  - Severe cardiovascular disease making the patient unfit for general anesthesia
  - Cirrhosis with portal hypertension
  - Active hemorrhage with hemodynamic instability (A relative contraindication)
  - Pregnancy (A relative contraindication)

19.3 Open Splenectomy

- Open splenectomy is done through:
  - A midline incision
  - A left subcostal incision
  - A left upper transvers incision
- The type of incision depends on:
  - The indication for splenectomy, whether it is elective or emergency
  - The size of the spleen
  - The surgeon preference
- Generally, an upper midline incision is preferable in traumatic exploration or if splenectomy is done as an emergency.
- In children, a left upper transverse incision is preferable.
- An upper midline incision has several advantages:
  - Provides an excellent exposure of the abdominal cavity
  - It can be performed quickly which is an advantage in patients with intraabdominal bleeding following trauma
  - It provides access for all abdominal organs including evaluation and management of other associated abdominal injuries
  - It is easily extendable

- Once the abdominal cavity is entered, the splenic peritoneal attachments and ligaments are divided using both sharp and blunt dissection.
- There are three main ligaments that hold the spleen in place:
  - The gastrosplenic ligament
  - The splenorenal ligament
  - The splenocolic ligament
- The peritoneal attachments are avascular and ligaments are incised with an electrocautery or scissors. Electrocautery is preferable to avoid any bleeding specially in patients with portal hypertension where the ligaments may have blood vessels and these can be large and should be ligated to avoid further bleeding.
- It is also important to take good care not to divide the posterior attachment of the spleen too far medially as this may result in injury to the splenic vein.
- The gastrosplenic ligament contains the short gastric vessels. These should be identified and ligated. The short gastric vessels are usually short and care must be taken at the time of ligation as they may slip and cause bleeding.
- Once these attachments are released, the spleen is mobilized to expose the hilum where the splenic artery and vein can be identified.
- The splenic artery and veins are identified in the splenic hilum.
- The splenic vessels are dissected in close proximity to the spleen to avoid injury to the pancreas. The pancreatic tail and the splenic hilum are close to each other and care should be taken at this stage as pancreatic injury can occur. This is especially in the presence of severe inflammation and/or abscess. The majority of pancreatic injuries resolve spontaneously with nonoperative management, which includes:
  - Placement of a wide drainage in close proximity to the site of pancreatic injury.
  - The use of a somatostatin analogue to decrease exocrine pancreatic secretions.
  - Total parenteral nutrition (TPN) or enteral feeding using a gastojejunal feeding tube.
- Once the splenic vessels are identified, they are carefully dissected, doubly ligated with nonabsorbable suture. It is also preferable to transfix these vessels prior to their division to avoid accidental slippage of the sutures and intraoperative bleeding. It is generally preferable to individually ligate the splenic artery or arterial branches and the splenic vein or venous branches.
- Following this step, the short gastric vessels are identified. These are short and care should be taking during their ligation and division. This is to avoid intraoperative bleeding once the tie slip and in order to stop bleeding it is important to place a Lembert suture in the gastric wall in a seromuscular fashion to avoid development of gastric fistula.
- Once the short gastric vessels are divided, the spleen is removed and hemostasis is secured.
- Drains are not routinely placed following splenectomy except in cases where there is an injury of the tail of the pancreas or splenic abscess. Drains are known to be associated with an increased risk of local infection.
- When splenectomy is performed for hematologic disease, a thorough abdominal exploration should be performed to look for any accessory spleens. This is very

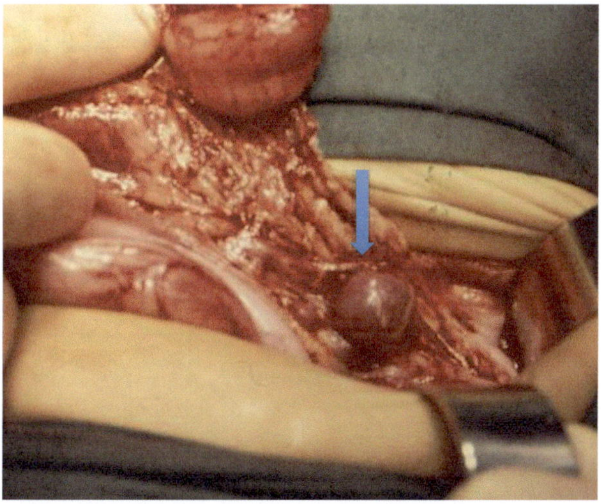

Fig. 19.1 A clinical intraoperative photograph showing an accessory spleen

important for patients with idiopathic thrombocytopenic purpura (ITP) to avoid recurrence. Patients with ITP who require intraoperative platelets transfusion, platelets should be administered after ligation of the splenic artery.

- There are common locations of accessory spleens and these include (Fig. 19.1):
  - The hilum of the spleen
  - The gastrocolic ligament
  - The gastrosplenic ligaments
  - The greater omentum
  - The mesentery
  - The pelvis
- Once the spleen is removed and hemostasis is secured, the abdominal incision is closed.

19.4 Laparoscopic Splenectomy

- Laparoscopic splenectomy is currently the procedure of choice to remove the spleen.
- The indications for laparoscopic splenectomy are the similar to those for open splenectomy.
- Laparoscopic splenectomy is indicated for:
  - Various benign hematologic diseases
  - Malignant hematologic diseases
  - Secondary hypersplenism
  - Anatomic abnormalities of the spleen
- The most common benign hematologic disease requiring laparoscopic splenectomy is immune thrombocytopenic purpura (ITP).
- Laparoscopic splenectomy for ITP is recommended when:

- The medical therapy, including steroids and intravenous gamma globulin, fails
- When long-term steroids are needed

- The presence of splenic infarct or abscess is currently not a contraindication for laparoscopic splenectomy.
- Laparoscopic splenectomy is contraindicated in emergency situations, such as hemorrhagic shock and spontaneous splenic rupture.
- The size of the spleen is a relative contraindication for laparoscopic splenectomy as massive splenomegaly may be difficult to remove laparoscopically.
- The following laparoscopic equipment are need for laparoscopic splenectomy:
  - Laparoscopic scissors and atraumatic graspers
  - Telescopes, including a 30° or 45° 5- or 10-mm laparoscope
  - Three or four 5-mm trocars
  - One 12-mm port for laparoscopic stapler and ligasure introduction and specimen retrieval bag.
  - Electrosurgical devices such as Ligasure or Harmonic knife can be used to assist with splenic dissection, vessels cautery and mobilization.
  - An endoscopic stapling device with a vascular load can be used to divide the hilar splenic vessels.
  - A retrieval bag is needed for removal of the specimen.
  - A suction/irrigation
  - A fan or snake retractors may be needed for elevation and retraction of the spleen.
  - Ring forceps are useful for morcellating the spleen in the retrieval sac.
- Depending on the surgeon's experience and preference, laparoscopic splenectomy can be done in one of three positions (Figs. 19.2 and 19.3):
  - The "hanging spleen" technique:
    The patient is placed in the lateral decubitus position.
  - The "leaning spleen" technique:
    The patient is placed at a 45° tilt.

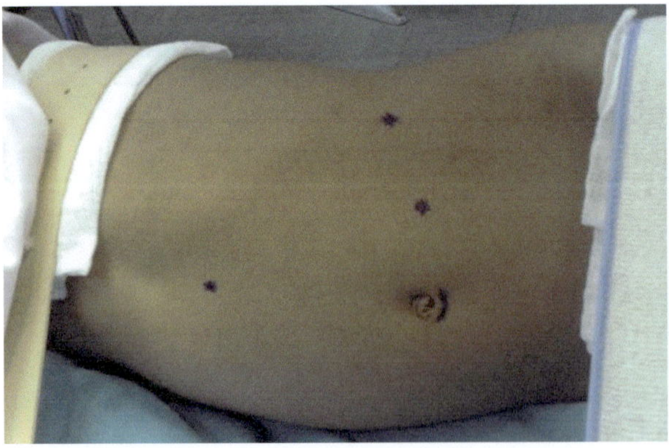

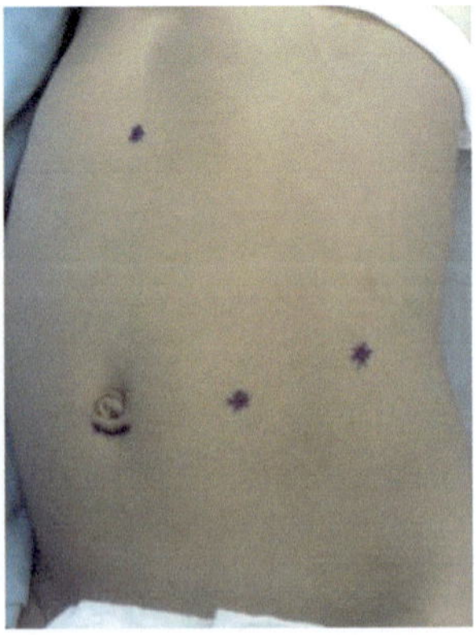

Figs. 19.2 and 19.3 Clinical photographs showing the position of the patient and placement of the trocars for laparoscopic splenectomy

  - The supine approach
  - The anterior approach was the first technique described for laparoscopic splenectomy; however, it is seldom used today, except when the spleen is very large. For the anterior approach, the patient is placed in the lithotomy position to allow the surgeon to operate while standing between the patient's legs with the assistants on either side of the patient.
- The surgeon and camera operator stand on the patient's right side, with the video monitors above and lateral to the patient's left shoulder.
- Once pneumoperitoneum is established, the peritoneal cavity is inspected.
- It is very important to look for accessory spleens, which are present in 12–16% of patients and as many as 32% of patients with immune (idiopathic) thrombocytopenia purpura (ITP).
- The splenocolic ligament and splenic attachments are divided with ultrasonic shears or ligasure.
  - Although these planes are relatively avascular, the use of an ultrasonic coagulation shears or an electrothermal bipolar sealing device can help in hemostasis.
  - This mobilization will open the lesser sac.
  - The spleen is rotated laterally, and the gastrosplenic ligament containing the short gastric vessels is carefully divided.
  - The splenophrenic attachment of the spleen is usually left undivided. This will keep the spleen attached superiorly and prevent it from falling down which makes dissection of the splenic vessels at the hilum more difficult.
- Once the gastrosplenic ligament and the short gastric vessels are divided, the spleen is elevated to expose the hilum. This can be done using a fan retractor, snake retractor, or long atraumatic bowel grasper.
- When elevating the spleen, it is important not to injure the splenic parenchyma as this will cause intraoperative bleeding.
- Once the spleen is elevated, the hilum and the tail of the pancreas become more visible and accessible.
  - When the hilum of the spleen is not adequately visualized, the splenophrenic ligament can be divided superiorly to facilitate splenic mobilization.
  - After the splenic hilum and the tail of the pancreas are well visualized, an endoscopic stapler with a vascular load can be used to ligate and divide the splenic vasculature.
  - The splenic artery and vein are dissected along the superior border of the pancreas and divided with the vascular stapler or using ligasure.
  - Some surgeons prefer to first dissect and expose the splenic vein and artery 1 cm proximal to the hilum, and then divide them with the vascular stapler or using ligasure.
  - Then, the remaining peritoneal attachments of the spleen are divided with the ultrasonic scalpel or ligasure.
- Others prefer to divide the splenic attachments first and the spleen is then mobilized to expose the hilar vessels. These vessels are then transected with the laparoscopic vascular stapler or divided by ligasure.
- Care should be taken to avoid injury to the tail of the pancreas.

- Following division of the splenic hilum, hemostasis is secured.
- Sometimes there is bleeding from the staple-line or the ligasure dividing line which can be controlled with clips, hemostatic agents or the ligasure itself.
- It is also important to keep the stapling device or ligasure jaws open after dividing the splenic vessels as this can be closed again to control any bleeding from the divided blood vessels.
- Another dissection technique is to divide the gastrosplenic ligament first, then the splenocolic and splenorenal ligaments. This allows the spleen to be suspended by the lateral and posterior attachments.
- The spleen once become free from all the attachments it is placed into an EndoCatch bag and is divided into pieces and extracted through the largest trocar port.
- The process of placing the spleen into the EndoCatch bag can sometimes be difficult and frustrating. The spleen can be grasped by the remnants of the splenocolic ligament left on the inferior border of the spleen and flipped onto its ventral surface with the hilum facing up. This will require some maneuvers.
- A retrieval bag is introduced through the 12-mm port and unfolded in the left upper quadrant. It is important to open the bag widely and having an atraumatic grasper can greatly facilitate placement of the spleen into the retrieval bag.
- After the spleen is placed totally in the retrieval bag, the purse-string suture is pulled tight and brought up through the 12-mm trocar. The 12-mm trocar is then removed, pulling the neck of the bag up through the abdominal wall (Fig. 19.4).
- Morcellation or piecemeal extraction of the spleen is then undertaken.
- A finger-fracture technique can also be used to break the spleen into smaller pieces but a ring forceps also work well. Suction may also be used to aspirate blood and aid in the removal of the spleen.
- Occasionally, depending on the size of the spleen, the 12-mm fascial incision must be enlarged to allow the removal of the morcellated spleen in the bag. In

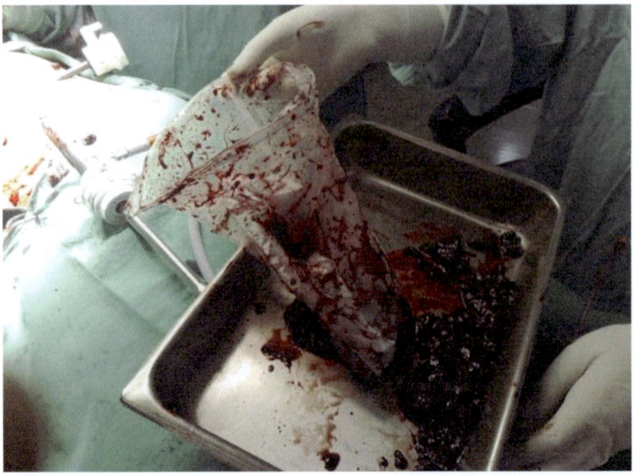

Fig. 19.4 A clinical intraoperative photograph showing morcellation of the spleen and extracting it via the endobag

morcellating the spleen, care must be taken not to rupture the bag and spill splenic tissue in the peritoneal cavity.
- Once the spleen is completely removed, the peritoneal cavity is insufflated again, and inspected to insure hemostasis.
- The ports are also removed under direct vision to ensure hemostasis, and the 12-mm fascial incisions are closed, followed by routine skin closure.

19.5 Hand-assisted Laparoscopic Splenectomy

- Hand-assisted laparoscopic surgery (HALS) is another technique for laparoscopic splenectomy that offers benefits of both open and laparoscopic techniques.
- HALS can be used with the anterior or lateral approach.
- It is generally agreed that the nondominant hand should be placed into the abdomen.
- The hand-assist device can be placed in the midline at or slightly below the inferior pole of the spleen.
- The incision should be 7–8 cm (or 1 cm less than the surgeon's glove size) and should be located 2–4 cm caudal to the inferior pole of the enlarged spleen.
- The surgeon stands on the patient's right side, and the nondominant hand is inserted through the hand-assist device, allowing medial retraction, rotation, and elevation of the spleen.
- Laparoscopic ports are placed normally as for a laparoscopic splenectomy.
- When the spleen is extremely large, the trocars must be placed more inferiorly than normal.
- After dividing all of the anterior and posterior attachments of the spleen, the hilum is ligated with an endoscopic stapling device and divided.
- The spleen is placed in a retrieval sac, brought up through the hand-assist incision, and, if necessary, morcellated.

19.6 Complications of Splenectomy

- Intraoperative complications include:
  - Bleeding
  - Pancreatic injury
  - Vascular injury
  - Colonic injury
  - Gastric injury
  - Diaphragmatic injury
- Postoperative complications include:
  - Pulmonary complications including atelectasis and pneumonia
  - Subphrenic abscess
  - Postoperative ileus
  - Portal vein thrombosis

 - Thrombocytosis and if left untreated may lead to thrombotic complications. Antiplatelet therapy is initiated when the platelet counts is higher than 1,000,000/μL.
 - Wound hematoma, seroma, and wound infections
- Late postoperative complications include:
 - Splenosis
 - Overwhelming post splenectomy sepsis
 - Chronic thromboembolic pulmonary hypertension
- Intraoperative and postoperative hemorrhage
 - Bleeding is one of the most common and feared complications of laparoscopic splenectomy.
 - Postoperative bleeding following laparoscopic splenectomy occurs in approximately 3% of patients.
 - Bleeding is the most common reason for conversion from laparoscopic to an open approach.
 - Bleeding is commonly secondary to a tear in the splenic capsule or failing to achieve adequate control of the splenic hilar vessels.
 - Treatment of postoperative bleeding from the staple line is more difficult and may require relaparoscopy or if bleeding is severe then open exploration becomes necessary. Postoperative splenic artery embolization has been described with success to control post splenectomy bleeding.
- Infection, including wound infection, pneumonia, and overwhelming post splenectomy infection (OPSI)
 - OPSI is a well-known major long-term risk following splenectomy.
 - Patients are at lifelong risk for the development of OPSI; however, the highest risk is in the first 2 years after surgery.
 - Although the reported risk of OPSI is relatively low (3.2%), it is known to be associated with a mortality rate as high as 40–50%.
 - Patients undergoing elective splenectomy should be vaccinated against meningococcal, pneumococcal, and Haemophilus influenzae type B infections at least 15 days before splenectomy.
 - In patients undergoing emergency splenectomy, vaccination is recommended within 30 days after the procedure or prior to their discharge from the hospital.
 - The pneumococcal vaccine should be repeated every 5 years, and patients should also receive an influenza vaccine annually.
- Injury to other structures, such as the colon, stomach and the tail of the pancreas.
 - Injury to the tail of the pancreas is a serious complication.
 - This can be avoided by careful dissection of the splenic hilum and adequate visualization of the pancreatic tail.
 - Drains are rarely used following laparoscopic splenectomy; however, if there is concern about a possible pancreatic tail injury, a closed suction drain should be used.
- Risk of missed accessory spleens
 - Accessory spleens are present in as many as 12–32% of patients.

- A thorough exploration for accessory spleens should be made after initial trocar placement.
- Accessory spleens are typically found:
  - In the splenic hilum
  - Along the splenic vessels
  - In the greater omentum
  - Along the splenorenal ligament
- It is important to look for accessory spleens specially in patients having splenectomy for idiopathic thrombocytopenic purpura as failure to identify and remove accessory spleens may result in recurrence.

- Portal vein thrombosis
  - Portal vein thrombosis is increasingly being recognized as a postoperative complication of splenectomy.
  - Portal vein thrombosis should be considered in post splenectomy patients if they have the following symptoms and signs:
    - Anorexia
    - Abdominal pain
    - Postoperative ileus
    - Low-grade fever
    - Elevated platelet and leukocyte count
  - Portal vein thrombosis has been reported to occur in 0.7–14% of patients.
  - There are several factors known to be associated with portal vein thrombosis. These include:
    - Splenomegaly
    - Myeloproliferative disorder
    - Hemolysis
    - A postoperative platelet count increasing to more than eight times the baseline level before splenectomy.
    - A preoperative splenic vein diameter of 8 mm or greater has also been suggested as a risk factor for portal or splenic vein thrombosis.
- Sepsis:
  - Splenectomy carries an increased risk of sepsis, particularly overwhelming post-splenectomy sepsis due to encapsulated organisms such as S. pneumoniae, Haemophilus influenzae and Neisseria Meningitides.
  - Splenectomized patients have a 10–20 time's higher risk of acquiring sepsis when compared to non-splenectomized patients.
  - Splenectomy also increases the severity of babesiosis.
  - Parasitemia levels can reach up to 85% in patients without spleens, compared to 1–10% in individuals with spleens and effective immune systems.
- Leukocytosis:
  - Splenectomy can be followed by an increase in blood leukocytes.
- Thrombocytosis:
  - The post-splenectomy platelet count may rise to abnormally high levels (thrombocytosis). This can lead to an increased risk of potentially fatal thrombosis.

  - Mild thrombocytosis can be observed and treated conservatively after a splenectomy.
- Normally, erythrocytes are stored and removed from the circulating blood by the spleen, including the removal of damaged erythrocytes.
  - In splenectomized patients damaged erythrocytes will continue to circulate in the blood.
  - Damaged erythrocytes have a procoagulant activity and this can cause thromboembolic events e.g., pulmonary embolism, portal vein thrombosis and deep vein thrombosis.
- Splenectomized patients may be at a higher risk of subsequently developing diabetes.
- Splenectomy patients typically have Howell-Jolly bodies and less commonly Heinz bodies in their blood smears.

19.7 Immunization and Post Splenectomy Sepsis

- The spleen performs a variety of immunological and hematological functions.
- The spleen plays an important role in both the innate and adaptive immune systems, thereby protecting the body from invading organisms.
- In 1911, Emil T Kocher stated that splenic injuries required the removal of the whole organ, and it had no adverse effects (Fig. 19.5).
- This led to the belief that splenectomy could be performed in all cases without adverse effects.
- In 1919, Morris and Bullock established that splenectomy was associated with an increased risk of infections.
- This finding was emphasized and further stressed by King and Schumacher in 1952 when they reported severe sepsis in five children who had undergone splenectomy for congenital hemolytic anemia.
- Overwhelming post-splenectomy infection (OPSS) occurs in all age groups regardless of the indication for splenectomy.
- The incidence of fatal infection in splenectomized children is 10 times higher than the control group.
- Splenectomized patients are more susceptible to a variety of different infections caused by organisms such as *Staphylococcus aureus*, *Haemophilus influenzae*, *Streptococcus pneumoniae*, and malarial parasites.

- **IN 1919, MORRIS AND BULLOCK STATED THAT SPLENECTOMY WAS ASSOCIATED WITH AN INCREASED RISK OF INFECTIONS**
- **IN 1952, KING AND SHUMACKER EMPHASID THAT SPLENECTOMY IS ASSOCIATED WITH AN INCREASED RISK OF OVERWHELMING INFECTION**

Fig. 19.5 Emil Theodor Kocher (25 August 1841–27 July 1917) was a Swiss physician and medical researcher who received the 1909 Nobel Prize in Physiology or Medicine for his work in the physiology, pathology and surgery of the thyroid. Among his many accomplishments are the introduction and promotion of aseptic surgery and scientific methods in surgery, specifically reducing the mortality of thyroidectomies below 1% in his operations. In 1911, Emil T Kocher stated that splenic injuries required the removal of the whole organ, and this had no adverse effects (Wikipedia, the free encyclopedia)

- The most severe complication of splenectomy is overwhelming post-splenectomy infection (OPSI). This is known to be associated with significant morbidity and mortality.
- Every effort should be done to protect splenectomized patients against OPSI including:
  - These patients and their families should be educated about the risks and complications of OPSI.
  - These patients should be vaccinated.
  - These patients should receive antibiotics prophylaxis.
- Splenectomy is known to be associated with an increased risk of sepsis due to encapsulated organisms such as:
  - S. pneumoniae
  - Haemophilus influenzae
  - Meningococcal

- These patients especially children should receive immunization prior to splenectomy.
- These vaccines include:
  - Pneumococcal vaccine
  - Meningococcal vaccines
  - Haemophilus influenza type b vaccine
- These vaccines should be given at least 2 weeks prior to splenectomy and followed subsequently by poster doses.
- If splenectomy is done for an emergency reason these vaccines can be given postoperatively.
- Children who undergo splenectomy should also receive prophylactic antibiotics for a minimum of 2 years depending on their age at the time of splenectomy. This is in the form of penicillin prophylaxis either orally or intramuscularly.
- This is because overwhelming post-splenectomy sepsis is commonly seen within the first 2 years post splenectomy.

19.8 Prevention of Post Splenectomy Sepsis

- Splenectomized patients are at a significantly increased risk of infection.
- The estimated incidence of OPSI is around 0.23–0.42% per year, and a lifetime risk of 5%.
- Splenectomy should be avoided whenever possible.
- Alternatives to splenectomy should be considered.
- Elective splenectomy should be deferred as long as possible.
- Asplenic patients should not only be vaccinated, but also educated about the risks of OPSI.
- Vaccinations against S. pneumoniae, Haemophilus influenza type b, N. meningitides and influenza virus are strongly recommended and should be given at least 2 weeks before splenectomy in elective cases or at least 2 weeks after splenectomy in emergency cases.
- Pneumococcal Vaccine:
  - This helps will help protect against pneumococcal disease, which is caused by Streptococcus pneumoniae bacteria.
  - There are two kinds of pneumococcal vaccines available:
    - Pneumococcal conjugate vaccines (PCV13, PCV15, and PCV20)
    - Pneumococcal polysaccharide vaccine (PPSV23)
    - PCV or PCV13 = Pneumococcal conjugate vaccine
    - PPSV or PPSV23 = Pneumococcal polysaccharide vaccine
  - CDC (Centers for Disease Control and prevention) recommends giving PCV13 or PCV15 for all children younger than 5 years old and children 5–18 years old with certain medical conditions that increase their risk of pneumococcal infections.
  - CDC recommends PCV15 or PCV20 for:
    - Adults 65 years or older

Adults 19–64 years old with certain medical conditions or risk factors

- CDC recommends PPSV23 for:
  - Children 2–18 years old with certain medical conditions
  - Adults 19 years or older who received PCV15
- For those who have never received any pneumococcal conjugate vaccine, CDC recommends PCV15 or PCV20 for adults 65 years or older and adults 19–64 years old with certain medical conditions or risk factors. If PCV15 is used, this should be followed by a dose of PPSV23.
- These serotypes account for over 80% of pneumococcal infections.
- Significant antibodies develop when the vaccine is administered to healthy individuals. The antibody titers decline in 6 months to 3 years. These patients should receive booster doses.
- Polyvalent vaccine should be given at least 2–4 weeks before splenectomy to achieve an optimal antibody response.
- The Centers for Disease Control and Prevention recommends that patients with asplenia receive the 23-valent pneumococcal vaccine to prevent OPSI, as this covers about 73–90% of strains causing OPSI.
- Asplenic children should be revaccinated every 3–6 years, as levels of circulating antibodies to the pneumococcal antigens decline and they reach a low level 5 years after vaccination in asplenic children.
- These patients should also receive:
  - Meningococcal vaccines
  - Haemophilus influenza type b vaccine
- These patients should be vaccinated as recommended unless:
  - They are actively on vasopressors
  - They are on steroids
  - They are undergoing a major procedure likely to reduce the ability of the immune system to respond to the vaccine appropriately.
- Patients who have received the Pneumovax 23 vaccine in the past year should wait at least 1 year to receive the Prevnar 13 vaccine, followed by a second Pneumovax 23 vaccine at least 8 weeks later.
- If 2 doses of the Pneumovax 23 vaccine have been given in the past year, Prevnar 13 should be given 1 year after the last Pneumovax 23.
- If Prevnar 13 has been given to the patient previously, a Pneumovax 23 dose may be given 8 weeks after Prevnar 13.

• Penicillin Prophylaxis
  - Oral penicillin has been used to prevent OPSI in splenectomized children.
  - One of the main concerns of oral penicillin is patient's noncompliance.
  - Add to this the emergence of penicillin-resistant pneumococci.
  - Patients sensitive to penicillin may be given erythromycin.

Splenectomy Vaccination Guidelines
INITIAL VACCINATIONS (2 WEEKS PRIOR TO SPLENECTOMY OR 2 WEEKS AFTER SPLENECTOMY

- **Pneumococcal 13-valent conjugate (PCV13—Prevnar 13) 0.5 mL IM**
- **Haemophilus influenza type b vaccine (Hib—ActHIB) 0.5 mL IM**
- **Meningococcal vaccine (Menactra) 0.5 mL IM**
- **Meningococcal serogroup B (Bexsero) 0.5 mL IM**

TWO MONTHS FOLLOW-UP VACCINATIONS AFTER THE INITIAL VACCINATIONS

- **Pneumococcal polysaccharide (PPSV23—Pneumovax 23) 0.5 mL IM**
- **Meningococcal vaccine 0.5 mL IM**
- **Meningococcal serogroup B 0.5 mL IM (>1 month after first dose)**

LONG-TERM FOLLOW-UP VACCINATIONS

- **Pneumococcal polysaccharide 0.5 mL IM 5 years after the first dose of this vaccine**
- **Meningococcal vaccine 0.5 mL IM recommended every 5 years**
- **No additional haemophilus vaccine is needed**
- **Seasonal influenza vaccine is indicated annually**

- Sepsis in splenectomized patients can occur with any organism including bacteria, viruses, fungi, or protozoan, but encapsulated organisms are the most frequently causative organisms.
- The most important pathogen is Streptococcus pneumoniae, but Haemophilus influenza and Neisseria meningitidis are also significant.
- Other causative organisms include:
  - Plasmodium falciparum malaria
  - Babesiosis (Babesia microti)
  - Capnocytophaga canimorsus
- Overwhelming post splenectomy infection (OPSI) is a serious fulminant process that carries a high mortality rate.
- The rates of OPSI in splenectomized patients is variable and depends on several factors including:
  - Age at splenectomy
  - Indication for splenectomy
  - Ongoing immunosuppression
  - Splenectomy performed for a hematological disorder including:
    - Thalassemia

    Hereditary spherocytosis
    Autoimmune hemolytic anemia
    Immune thrombocytopenic purpura
    Lymphoma
  - Splenectomy for these conditions appears to carry a higher risk of OPSI than splenectomy performed for trauma.
  - The greatest risk for OPSI is within the first 2 years after splenectomy.
- The mortality rate of OPSI is about 50%–70% despite aggressive medical therapy that includes:
  - Intravenous fluids
  - Broad spectrum antibiotics should be started after taking a blood sample for culture and sensitivity.
    A combination of I.V. vancomycin and ceftriaxone are recommended initially and these can be modified based on the result of culture and sensitivity.
  - Vasopressors
  - Steroids
  - Heparin
  - Packed red blood cells
  - Platelets
  - Cryoprecipitates
  - Fresh frozen plasma

Further Reading

1. Chahine AH, Gilyard S, Hanna TN, Fan S, Risk B, Johnson JO, et al. Management of splenic trauma in contemporary clinical practice: a National Trauma Data Bank Study. Acad Radiol. 2021;28(Suppl 1):S138–47.
2. Scarborough JE, Ingraham AM, Liepert AE, Jung HS, O'Rourke AP, Agarwal SK. Nonoperative management is as effective as immediate splenectomy for adult patients with high-grade blunt splenic injury. J Am Coll Surg. 2016;223(2):249–58.
3. Maurus CF, Schäfer M, Müller MK, Clavien PA, Weber M. Laparoscopic versus open splenectomy for nontraumatic diseases. World J Surg. 2008;32(11):2444–9.
4. Chen J, Ma R, Yang S, Lin S, He S, Cai X. Perioperative outcomes of laparoscopic versus open splenectomy for nontraumatic diseases: a meta-analysis. Chin Med J. 2014;127(13):2504–10.
5. Poulin EC, Schlachta CM, Mamazza J. Splenectomy. In: Ashley SW, et al., editors. ACS surgery: principles and practice. Hamilton, ON: BC Decker Inc.; 2010.
6. Hassan ME, Al AK. Massive splenomegaly in children: laparoscopic versus open splenectomy. JSLS. 2014;18(3)
7. Liu Y, Zhao L, Tang Y, Zhang Y, Shi SC, Xie FX, et al. Laparoscopic versus open splenectomy and devascularization for massive splenomegaly due to portal hypertension. J Huazhong Univ Sci Technolog Med Sci. 2016;36(6):876–80.
8. Casaccia M, Sormani MP, Palombo D, Dellepiane C, Ibatici A. Laparoscopic splenectomy versus open splenectomy in massive and Giant spleens: should we update the 2008 EAES guidelines? Surg Laparosc Endosc Percutan Tech. 2019;29(3):178–81.
9. Wang D, Chen X, Lv L, Yang T, Huang B, Cao Y, et al. Laparoscopic splenectomy and devascularization for massive splenomegaly in portal hypertensive patients: a retrospective

study of a single surgical team's experience with 6-year follow-up data. Ann Transl Med. 2022;10(4):207.
10. Shin RD, Lis R, Levergood NR, Brooks DC, Shoji BT, Tavakkoli A. Laparoscopic versus open splenectomy for splenomegaly: the verdict is unclear. Surg Endosc. 2019;33(4):1298–303.
11. Rodríguez-Luna MR, Balagué C, Fernández-Ananín S, Vilallonga R, Targarona Soler EM. Outcomes of laparoscopic splenectomy for treatment of splenomegaly: a systematic review and meta-analysis. World J Surg. 2021;45(2):465–79.
12. Shoultz TH, Minei JP. Spleen. In: Feliciano DV, Mattox KL, Moore EE, editors. Trauma. 9th ed. New York: McGraw Hill; 2021. p. 679–98.
13. Theodoropoulos J, Krecioch P, Myrick S, Atkins R. Delayed presentation of a splenic injury after colonoscopy: a diagnostic challenge. Int J Color Dis. 2010;25(8):1033–4.
14. Crawford RS, Tabbara M, Sheridan R, Spaniolas K, Velmahos GC. Early discharge after nonoperative management for splenic injuries: increased patient risk caused by late failure? Surgery. 2007;142(3):337–42.
15. Wang C, Tu X, Li S, Luo G, Norwitz ER. Spontaneous rupture of the spleen: a rare but serious case of acute abdominal pain in pregnancy. J Emerg Med. 2011;41(5):503–6.
16. Renzulli P, Hostettler A, Schoepfer AM, Gloor B, Candinas D. Systematic review of atraumatic splenic rupture. Br J Surg. 2009;96(10):1114–21.
17. Watson GA, Rosengart MR, Zenati MS, Tsung A, Forsythe RM, Peitzman AB, et al. Nonoperative management of severe blunt splenic injury: are we getting better?*J*. Trauma. 2006;61(5):1113–8; discussion 1118–9.
18. McIntyre LK, Schiff M, Jurkovich GJ. Failure of nonoperative management of splenic injuries: causes and consequences. Arch Surg. 2005;140(6):563–8; discussion 568–9.
19. Li Y, Zhang D, Hua F, Gao S, Wu Y, Xu J. Factors associated with the effect of open splenectomy for immune thrombocytopenic purpura. Eur J Haematol. 2017;98(1):44–51.
20. Cáp J. Long-term clinical experience with spherocytic hemolytic anemia in children. Cesk Pediatr. 1992;47(10):592–5. [QxMD MEDLINE Link].
21. Meekes I, van der Staak F, van Oostrom C. Results of splenectomy performed on a group of 91 children. Eur J Pediatr Surg. 1995;5(1):19–22.
22. Alizai NK, Richards EM, Stringer MD. Is cholecystectomy really an indication for concomitant splenectomy in mild hereditary spherocytosis? Arch Dis Child. 2010;95(8):596–9.
23. Sandler A, Winkel G, Kimura K, Soper R. The role of prophylactic cholecystectomy during splenectomy in children with hereditary spherocytosis. J Pediatr Surg. 1999;34(7):1077–8.
24. Luo HP, Zhang ZG, Long X, Liu FL, Chen XP, Zhang L, et al. Combined laparoscopic splenectomy and esophagogastric devascularization versus open splenectomy and esophagogastric devascularization for portal hypertension due to liver cirrhosis. Curr Med Sci. 2020;40(1):117–22.
25. Cines DB, Blanchette VS. Immune thrombocytopenic purpura. N Engl J Med. 2002;346(13):995–1008.
26. Demetriades D, Lam L. Splenectomy and splenorrhaphy. In: Fischer JE, Jones DB, Pomposelli FB, Upchurch Jr GR, Klimberg VS, Schwaitzberg SD, et al., editors. Fischer's mastery of surgery, vol. 2. 6th ed. Philadelphia: Wolters Kluwer; 2012: Chap 175.
27. Winslow ER, Brunt LM, Drebin JA, Soper NJ, Klingensmith ME. Portal vein thrombosis after splenectomy. Am J Surg. 2002;184(6):631–5; discussion 635–6.
28. Szasz P, Ardestani A, Shoji BT, Brooks DC, Tavakkoli A. Predicting venous thrombosis in patients undergoing elective splenectomy. Surg Endosc. 2020;34(5):2191–6.
29. Bisharat N, Omari H, Lavi I, Raz R. Risk of infection and death among post-splenectomy patients. J Infect. 2001;43(3):182–6.
30. Okabayashi T, Hanazaki K. Overwhelming postsplenectomy infection syndrome in adults - a clinically preventable disease. World J Gastroenterol. 2008;14(2):176–9.
31. Kimmig LM, Palevsky HI. Review of the association between splenectomy and chronic thromboembolic pulmonary hypertension. Ann Am Thorac Soc. 2016;13(6):945–54.
32. Coon WW. The spleen and splenectomy. Surg Gynecol Obstet. 1991;173(5):407–14.

33. Uranues S, Alimoglu O. Laparoscopic surgery of the spleen. Surg Clin North Am. 2005;85(1):75–90. ix
34. Misiakos EP, Bagias G, Liakakos T, Machairas A. Laparoscopic splenectomy: current concepts. World J Gastrointest Endosc. 2017;9(9):428–37.
35. Delaitre B, Maignien B. Splenectomy by the laparoscopic approach. Report of a case. Presse Med. 1991;20(44):2263.
36. Habermalz B, Sauerland S, Decker G, Delaitre B, Gigot JF, Leandros E, et al. Laparoscopic splenectomy: the clinical practice guidelines of the European Association for Endoscopic Surgery (EAES). Surg Endosc. 2008;22(4):821–48.
37. Huscher CG, Mingoli A, Sgarzini G, Brachini G, Ponzano C, Di Paola M, et al. Laparoscopic treatment of blunt splenic injuries: initial experience with 11 patients. Surg Endosc. 2006;20(9):1423–6.
38. Shamim AA, Zafar SN, Nizam W, Zeineddin A, Ortega G, Fullum TM, et al. Laparoscopic splenectomy for trauma. JSLS. 2018;22(4)
39. Fransvea P, Costa G, Serao A, Cortese F, Balducci G, Sganga G, et al. Laparoscopic splenectomy after trauma: who, when and how. A systematic review. J Minim Access Surg. 2021;17(2):141–6.
40. Hao X, Dai K, He Y, Tao L, Yu H. Laparoscopic splenectomy via the spleen bed in combination with selective esophagogastric devascularization for patients with cirrhotic portal hypertension: a single-institution experience. Wideochir Inne Tech Maloinwazyjne. 2020;15(3):462–8.
41. Zhu JH, Wang YD, Ye ZY, Zhao T, Zhu YW, Xie ZJ, et al. Laparoscopic versus open splenectomy for hypersplenism secondary to liver cirrhosis. Surg Laparosc Endosc Percutan Tech. 2009;19(3):258–62.
42. Wang Y, Zhan X, Zhu Y, Xie Z, Zhu J, Ye Z. Laparoscopic splenectomy in portal hypertension: a single-surgeon 13-year experience. Surg Endosc. 2010;24(5):1164–9.
43. Cobb WS, Heniford BT, Burns JM, Carbonell AM, Matthews BD, Kercher KW. Cirrhosis is not a contraindication to laparoscopic surgery. Surg Endosc. 2005;19(3):418–23.
44. Kawanaka H, Akahoshi T, Kinjo N, Harimoto N, Itoh S, Tsutsumi N, et al. Laparoscopic splenectomy with technical standardization and selection criteria for standard or hand-assisted approach in 390 patients with liver cirrhosis and portal hypertension. J Am Coll Surg. 2015;221(2):354–66.
45. Somasundaram SK, Massey L, Gooch D, Reed J, Menzies D. Laparoscopic splenectomy is emerging 'gold standard' treatment even for massive spleens. Ann R Coll Surg Engl. 2015;97(5):345–8.
46. Tsamalaidze L, Stauffer JA, Permenter SL, Asbun HJ. Laparoscopic splenectomy for massive splenomegaly: does size matter? J Laparoendosc Adv Surg Tech A. 2017;27(10):1009–14.
47. Casaccia M, Sormani MP, Palombo D, Dellepiane C, Ibatici A. Laparoscopic splenectomy versus open splenectomy in massive and giant spleens: should we update the 2008 EAES guidelines? Surg Laparosc Endosc Percutan Tech. 2019;29(3):178–81.
48. Misawa T, Yoshida K, Iida T, Sakamoto T, Gocho T, Hirohara S, et al. Minimizing intraoperative bleeding using a vessel-sealing system and splenic hilum hanging maneuver in laparoscopic splenectomy. J Hepato-Biliary-Pancreat Surg. 2009;16(6):786–91.
49. Canda AE, Ozsoy Y, Yuksel S. Laparoscopic splenectomy using LigaSure in benign hematologic diseases. Surg Laparosc Endosc Percutan Tech. 2009;19(1):69–71.
50. Romano F, Gelmini R, Caprotti R, Andreotti A, Guaglio M, Franzoni C, et al. Laparoscopic splenectomy: ligasure versus EndoGIA: a comparative study. J Laparoendosc Adv Surg Tech A. 2007;17(6):763–7.
51. Gelmini R, Romano F, Quaranta N, Caprotti R, Tazzioli G, Colombo G, et al. Sutureless and stapleless laparoscopic splenectomy using radiofrequency: LigaSure device. Surg Endosc. 2006;20(6):991–4.
52. Mortelé KJ, Mortelé B, Silverman SG. CT features of the accessory spleen. AJR Am J Roentgenol. 2004;183(6):1653–7.

53. Park AE, Birgisson G, Mastrangelo MJ, Marcaccio MJ, Witzke DB. Laparoscopic splenectomy: outcomes and lessons learned from over 200 cases. Surgery. 2000;128(4):660–7.
54. Pace DE, Chiasson PM, Schlachta CM, Mamazza J, Poulin EC. Laparoscopic splenectomy for idiopathic thrombocytopenic purpura (ITP). Surg Endosc. 2003;17(1):95–8.
55. Wang KX, Hu SY, Zhang GY, Chen B, Zhang HF. Hand-assisted laparoscopic splenectomy for splenomegaly: a comparative study with conventional laparoscopic splenectomy. Chin Med J. 2007;120(1):41–5.
56. Han ES, You YK, Kim DG, Lee JS, Kim EY, Lee SH, et al. Clinical significance of single-port laparoscopic splenectomy: comparison of single-port and multiport laparoscopic procedure. Ann Surg Treat Res. 2015;89(2):55–60.
57. Barbaros U, Aksakal N, Tukenmez M, Agcaoglu O, Bostan MS, Kilic B, et al. Comparison of single port and three port laparoscopic splenectomy in patients with immune thrombocytopenic purpura: clinical comparative study. J Minim Access Surg. 2015;11(3):172–6.
58. Wu S, Lai H, Zhao J, Deng X, Wei J, Liang J, et al. Systematic review and meta-analysis of single-incision versus conventional multiport laparoscopic splenectomy. J Minim Access Surg. 2018;14(1):1–8.
59. Bai DS, Jin SJ, Xiang XX, Qian JJ, Zhang C, Zhou BH, et al. Robotic-assisted versus laparoscopic approach of Bai-Jiang-style vagus nerve-preserving splenectomy and azygoportal disconnection. Updat Surg. 2022;74:1773–80.
60. Babala J, Pini Prato A, Zahradnikova P, Beder I. Minimally invasive partial spleen resection preserving the lower pole in children: a technical report on feasibility and safety. J Laparoendosc Adv Surg Tech A. 2021; https://doi.org/10.1089/lap.2021.0232.
61. Makansi M, Hutter M, Theilen TM, Fiegel HC, Rolle U, Gfroerer S. Comparison of perioperative outcomes between laparoscopic and open partial splenectomy in children and adolescents. World J Gastrointest Surg. 2021;13(9):979–87.
62. Schaarschmidt K, Kolberg-Schwerdt A, Lempe M, Saxena A. Ultrasonic shear coagulation of main hilar vessels: a 4-year experience of 23 pediatric laparoscopic splenectomies without staples. J Pediatr Surg. 2002;37(4):614–6.
63. Rosen M, Brody F, Walsh RM, Ponsky J. Hand-assisted laparoscopic splenectomy vs conventional laparoscopic splenectomy in cases of splenomegaly. Arch Surg. 2002;137(12):1348–52.
64. Kercher KW, Matthews BD, Walsh RM, Sing RF, Backus CL, Heniford BT. Laparoscopic splenectomy for massive splenomegaly. Am J Surg. 2002;183(2):192–6.
65. Kaban GK, Czerniach DR, Cohen R, Novitsky YW, Yood SM, Perugini RA, et al. Hand-assisted laparoscopic splenectomy in the setting of splenomegaly. Surg Endosc. 2004;18(9):1340–3.
66. Glasgow RE, Mulvihill SJ. Laparoscopic splenectomy. World J Surg. 1999;23(4):384–8.
67. Brodsky JA, Brody FJ, Walsh RM, Malm JA, Ponsky JL. Laparoscopic splenectomy. Surg Endosc. 2002;16(5):851–4.
68. Targarona EM, Espert JJ, Bombuy E, Vidal O, Cerdán G, Artigas V, et al. Complications of laparoscopic splenectomy. Arch Surg. 2000;135(10):1137–40.
69. Kercher KW, Novitsky YW, Czerniach DR, Litwin DE. Staple line bleeding following laparoscopic splenectomy: intraoperative prevention and postoperative management with splenic artery embolization. Surg Laparosc Endosc Percutan Tech. 2003;13(5):353–6.
70. Legrand A, Bignon A, Borel M, Zerbib P, Langlois J, Chambon JP, et al. Perioperative management of asplenic patients. Ann Fr Anesth Reanim. 2005;24(7):807–13.
71. Working Party of the British Committee for Standards in Haematology Clinical Haematology Task Force. Guidelines for the prevention and treatment of infection in patients with an absent or dysfunctional spleen. BMJ. 1996;312(7028):430–4.
72. Stanek A, Stefaniak T, Makarewicz W, Kaska L, Podgórczyk H, Hellman A, et al. Accessory spleens: preoperative diagnostics limitations and operational strategy in laparoscopic approach to splenectomy in idiopathic thrombocytopenic purpura patients. Langenbeck's Arch Surg. 2005;390(1):47–51.
73. Gigot JF, Jamar F, Ferrant A, van Beers BE, Lengele B, Pauwels S, et al. Inadequate detection of accessory spleens and splenosis with laparoscopic splenectomy. A shortcoming of the laparoscopic approach in hematologic diseases. Surg Endosc. 1998;12(2):101–6.

74. Katkhouda N, Manhas S, Umbach TW, Kaiser AM. Laparoscopic splenectomy. J Laparoendosc Adv Surg Tech A. 2001;11(6):383–90.
75. Velanovich V, Shurafa M. Laparoscopic excision of accessory spleen. Am J Surg. 2000;180(1):62–4.
76. Rattner DW, Ellman L, Warshaw AL. Portal vein thrombosis after elective splenectomy. An underappreciated, potentially lethal syndrome. Arch Surg. 1993;128(5):565–9; discussion 569–70.
77. Harris W, Marcaccio M. Incidence of portal vein thrombosis after laparoscopic splenectomy. Can J Surg. 2005;48(5):352–4.
78. Romano F, Caprotti R, Conti M, Piacentini MG, Uggeri F, Motta V, et al. Thrombosis of the splenoportal axis after splenectomy. Langenbeck's Arch Surg. 2006;391(5):483–8.
79. Huang D, Tao M, Cao L, Wang X, Zheng S, Cao Y. Risk factors and anticoagulation effects of portal vein system thrombosis after laparoscopic splenectomy in patients with or without cirrhosis. Surg Laparosc Endosc Percutan Tech. 2019;29(6):498–502.
80. De'Angelis N, Abdalla S, Lizzi V, Esposito F, Genova P, Roy L, et al. Incidence and predictors of portal and splenic vein thrombosis after pure laparoscopic splenectomy. Surgery. 2017;162(6):1219–30.
81. Swinson B, Waters PS, Webber L, Nathanson L, Cavallucci DJ, O'Rourke N, et al. Portal vein thrombosis following elective laparoscopic splenectomy: incidence and analysis of risk factors. Surg Endosc. 2021;36:3332–9.
82. Katz SC, Pachter HL. Indications for splenectomy. Am Surg. 2006;72:565–80.
83. Cappellini MD, Taher AT. The use of luspatercept for thalassemia in adults. Blood Adv. 2021;5:326–33.
84. Locatelli F, Thompson AA, Kwiatkowski JL, et al. Betibeglogene Autotemcel gene therapy for non-β0/β0 genotype β-thalassemia. N Engl J Med. 2022;386:415–27.
85. Owusu-Ofori S, Remmington T. Splenectomy versus conservative management for acute sequestration crises in people with sickle cell disease. Cochrane Database Syst Rev. 2017;11:CD003425.
86. Grace RF, Barcellini W. Management of pyruvate kinase deficiency in children and adults. Blood. 2020;136:1241–9.
87. Grace RF, Bianchi P, van Beers EJ, et al. Clinical spectrum of pyruvate kinase deficiency: data from the pyruvate kinase deficiency natural history study. Blood. 2018;131:2183–92.
88. Reisner DC, Burgan CM. Wandering spleen: an overview. Curr Probl Diagn Radiol. 2018;47:68–70.
89. Cohen O, Baazov A, Samuk I, et al. Emergencies in the treatment of wandering spleen. Isr Med Assoc J. 2018;20:354–7.
90. Makrin V, Avital S, White I, et al. Laparoscopic splenectomy for solitary splenic tumors. Surg Endosc. 2008;22:2009–12.
91. Kraus MD, Fleming MD, Vonderheide RH. The spleen as a diagnostic specimen: a review of 10 years' experience at two tertiary care institutions. Cancer. 2001;91:2001–9.
92. Carr JA, Shurafa M, Velanovich V. Surgical indications in idiopathic splenomegaly. Arch Surg. 2002;137:64–8.
93. Pottakkat B, Kashyap R, Kumar A, et al. Redefining the role of splenectomy in patients with idiopathic splenomegaly. ANZ J Surg. 2006;76:679–82.
94. Mistry PK, Batista JL, Andersson HC, et al. Transformation in pretreatment manifestations of Gaucher disease type 1 during two decades of alglucerase/imiglucerase enzyme replacement therapy in the International Collaborative Gaucher Group (ICGG) Gaucher Registry. Am J Hematol. 2017;92:929–39.
95. Gochhait D, Dey P, Rajwanshi A, et al. Role of fine needle aspiration cytology of spleen. APMIS. 2015;123:190–3.
96. Lisotti A, Crinò SF, Mangiavillano B, et al. Diagnostic performance of endoscopic ultrasound-guided tissue acquisition of splenic lesions: systematic review with pooled analysis. Gastroenterol Rep (Oxf). 2022;10:goac022.

97. John S, Shabana W, Salameh JP, McInnes MDF. Percutaneous image-guided biopsy of the spleen: experience at a single tertiary care center. Can Assoc Radiol J. 2021;72:311–6.
98. Terrault N, Chen YC, Izumi N, et al. Avatrombopag before procedures reduces need for platelet transfusion in patients with chronic liver disease and thrombocytopenia. Gastroenterology. 2018;155:705–18.
99. Nilles KM, Caldwell SH, Flamm SL. Thrombocytopenia and procedural prophylaxis in the era of thrombopoietin receptor agonists. Hepatol Commun. 2019;3:1423–34.
100. Quah C, Ayiomamitis GD, Shah A, Ammori BJ. Computed tomography to detect accessory spleens before laparoscopic splenectomy: is it necessary? Surg Endosc. 2011;25:261–5.
101. Bickenbach KA, Gonen M, Labow DM, et al. Indications for and efficacy of splenectomy for haematological disorders. Br J Surg. 2013;100:794–800.
102. MacRae HM, Yakimets WW, Reynolds T. Perioperative complications of splenectomy for hematologic disease. Can J Surg. 1992;35:432.
103. Tefferi A, Pardanani A. Serious adverse events during ruxolitinib treatment discontinuation in patients with myelofibrosis. Mayo Clin Proc. 2011;86:1188–91.
104. Mukherjee D, Lidor AO, Chu KM, et al. Postoperative venous thromboembolism rates vary significantly after different types of major abdominal operations. J Gastrointest Surg. 2008;12:2015–22.
105. Rettenmaier MA, Abaid LN, Brown JV 3rd, et al. Dramatically reduced incidence of vaginal cuff dehiscence in gynecologic patients undergoing endoscopic closure with barbed sutures: a retrospective cohort study. Int J Surg. 2015;19:27–30.
106. Radkowiak D, Zychowicz A, Lasek A, et al. 20 years' experience with laparoscopic splenectomy. Single center outcomes of a cohort study of 500 cases. Int J Surg. 2018;52:285–92.
107. Ardestani A, Tavakkoli A. Laparoscopic versus open splenectomy: the impact of spleen size on outcomes. J Laparoendosc Adv Surg Tech A. 2013;23:760–4.
108. Kojouri K, Vesely SK, Terrell DR, George JN. Splenectomy for adult patients with idiopathic thrombocytopenic purpura: a systematic review to assess long-term platelet count responses, prediction of response, and surgical complications. Blood. 2004;104:2623–34.
109. Musallam KM, Khalife M, Sfeir PM, et al. Postoperative outcomes after laparoscopic splenectomy compared with open splenectomy. Ann Surg. 2013;257:1116–23.
110. Deeb AP, Kim MJ, Fleming FJ, et al. The impact of operative approach in elective splenectomy: a multivariate analysis of outcomes from the NSQIP database. Surg Laparosc Endosc Percutan Tech. 2012;22:415–9.
111. Rodríguez-Luna MR, Balagué C, Fernández-Ananín S, et al. Outcomes of laparoscopic splenectomy for treatment of splenomegaly: a systematic review and meta-analysis. World J Surg. 2021;45:465–79.
112. Shin RD, Lis R, Levergood NR, et al. Laparoscopic versus open splenectomy for splenomegaly: the verdict is unclear. Surg Endosc. 2019;33:1298–303.
113. Costi R, Castro Ruiz C, Romboli A, et al. Partial splenectomy: who, when and how. A systematic review of the 2130 published cases. J Pediatr Surg. 2019;54:1527–38.
114. Vikse J, Sanna B, Henry BM, et al. The prevalence and morphometry of an accessory spleen: a meta-analysis and systematic review of 22,487 patients. Int J Surg. 2017;45:18–28.
115. Qu Y, Ren S, Li C, et al. Management of postoperative complications following splenectomy. Int Surg. 2013;98:55–60.
116. Kristinsson SY, Gridley G, Hoover RN, et al. Long-term risks after splenectomy among 8,149 cancer-free American veterans: a cohort study with up to 27 years follow-up. Haematologica. 2014;99:392–8.
117. Musser G, Lazar G, Hocking W, Busuttil RW. Splenectomy for hematologic disease. The UCLA experience with 306 patients. Ann Surg. 1984;200:40–5.
118. Horowitz J, Smith JL, Weber TK, et al. Postoperative complications after splenectomy for hematologic malignancies. Ann Surg. 1996;223:290–6.
119. Nelson EW, Mone MC. Splenectomy in high-risk patients with splenomegaly. Am J Surg. 1999;178:581–5.

120. Crary SE, Buchanan GR. Vascular complications after splenectomy for hematologic disorders. Blood. 2009;114:2861–8.
121. Ho G, Brunson A, Keegan THM, Wun T. Splenectomy and the incidence of venous thromboembolism and sepsis in patients with autoimmune hemolytic anemia. Blood Cells Mol Dis. 2020;81:102388.
122. Swinson B, Waters PS, Webber L, et al. Portal vein thrombosis following elective laparoscopic splenectomy: incidence and analysis of risk factors. Surg Endosc. 2022;36:3332–9.
123. Szasz P, Ardestani A, Shoji BT, et al. Predicting venous thrombosis in patients undergoing elective splenectomy. Surg Endosc. 2020;34:2191–6.
124. Dendle C, Spelman T, Sundararajan V, et al. An analysis of the thromboembolic outcomes of 2472 splenectomized individuals. Blood. 2015;125:1681–2.
125. Stewart GW, Amess JA, Eber SW, et al. Thrombo-embolic disease after splenectomy for hereditary stomatocytosis. Br J Haematol. 1996;93:303–10.
126. Zhang L, Yan P, Yang K, et al. Association between splenectomy and chronic thromboembolic pulmonary hypertension: a systematic review and meta-analysis. BMJ Open. 2021;11:e038385.
127. Gubergrits NB, Zubov A, Golubova O, et al. Clinical observation of macroamylasemia in splenosis due to the posttraumatic splenectomy. JOP. 2014;15:394.
128. Ferreira H, Maciel C, Morgado A, Pereira A. Rectovaginal splenosis: an unexpected cause of dyspareunia approached by laparoscopy. J Minim Invasive Gynecol. 2017;24:715–6.
129. Lemos AA, Crespi S, Costa S, Marini A. Splenosis of the abdomen and pelvis complicated by torsion of a splenic implant clinically mimicking an acute bowel ischemia. BJR Case Rep. 2018;4:20180024.
130. Leitz EM, Kwan SW. Splenosis: a rare cause of gastrointestinal bleeding successfully treated with transarterial embolization. Clin J Gastroenterol. 2015;8:126–9.
131. Sun LM, Chen HJ, Jeng LB, et al. Splenectomy and increased subsequent cancer risk: a nationwide population-based cohort study. Am J Surg. 2015;210:243–51.
132. Linet MS, Nyrén O, Gridley G, et al. Causes of death among patients surviving at least one year following splenectomy. Am J Surg. 1996;172:320–3.
133. Singer DB. Post-splenectomy sepsis. Perspect Pediatr Pathol. 1973;1:285.
134. Morris DH, Bullock FD. The importance of the spleen in resistance to infection. Ann Surg. 1919;70:513.
135. King H, Schumacker HB. Splenic studies: I. Susceptibility to infection after splenectomy performed in infancy. Ann Surg. 1952;136:239–42.
136. Perla D, Marmorston J. The spleen and resistance. Baltimore: Williams and Wilkins Co.; 1935. p. 2–50.
137. Marmorston J. The effect of splenectomy on acute bacterium enteritidis infections in white mice. Proc Soc Exp Biol Med. 1935;32:981–5.
138. Rowley DA. The effect of splenectomy on the formation of circulating antibiotics in the adult male albino rat. J Immunol. 1950;64:289.
139. Huntley CG. Infection following splenectomy in infants and children: a review of the experience at Duke Hospital in infants and children during a 22-year period (1933–1954). Am J Dis Child. 1958;95:477.
140. Horan M, Colebatch JH. Relation between splenectomy and subsequent infection: a clinical study. Arch Dis Child. 1962;37:398–414.
141. Kiesewetter WB. Pediatric splenectomy: indications, technique, complications, and mortality. Surg Clin North Am. 1975;55:449–60.
142. Pearson HA, Spencer RP, Cornelius EA. Functional asplenia in sickle cell disease. N Engl J Med. 1969;281:923–6.
143. Grosfeld JL, Ranochak JE. Are hemisplenectomy and/or primary splenic repair feasible? J Pediatr Surg. 1976;11:419–24.
144. Robinson MG, Watson RJ. Pneumococcal meningitis in sickle cell anemia. N Engl J Med. 1966;274:1006–8.

145. Waldman JD, Rosenthal A, Smith AL, et al. Sepsis and congenital asplenia. J Pediatr. 1977;90:555–9.
146. Witte MH, Wittle CL, VanWyck DB. Preservation of the spleen. Lymph. 1983;16:128.
147. Eraklis AJ, Kewy SV, Diamond LK. Hazard of overwhelming infection after splenectomy in childhood. N Engl J Med. 1967;276:1225–9.
148. Eraklis AJ, Filler RM. Splenectomy in childhood: a review of 1413 cases. J Pediatr Surg. 1972;4:382.
149. Grosfeld JL, Malangoni MA. Blunt splenic trauma: a reassessment of surgical therapy based on laboratory and clinical observations. In: Nyhus LM, editor. Surgery annual, vol. 12. Appleton-Century-Crofts; 1980. p. 123–8.
150. Gopal V, Bisno AL. Fulminant pneumococcal infections in normal asplenic hosts. Arch Intern Med. 1977;1937:1526.
151. Robinette CD, Fraumeni JF. Splenectomy and subsequent mortality in veterans of the 1939–45 war. Lancet. 1977;2:127.
152. Balfanz JR, Nesbit ME, Jarvis C, Krivit W. Overwhelming sepsis following splenectomy for trauma. J Pediatr. 1976;88:458–60.
153. Oakes D. Splenic trauma. Curr Probl Surg. 1981;18:346–401.
154. O'Neal BJ, McDonald JC. The risk of sepsis in the asplenic adult. Ann Surg. 1981;194:775–8.
155. Chilcote RR, Baehner RL, Hammond D, The investigators of the Children's Cancer Study Group. Septicemia and meningitis in children splenectomized for Hodgkins disease. N Engl J Med. 1976;295:798–800.
156. Coil JA, Dickerman JD, Boultan E. Increased susceptibility of splenectomized mice to infection after exposure to an aerosolized suspension of type III streptococcus pneumonia. Infect Immunol. 1978;21:42.
157. Schulkin ML, Ellis EF, Smith RT. Effect of antibody upon clearance of I^{125} labelled pneumococci by spleen and liver. Pediatr Res. 1967;1:178–84.
158. Chaundry IH, Tobato Y, Schleck S. Effect of splenectomy on reticuloendothelial function and survival following sepsis. J. Trauma. 1980;20:649.
159. Claret I, Morales L, Montaner A. Immunological studies in the postsplenectomy syndrome. J Pediatr Surg. 1975;10:59–64.
160. Schumacher MJ. Serum immunoglobulin and transferrin levels after childhood splenectomy. Arch Dis Child. 1970;45:114–7.
161. Winkelstein JA, Lambert GH. Pneumococcal serum opsonizing activity in splenectomized children. J Pediatr. 1975;87:430–3.
162. Chainoff O, Douer D, Pick IA. Serum immunoglobulin changes after accidental splenectomy in adults. Am J Surg. 1978;136:332–3.
163. Hobles JR, Melner RD, Watt P. Gamma-M deficiency predisposing to meningoccal septicemia. Br Med J. 1967;4:583.
164. Sullivan JL, Ochs HD. Immune response after splenectomy. Lancet. 1978;1:178.
165. Najjar VA, Nishioka K. Tuftsin: a physiological phagocytosis stimulating peptide. Nature. 1970;228:612.
166. Constantopoulos A, Najjar VA, Smith JW. Tuftsin deficiency: a new syndrome with defective phagocytosis. J Pediatr. 1972;80:564–72.
167. Wright AE, Douglas SR. Experimental investigation of role of blood fluids in connection with phagocytosis. Proc R Soc London. 1903;72:357.
168. Merikanto J, Eskola J. Immunologic consequences of neonatal splenectomy. J Pediatr Surg. 1980;15:651–4.
169. Skin HS, Smith MR, Wood WB. Heat labile opsonins to pneumococcus II. Involvement of C_3 and C_5. J Exp Med. 1969;130:1229.
170. Carlisle HN, Seslaw S. Properdin levels in splenectomized persons (25173). Proc Soc Exp Biol Med. 1959;102:150–4.
171. Amsbaugh DG, Prescott B, Baker PJ. Effect of splenectomy on the expression of regulating T-cell activities. J Immunol. 1978;121:1483–5.

172. Treng G, Miller C, Lim RC, Fisher E. Effect of splenectomy on patients immunocompetent leukocytes. Surg Forum. 1979;28:337.
173. Miller CL, Baker CL. Development of inhibitory macrophages after splenectomy. Transplant Proc. 1979;11:1460.
174. Ernstrom U, Sandberg G. Influence of splenectomy on thymic release of lymphocytes into the blood. Scand J Haematol. 1970;7:342–8.
175. Yamagishi H, Pellis RN, Kahar BD. Surgical immunodpression: role of splenic suppressor cells. Surg Forum. 1978;29:465–8.
176. Leonard AS, Grebink S, Baesl TJ. The overwhelming postsplenectomy problems. World J Surg. 1980;4:423–7.
177. Diamond LK. Splenectomy in childhood and the hazard of overwhelming infection. Pediatrics. 1969;45:886.
178. Bisno A, Freeman JK. The syndrome of asplenia, pneumoccal sepsis, and disseminated intravascular coagulation. Ann Intern Med. 1970;72:389–93.
179. Browder W, Rakinic BS, McNamee R. Protective effect of nonspecific immunostimulation in postsplenectomy sepsis. J Surg Res. 1983;35:474.
180. Cooney DR, Dearth JC, Swanson SE. Relative merits of partial splenectomy, splenic reimplantation, and immunization in preventing postsplenectomy sepsis. Surgery. 1979;86:561.
181. Leung LE, Szal GJ, Frachman RH. Increased susceptibility of splenectomized rats to infection with *S. Pneumoniae*. J Infect Dis. 1972;126:507–13.
182. Mufson MA, Kruss DM, Wasil RE, et al. Capsular types and outcome of bacteremic pneumococcal disease in the antibiotic era. Arch Intern Med. 1974;134:505–10.
183. Rice HM, James PD. Ectopic splenic tissue failed to prevent fatal pneumococcal septicemia after splenectomy for trauma. Lancet. 1980;1:565.
184. Cooney DR, Michalak WA, Michalak DM. Comparative methods of splenic preservation. J Pediatr Surg. 1981;16:327–38.
185. Malangoni MA, West KW, Grosfeld JL. Response to pneumococcal challenge after ligation of the splenic artery. Surg Forum. 1980;31:44.
186. Van Wyck DB, Witte MH, Witte CL. Critical splenic mass for survival from experimental pneumococcemia. J Surg Res. 1980;21:14.
187. Coil JA, Dickerman JD, Homer SR, Chalmer BJ. Pulmonary infection in splenectomized mice: protection by splenic remnant. J Surg Res. 1980;28:18–22.

Partial Splenectomy and Splenorrhaphy 20

20.1 Introduction

- Splenectomy is usually done through an open incision either:
  - An upper midline incision
  - A left upper transverse incision
  - A left subcostal incision
- Currently and as a result of the recent advances in minimal invasive surgery, laparoscopic splenectomy is the preferred procedure.
- This is in cases where the spleen is not too large and when splenectomy is done electively.
- Total splenectomy should however be avoided whenever possible.
- This is to obviate the increased risk of post-splenectomy sepsis.
- Vaccinations and antibiotics provide good protection against the risks of asplenia but this is not 100% effective and these are not always available especially in poorer countries.
- Add to this the poor compliance of patients specially children in taking these antibiotics.
- To overcome this, partial splenectomy or partial splenic arterial embolization has been advocated.
- It was shown that the spleen's protective functions can be maintained if a small part of the spleen can be left behind.
- It was also shown that up to 1/3 of the size of the normal spleen can provide protection against invading organisms.
- Currently and where clinically appropriate, attempts are made to perform either:
  - Surgical subtotal (partial) splenectomy
  - Or partial splenic embolization (Figs. 20.1, 20.2, 20.3, 20.4, 20.5, 20.6 and 20.7).
  - Partial splenectomy is usually performed through an open technique but recently, laparoscopic partial splenectomy was shown to be feasible and safe both in children and adults.

A. H. Al-Salem, *The Spleen*, https://doi.org/10.1007/978-981-99-6191-7_20

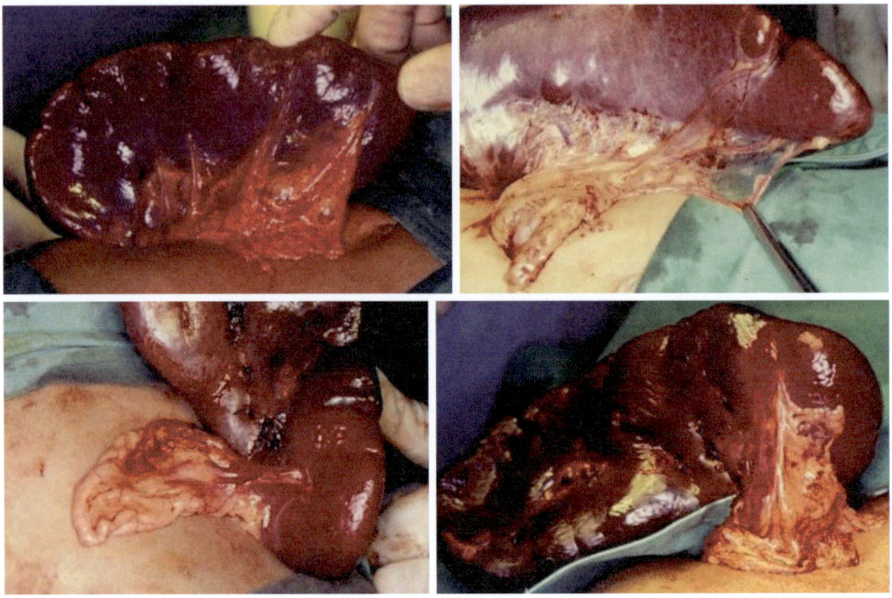

Figs. 20.1–20.4 Clinical intraoperative photographs showing partial splenectomy in a patient with beta thalassemia major. Note the identification of the lower polar vessels and demarcation after division of the remaining blood vessels

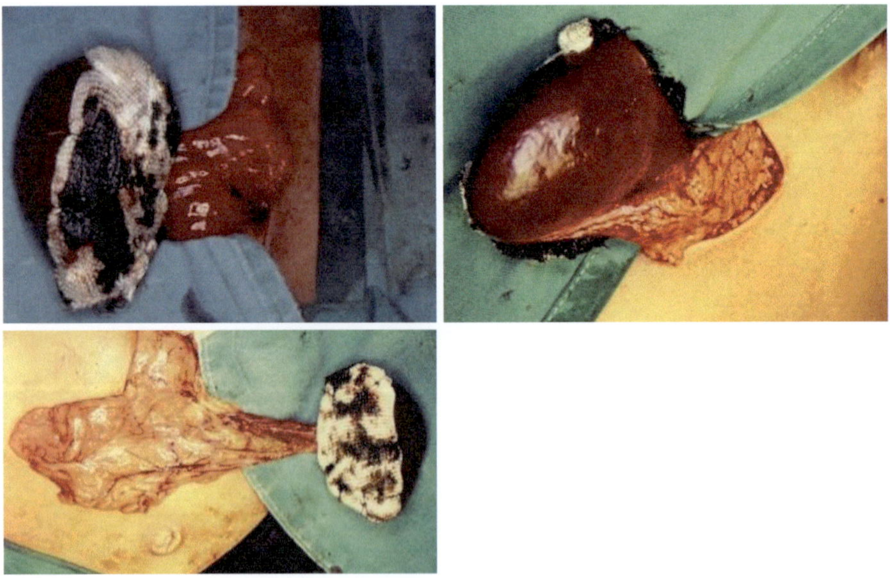

Figs. 20.5–20.7 Clinical intra-operative photographs showing the end result of partial splenectomy. Note the size of the remaining part of the spleen. This must be fixed to avoid torsion. The cut surface is covered with surgicele for hemostasis

  - As it may take some time for the preserved splenic tissue to provide full immunological protection, it is advisable to immunize these patients preoperatively and cover them with antibiotics.
- Splenectomy is associated with severe peri-and post-operative complications such as:
  - Infections
  - Intra-abdominal abscess
  - Portal vein thrombosis
  - Pulmonary hypertension
  - Thrombocytosis
  - Venous thromboembolism
- Patients with anatomical or functional asplenia are at high risk of developing severe and potentially fatal infections caused by encapsulated bacteria (OPSI).
- Splenorrhaphy:
  - This technique is still used to manage small lacerations or other injuries the spleen.
  - This will help preserve the spleen and its protective functions.
  - Horizontal mattress sutures placed over pledges are commonly used.
  - Omentum or a local hemostatic agent (e.g., fibrin glue) may be used as an adjuvant to control bleeding and achieving hemostasis.
  - The advantage of partial splenectomy is the removal of the etiology and at the same time, preservation of the splenic function.
  - Add to this, partial splenectomy avoids all the known complications of total splenectomy, such as:

    Overwhelming post-splenectomy infection (OPSI)
    Intra-abdominal abscesses
    Thrombocytosis
    Portal hypertension
    Thromboembolism
    Pulmonary hypertension
- Morgenstern and Shapiro were the first to describe partial splenectomy in 1980.
- Laparoscopic partial splenectomy was performed by Poulin in 1995.
- Partial splenectomy has become one of the most widely used surgical procedures in the treatment of:
  - Hematologic disorders
  - Trauma
  - Benign, solid and cystic diseases of the spleen
- Partial splenectomy should be preferred in selected cases, especially in younger patients.

- **PARTIAL SPLENECTOMY WAS FIRST DESCRIBED BY MORGENSTEN AND SHAPIRO IN 1980**
- **LAPAROSCOPIC PARTIAL SPLENECTOMY WAS FIRST PERFORMED BY POULIN IN 1995**

- Laparoscopic partial splenectomy still represents an important surgical challenge, because of intraoperative and postoperative risk of bleeding.
- During partial splenectomy, a variety of techniques has been used for parenchyma transection and hemostasis.
  - Initially, the splenic transection was mostly performed by knife and hemostasis was accomplished by 5-0 silk suturing the visible vessels and methyl 2-cyanoacrylate monomer.
  - The use of various collagen products and omentoplasty.
  - Currently resection of splenic parenchyma and hemostasis can be performed by various methods including:
    - Electrocoagulation
    - Ultrasonic dissectors
    - Mechanical staplers
    - Reinforcing the raw surface of the spleen with interrupted or mattress sutures.
    - The use of new generation hemostatic agents, including Surgicel®, Spongostan®, and various other cellulose products.
    - The use of Argon beam and Microwave coagulation.
    - The use of new energies' cutting/coagulating devices (harmonic scalpel, radiofrequency) for parenchymal section.
    - The use of newly introduced hemostatic agents (TachoComb®, Tachosil® and FloSeal®).
- The size of splenic remnant "optimal size" to reduce the risk of hematological recurrence to a minimum without increasing the risk of overwhelming sepsis is not exactly known.
- Preservation of 25–30% of splenic parenchyma was shown to be adequate to ensure a satisfactory immunological response to an antigenic stimulus.

20.2 Partial Splenectomy Procedure

- Splenectomy is a common surgical procedure performed for various hematological and non-hematological indications.
- The surgical approach can range from conventional open splenectomy to laparoscopic or robotic approach based on indication and experience of the operating surgeon.
- Total splenectomy is known to be associated with several peri and post-operative complications.
  - Asplenic patients are at increased risk of developing fatal infections with encapsulated organisms.
  - Overwhelming post splenectomy infections (OPSI) remains a threat for these patients. This is despite vaccination and antimicrobial prophylaxis.
  - To avoid this life-threatening complication, spleen-preserving surgical techniques have become increasingly common.
  - Partial Splenectomy was first described by Morgenstern and Shapiro in 1980.
  - Poulin performed the first laparoscopic partial splenectomy in 1995.

- It is well known that patients with asplenia are at high risk of developing severe life-threatening infections. These are usually caused by encapsulated bacteria and although this risk is throughout their lifespan, it is highest during the first three years post splenectomy.
- Post-splenectomy prevalence of OPSI is 0.1–0.5% and has a mortality rate of up to 50%.
- Splenectomy is also associated with an increase in intravascular hemolysis and subsequent long-term risks of:
  - Hypertension
  - Vascular thrombosis
  - Pulmonary hypertension
  - Cardiovascular disease
- To overcome these complications and to preserve splenic function, alternatives to total splenectomy became desirable.
- Partial splenectomy can be done via a laparotomy or laparoscopically (Figs. 20.8, 20.9, 20.10, 20.11, 20.12, 20.13, 20.14 and 20.15).
  - Laparoscopic splenectomy takes a slightly longer operative time than the open technique.
  - Laparoscopic partial splenectomy is associated with a shorter duration of hospitalization and less postoperative analgesia.
  - The patients who had laparoscopic partial splenectomy had a lower percentage of postoperative transfusions and a lower blood loss.
  - The patients who had laparoscopic partial splenectomy had a lower rate of postoperative complications and better cosmetic results.
- The recent advances in minimal invasive surgery and development of new devices contributed to minimizing blood loss during laparoscopic partial splenectomy.
- Add to this the value of preoperative embolization of the splenic vessels. This is important for the following reasons:
  - Reduction in intra- and postoperative bleeding
  - A better visualization of the ischemic demarcation area on the surface of the spleen.
  - A better control of the blood vessels during the division of splenic parenchyma.
  - The discovery of anatomical variants of the spleen.
- The time that must elapse between embolization and surgery is controversial. In the past surgery was performed about a month post embolization. Currently and with the development of laparoscopic techniques, laparoscopic partial splenectomy should be done immediately after splenic embolization. This is to avoid the known risks of splenic embolization which include:
  - Splenic abscesses
  - Sudden rupture of the spleen
  - Post-embolization syndrome
- Nowadays laparoscopic partial splenectomy is considered the gold standard for partial splenectomy.
- Partial splenectomy is a safe procedure and has low morbidity and mortality.

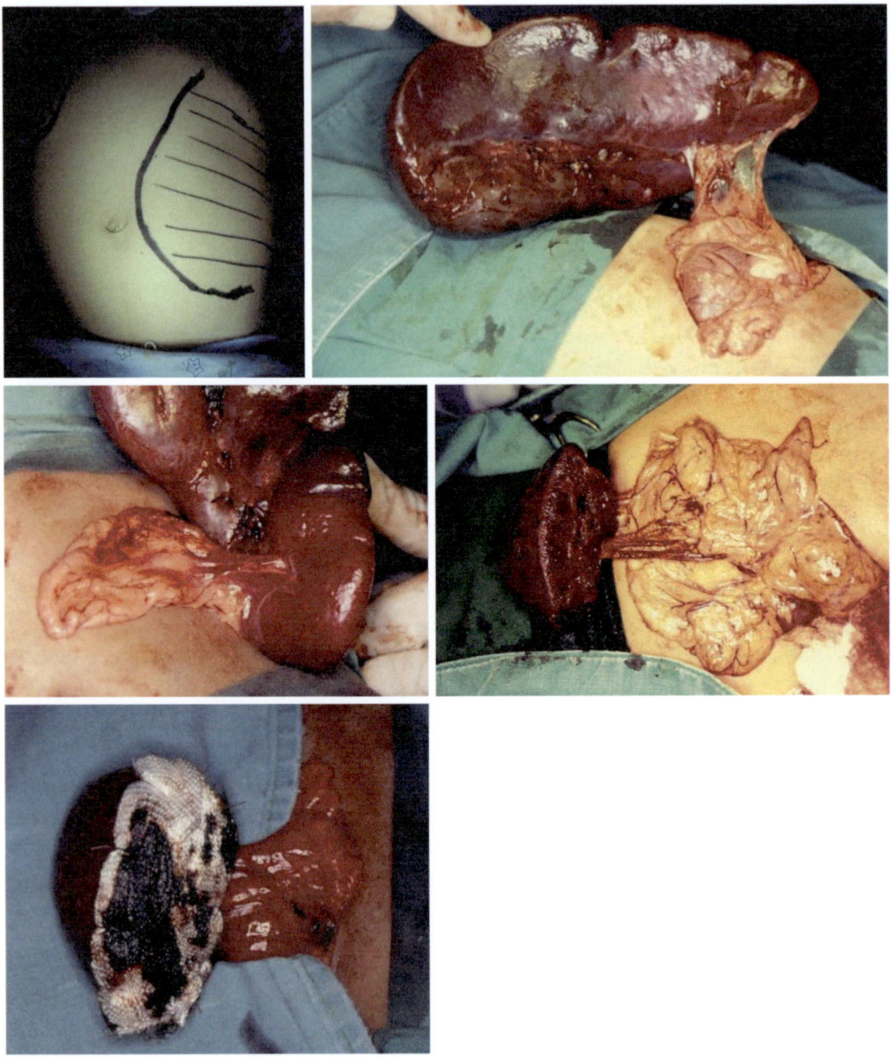

Figs. 20.8–20.12 Clinical photographs showing open partial splenectomy for a patient with hemplytic anemia. Nnote the decarcation line after division of the main blood vessels. Not also the preserved part of the spleen and the cut surface wrapped in surgicele for hemostasis

- The indications for partial splenectomy keep increasing and include:
  - Patients with traumatic splenic injuries
  - Hematological diseases
  - Portal hypertension
  - Parasitic and nonparasitic splenic cysts
  - Splenic tumors

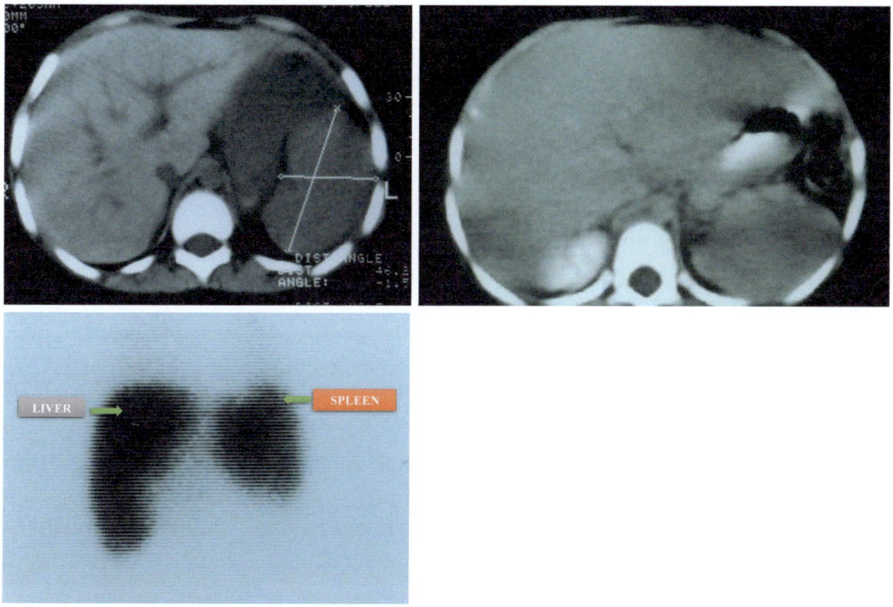

Figs. 20.13–20.15 Abdominal CT-scan and Isotope scan of a patient who had partial splenectomy. Note the splenic remnant which looks homogenous and functioning by the isotope scan

- Partial splenectomy is based on the selective devascularization of splenic parenchyma followed by the resection of the ischemic area along the line of demarcation.
- During partial splenectomy, two main approaches have been described:
  - The splenic artery and vein are ligated and the splenic remnant relies on blood supply from the short gastric vessels and phrenic collaterals or left gastroepiploic artery, depending on the preserved splenic pole.
  - The selective segmental splenic devascularization at the hilum, where only branches of the main splenic vessels are ligated and divided.
- Ligation of segmental vessels at the hilum is nowadays the most commonly used approach. This requires careful dissection of the splenic hilum in order to identify, ligate and divide the vessels tributary of the segment to be resected.
- From a technical point of view, the upper part of the spleen was more often preserved than the lower pole.
- This is for the following reasons:
  - Mobilization of colonic splenic flexure in order to dissect the lower pole was considered an easier step than dissecting the upper pole from the diaphragm and stomach.
  - The short gastric vessels are more effective a blood supply to the remnant spleen than the gastroepiploic vessels in preventing splenic ischemia and infarction of the preserved segment.

- Partial splenectomy is a technically challenging surgical procedure designed to resect enough spleen to achieve the desired effect based on the indication while at the same time preserving splenic immune function.
- The "optimal size of the splenic remnant", is not known exactly.
- Partial splenectomy with preservation of 25%–30% of the normal spleen is sufficiently enough and protective in preventing fatal complications of OPSI.
- There are several techniques described in the literature for performing partial splenectomy while minimizing blood loss.
- It is important to understand the anatomy, vascular supply of the spleen, and operative technique.
  - A Left subcostal, transverse or midline laparotomy incision is used.
  - Recently, laparoscopy, robotic or lap assisted techniques have been used.
  - Adequate mobilization of the spleen is performed initially, followed by clamping of the splenic artery with a vascular clamp.
  - Then the branch supplying the normal splenic remnant is isolated and preserved.
  - Liga Sure™ can be used to divide the spleen and ligate the smaller vessels.
  - After releasing the vascular clamp, various techniques can be used to control bleeding from the cut surface of the spleen. These include:

    Pledged sutures

    Surgicel®Fibrillar™ wrapping of splenic raw surface
- Technical challenges in performing partial splenectomy include:
  - Understanding the vascular anatomy of the spleen
  - Splenic vascular dissection may be performed in several ways, with the most common approach nowadays being the splenic artery/vein-sparing dissection of the splenic hilum to selectively revascularize the preserved splenic segment.
  - The other method is the division of splenic artery and vein and the blood supply of the preserved splenic remnant provided by the short gastric vessels or left gastroepiploic artery.
  - Temporary clamping by bulldog clamps or loops allows identification of the demarcation line and also helps reduce bleeding.
  - Preoperative arterial embolization:

    This will help in outlining the line of demarcation and plan parenchymal dissection which will reduce the requirement of peri-operative blood transfusions.
  - Detachable clips have also been safely used to temporarily clamp splenic vessels for demarcation and better control.
- Dividing the splenic parenchyma may be carried out in several ways:
  - Using sharp or blunt dissection
  - Electrocautery
  - Ultrasonic dissectors
  - Mechanical linear stapler
  - Harmonic scalpel
  - Radiofrequency ablator

- Hemostasis of the raw surface is achieved with:
  - Electrocautery devices
  - Pledges placement
  - Omentoplasty
  - Static agents such as Surgicel® or FloSeal® have been successfully used for hemostasis in partial splenectomy.
- Bleeding splenic vessels are controlled with:
  - Sutures
  - Metallic clips
  - Electrocautery
  - Argon beam
- To prevent torsion of the splenic remnant, it can be:
  - Fixed to the splenic bed
  - Fixed to the greater curvature of the stomach
  - Fixed to the abdominal wall
  - Placed in prosthesis or transposition in the retroperitoneal space
- Vaccination is recommended 2 weeks before elective splenectomy or 2 weeks after splenectomy in the emergency setting.

Further Reading

1. Di Sabatino A, Carsetti R, Corazza GR. Post-splenectomy and hyposplenic states. Lancet. 2011;378(9785):86–97.
2. Uranues S, Alimoglu O. Laparoscopic surgery of the spleen. Surg Clin North Am. 2005;85(1):75–90. ix
3. Redmond HP, Redmond JM, Rooney BP, et al. Surgical anatomy of the human spleen. Br J Surg. 1989;76(2):198–201.
4. Kristinsson SY, Gridley G, Hoover RN, et al. Long-term risks after splenectomy among 8,149 cancer-free American veterans: a cohort study with up to 27 years follow-up. Haematologica. 2014;99(2):392–8.
5. Guizzetti L. Total versus partial splenectomy in pediatric hereditary spherocytosis: a systematic review and meta-analysis. Pediatr Blood Cancer. 2016;63(10):1713–22.
6. Tubbs RR, Thomas F, Norris D, et al. Is hemisplenectomy a satisfactory option to total splenectomy in abdominal staging of Hodgkin's disease? J Pediatr Surg. 1987;22(8):727–9.
7. Rice HE, Oldham KT, Hillery CA, et al. Clinical and hematologic benefits of partial splenectomy for congenital hemolytic anemias in children. Ann Surg. 2003;237(2):281–8.
8. Tracy ET, Rice HE. Partial splenectomy for hereditary spherocytosis. Pediatr Clin North Am. 2008;55(2):503–19. x
9. Traub A, Giebink GS, Smith C, et al. Splenic reticuloendothelial function after splenectomy, spleen repair, and spleen autotransplantation. N Engl J Med. 1987;317(25):1559–64.
10. Resende V, Petroianu A. Functions of the splenic remnant after subtotal splenectomy for treatment of severe splenic injuries. Am J Surg. 2003;185(4):311–5.
11. Morgenstern L, Kahn FH, Weinstein IM. Subtotal splenectomy in myelofibrosis. Surgery. 1966;60(2):336–9.
12. White FN. Wound of abdominal wall-partial splenectomy: recovery. Ind Med Gaz. 1905;40(2):57–8.

13. Christo MC. Partial regulated splenectomies. Preliminary note on the first 3 cases operated on. Hospital (Rio J). 1959;56:645–50.
14. Hollingsworth CL, Rice HE. Hereditary spherocytosis and partial splenectomy in children: review of surgical technique and the role of imaging. Pediatr Radiol. 2010;40(7):1177–83.
15. Freud E, Cohen IJ, Mor C, et al. Splenic "regeneration" after partial splenectomy for Gaucher disease: histological features. Blood Cells Mol Dis. 1998;24(3):309–16.
16. Chu HB, Zhang TG, Zhao JH, et al. Assessment of immune cells and function of the residual spleen after subtotal splenectomy due to splenomegaly in cirrhotic patients. BMC Immunol. 2014;15:42.
17. Tchernia G, Gauthier F, Mielot F, et al. Initial assessment of the beneficial effect of partial splenectomy in hereditary spherocytosis. Blood. 1993;81(8):2014–20.
18. Todde G, Bagolan P, Fariello G, et al. Epidermoid cyst of the spleen in a newborn infant. Prenatal diagnosis and partial splenectomy. Chir Pediatr. 1989;30(3):172–4.
19. Petroianu A. Subtotal splenectomy for the treatment of chronic lymphocytic leukemia. Ann Hematol. 2003;82(11):708–9.
20. de la Villeon B, Zarzavadjian Le Bian A, Vuarnesson H, et al. Laparoscopic partial splenectomy: a technical tip. Surg Endosc. 2015;29(1):94–9.
21. Sheikha AK, Salih ZT, Kasnazan KH, et al. Prevention of overwhelming postsplenectomy infection in thalassemia patients by partial rather than total splenectomy. Can J Surg. 2007;50(5):382–6.
22. Rosman CWK, Broens PMA, Trzpis M, Tamminga RYJ. A long-term follow-up study of subtotal splenectomy in children with hereditary spherocytosis. Pediatr Blood Cancer. 2017;64:e26592.
23. Moorman DW, Evans DM, Wright DJ. Segmental splenectomy using the ultrasonic surgical aspirator. Am J Surg. 1988;155(2):266–7.
24. Sagar PM, McMahon MJ. Partial splenectomy for splenic cysts. Br J Surg. 1988;75(5):488.
25. Hall JG, Kurtzberg J, Szabolcs P, et al. Partial splenectomy before a hematopoietic stem cell transplantation in children. J Pediatr Surg. 2005;40(1):221–7.
26. Dutta S, Price VE, Blanchette V, et al. A laparoscopic approach to partial splenectomy for children with hereditary spherocytosis. Surg Endosc. 2006;20(11):1719–24.
27. Louis D, Duc ML, Reix P, et al. Partial splenectomy for portal hypertension in cystic fibrosis related liver disease. Pediatr Pulmonol. 2007;42(12):1173–80.
28. Morinis J, Dutta S, Blanchette V, et al. Laparoscopic partial vs total splenectomy in children with hereditary spherocytosis. J Pediatr Surg. 2008;43(9):1649–52.
29. Petroianu A, Cabezas-Andrade MA, Berindoague NR. Laparoscopic subtotal splenectomy. Surg Laparosc Endosc Percutan Tech. 2008;18(1):94–7.
30. Rezende-Neto JB, Petroianu A, Santana SK. Subtotal splenectomy and central splenorenal shunt for treatment of bleeding from Roux en Y jejunal loop varices secondary to portal hypertension. Dig Dis Sci. 2008;53(2):539–43.
31. Vasilescu C, Stanciulea O, Popa M, et al. Subtotal laparoscopic splenectomy and esophagogastric devascularization for the thrombocytopenia because of portal cavernoma–case report. J Pediatr Surg. 2008;43(7):1373–5.
32. Nagem RG, Petroianu A. Subtotal splenectomy for splenic abscess. Can J Surg. 2009;52(4):E91–2.
33. Slater BJ, Chan FP, Davis K, et al. Institutional experience with laparoscopic partial splenectomy for hereditary spherocytosis. J Pediatr Surg. 2010;45(8):1682–6.
34. Vasilescu C, Tudor S, Popa M, et al. Robotic partial splenectomy for hydatid cyst of the spleen. Langenbecks Arch Surg. 2010;395(8):1169–74.
35. Vasilescu C, Stanciulea O, Tudor S. Laparoscopic versus robotic subtotal splenectomy in hereditary spherocytosis. Potential advantages and limits of an expensive approach. Surg Endosc. 2012;26(10):2802–9.
36. Warshaw AL. Conservation of the spleen with distal pancreatectomy. Arch Surg. 1988;123(5):550–3.

37. Koyazounda A, Bradly R, Le Baron JC, et al. Partial splenectomy with the automatic stapler in the adult. Presse Med. 1984;13(21):1336.
38. Rosser SB. Partial splenectomy for a splenic pseudocyst. J Natl Med Assoc. 1984;76(10):1020–4.
39. Revillon Y, Girot R. Partial disarterialization of the spleen and partial splenectomy in children. Presse Med. 1985;14(7):423–5.
40. Rodgers BM, Tribble C, Joob A. Partial splenectomy for Gaucher's disease. Ann Surg. 1987;205(6):693–9.
41. Emery E, Houry S, Lacaine F, et al. Technic of partial splenectomy with linear stapler instrument (TA 90, U.S., Surgical Corps). J Chir. 1990;127(6-7):356–8.
42. Feliciano DV, Spjut-Patrinely V, Burch JM, et al. Splenorrhaphy. The alternative. Ann Surg. 1990;211(5):569–80.
43. Guzzetta PC, Ruley EJ, Merrick HF, et al. Elective subtotal splenectomy. Indications and results in 33 patients. Ann Surg. 1990;211(1):34–42.
44. Kehila M, Abderrahim T. Partial splenectomy requiring ligation of splenic vessels. Apropos of 40 cases. Ann Chir. 1993;47(5):433–5.
45. Louis D, Chazalette JP. Cystic fibrosis and portal hypertension interest of partial splenectomy. Eur J Pediatr Surg. 1993;3(1):22–4.
46. Castillo B, Cynober T, Bader-Meunier B, et al. Hereditary spherocytosis. Course and value of subtotal splenectomy. Arch Pediatr. 1997;4(6):515–20.
47. Sandoval C, Stringel G, Weisberger J, et al. Failure of partial splenectomy to ameliorate the anemia of pyruvate kinase deficiency. J Pediatr Surg. 1997;32(4):641–2.
48. Idowu O, Hayes-Jordan A. Partial splenectomy in children under 4 years of age with hemoglobinopathy. J Pediatr Surg. 1998;33(8):1251–3.
49. Kimber C, Pierro A, Drake D, et al. Hemisplenectomy for giant splenic cysts in children. Pediatr Surg Int. 1998;14(1-2):116–8.
50. Tulikoura I, Lassus J, Konttinen YT, et al. A safe surgical technique for the partial resection of the ruptured spleen. A clinical report. Injury. 1999;30(10):693–7.
51. Yamamoto H, Ohshiro H, Lee KJ, et al. Partial splenectomy for epidermoid cyst using TA-stapler. J Pediatr Surg. 1999;34(10):1582–3.
52. Bader-Meunier B, Gauthier F, Archambaud F, et al. Long-term evaluation of the beneficial effect of subtotal splenectomy for management of hereditary spherocytosis. Blood. 2001;97(2):399–403. 1536 R. Costi et al. J Pediatr Surg 54 (2019) 1527–1538.
53. Corcione F, Cuccurullo D, Caiazzo P, et al. Laparoscopic partial splenectomy for a splenic pseudocyst. Surg Endosc. 2003;17(11):1850.
54. Tatulli F, Ciani V, Caputi A, et al. Subtotal splenectomy for large splenic cyst. Chir Ital. 2003;55(5):761–4.
55. Thalhammer GH, Eber E, Uranus S, et al. Partial splenectomy in cystic fibrosis patients with hypersplenism. Arch Dis Child. 2003;88(2):143–6.
56. Velanovich V, Weaver M. Partial splenectomy using a coupled salineradiofrequency hemostatic device. Am J Surg. 2003;185(1):66–8.
57. Esposito C, Settimi A, Centonze A, et al. Enlarged wandering spleen treated with hemisplenectomy and fixation of the residual spleen. Pediatr Surg Int. 2005;21(6):488–90.
58. Stoehr GA, Stauffer UG, Eber SW. Near-total splenectomy: a new technique for the management of hereditary spherocytosis. Ann Surg. 2005;241(1):40–7.
59. Huscher CG, Mingoli A, Sgarzini G, et al. Laparoscopic treatment of blunt splenic injuries: initial experience with 11 patients. Surg Endosc. 2006;20(9):1423–6.
60. Itamoto T, Fukuda S, Tashiro H, et al. Radiofrequency-assisted partial splenectomy with a new and simple device. Am J Surg. 2006;192(2):252–4.
61. Khelif K, Maassarani F, Dassonville M, et al. Laparoscopic partial splenectomy using radiofrequency ablation for nonparasitic splenic cysts in two children. J Laparoendosc Adv Surg Tech A. 2006;16(4):414–7.
62. Breitenstein S, Scholz T, Schafer M, et al. Laparoscopic partial splenectomy. J Am Coll Surg. 2007;204(1):179–81.

63. Godiris-Petit G, Goasguen N, Munoz-Bongrand N, et al. Laparoscopic partial splenectomy using the harmonic scalpel. Two case reports. J Chir. 2007;144(4):339–41.
64. Rescorla FJ, West KW, Engum SA, et al. Laparoscopic splenic procedures in children: experience in 231 children. Ann Surg. 2007;246(4):683–7.
65. Uranues S, Grossman D, Ludwig L, et al. Laparoscopic partial splenectomy. Surg Endosc. 2007;21(1):57–60.
66. Wu SC, Wang CC, Yong CC. Partial splenectomy for benign splenic cysts with the aid of a Lin clamp: technical note. World J Surg. 2007;31(11):2144–7.
67. De Greef E, Hoffman I, Topal B, et al. Partial laparoscopic splenectomy for splenic abscess because of Salmonella infection: a case report. J Pediatr Surg. 2008;43(5):E35–8.
68. Ghuliani D, Agarwal S, Thomas S, et al. Giant cavernous haemangioma of the spleen presenting as massive splenomegaly and treated by partial splenectomy. Singapore Med J. 2008;49(12):e356–8.
69. Hery G, Becmeur F, Mefat L, et al. Laparoscopic partial splenectomy: indications and results of a multicenter retrospective study. Surg Endosc. 2008;22(1):45–9.
70. Jain P, Parelkar S, Shah H, et al. Laparoscopic partial splenectomy for splenic epidermoid cyst. J Laparoendosc Adv Surg Tech A. 2008;18(6):899–902.
71. Medejel N, Garcon L, Guitton C, et al. Effect of subtotal splenectomy for management of hereditary pyropoikilocytosis. Br J Haematol. 2008;142(2):315–7.
72. Stacey MJ, Rampaul RS, Rengaragan A, et al. Use of FloSeal matrix hemostatic agent in partial splenectomy after penetrating trauma. J Trauma. 2008;64(2):507–8.
73. Szczepanik AB, Meissner AJ. Partial splenectomy in the management of nonparasitic splenic cysts. World J Surg. 2009;33(4):852–6.
74. Costamagna D, Rizzi S, Zampogna A, Alonza A. Open partial splenectomy for trauma using GIA-Stapler and FloSeal matrix haemostatic agent. BMJ Case Rep. 2010;
75. Hong TH, Lee SK, You YK, et al. Single-port laparoscopic partial splenectomy: a case report. Surg Laparosc Endosc Percutan Tech. 2010;20(5):e164–6.
76. Bork U, Muller SA, Herpel E, et al. Partial splenectomy using a vascular stapler in a patient with a benign splenic cyst. Am Surg. 2011;77(1):118–9.
77. Budzynski A, Demczuk S, Kumiega B, et al. Sclerosing angiomatoid nodular transformation of the spleen treated by laparoscopic partial splenectomy. Wideochir Inne Tech Maloinwazyjne. 2011;6(4):249–55.
78. Buesing KL, Tracy ET, Kiernan C, et al. Partial splenectomy for hereditary spherocytosis: a multi-institutional review. J Pediatr Surg. 2011;46(1):178–83.
79. Fan H, Zhang D, Zhao X, et al. Laparoscopic partial splenectomy for large splenic epidermoid cyst. Chin Med J (Engl). 2011;124(11):1751–3.
80. Giulianotti PC, Buchs NC, Addeo P, et al. Robot-assisted partial and total splenectomy. Int J Med Robot. 2011;7(4):482–8.
81. Karadayi K, Turan M, Sen M. A new technique for partial splenectomy with radiofrequency technology. Surg Laparosc Endosc Percutan Tech. 2011;21(5):358–61.
82. Okano K, Kakinoki K, Suto H, et al. Hand-assisted laparoscopic partial splenectomy using an endopath monopolar sealer. Surg Laparosc Endosc Percutan Tech. 2011;21(6):e291–4.
83. Krl EA, Orhan D, Haliloglu M, et al. Invasive inflammatory myofibroblastic tumor of the spleen treated with partial splenectomy in a child. J Pediatr Hematol Oncol. 2012;34(4):e131–3.
84. Iimuro Y, Okada T, Sueoka H, et al. Laparoscopic management of giant splenic true cyst with partial splenectomy: a case report. Asian J Endosc Surg. 2013;6(3):226–30.
85. Sretenovic A, Perisic V, Krstic Z, et al. Warren shunt combined with partial splenectomy for children with extrahepatic portal hypertension, massive splenomegaly, and severe hypersplenism. Surg Today. 2013;43(5):521–5.
86. Cai YQ, Li CL, Zhang H, et al. Emergency laparoscopic partial splenectomy for ruptured spleen: a case report. World J Gastroenterol. 2014;20(46):17670–3.
87. Sretenovic A, Perisic V, Vujovic D, et al. Warren shunt combined with partial splenectomy in children with extra-hepatic portal hypertension, massive splenomegaly and severe hypersplenism. Srp Arh Celok Lek. 2014;142(7-8):419–23.

88. Wang X, Wang M, Zhang H, et al. Laparoscopic partial splenectomy is safe and effective in patients with focal benign splenic lesion. Surg Endosc. 2014;28(12):3273–8.
89. Zhang, et al. Treatment of hemolymphangioma of the spleen by laparoscopic partial splenectomy: a case report. World J Surg Oncol. 2014;12:60. https://doi.org/10.1186/1477-7819-12-60.
90. Costi R, Castro Ruiz C, Zarzavadjian le Bian A, et al. Spleen hydatidosis treated by hemi-splenectomy: a low-morbidity, cost-effective management by a recently improved surgical technique. Int J Surg. 2015;20:41–5.
91. Han XL, Zhao YP, Chen G, et al. Laparoscopic partial splenectomy for splenic hemangioma: experience of a single center in six cases. Chin Med J (Engl). 2015;128(5):694–7.
92. Hao F, Wang W, Yang X. Radiofrequency-assisted partial splenectomy plus splenopexy—a modified surgical procedure for wandering spleen: a case report. J Laparoendosc Adv Surg Tech A. 2015;25(9):760–2.
93. Lee SH, Lee JS, Yoon YC, et al. Role of laparoscopic partial splenectomy for tumorous lesions of the spleen. J Gastrointest Surg. 2015;19(6):1052–8.
94. Navas-Cuellar JA, Canete-Gomez J, Lopez-Bernal F, et al. Spleen-preserving surgery after blunt abdominal trauma with splenic hilum involvement. Cir Cir. 2015;83(6):516–21.
95. Petroianu A. Subtotal splenectomy for the treatment of chronic lymphocytic leukemia. Blood Cancer J. 2015;5:e289.
96. Garza-Serna U, Ovalle-Chao C, Martinez D, et al. Laparoscopic partial splenectomy for congenital splenic cyst in a pediatric patient: case report and review of literature. Int J Surg Case Rep. 2017;33:44–7.
97. Li H, Wei Y, Peng B, et al. Feasibility and safety of emergency laparoscopic partial splenectomy: a retrospective analysis. Medicine. 2017;96(16):e6450.
98. Mudan S, Kumar J, Mafalda NC, et al. Case report on the role of radiofrequencyassisted spleen-preserving surgery for splenic metastasis in the era of checkpoint inhibitors. Medicine. 2017;96(49):e9106.
99. Wang L, Xu J, Li F, et al. Partial splenectomy is superior to total splenectomy for selected patients with hemangiomas or cysts. World J Surg. 2017;41(5):1281–6.
100. Zarzavadjian Le Bian A, Cesaretti M, Costi R, et al. Laparoscopic partial splenectomy with selective pedicular approach (with video). J Visc Surg. 2017;154(1):63–4.
101. Kamel R, Dunn MA. Segmental splenectomy in schistosomiasis. Br J Surg. 1982;69(6):311–3.
102. Kamel R, Dunn MA, Skelly RR, et al. Clinical and immunological results of segmental splenectomy in schistosomiasis. Br J Surg. 1986;73(7):544–7.
103. Petroianu A. Treatment of portal hypertension by subtotal splenectomy and central splenorenal shunt. Postgrad Med J. 1988;64(747):38–41.
104. Golinsky D, Freud E, Steinberg R, et al. Vertical partial splenectomy for epidermoid cyst. J Pediatr Surg. 1995;30(12):1704–5.
105. Tsakayannis DE, Mitchell K, Kozakewich HP, et al. Splenic preservation in the management of splenic epidermoid cysts in children. J Pediatr Surg. 1995;30(10):1468–70.
106. Benetatos N, Filobbos R, Ammori BJ. Laparoscopic partial splenectomy for littoral cell angioma. JSCR. 2011;7:4.
107. Patrzyk M, Glitsch A, Hoene A, et al. Laparoscopic partial splenectomy using a detachable clamp with and without partial splenic embolisation. Langenbecks Arch Surg. 2011;396(3):397–402.
108. Abstracts of the 20th International Congress of the European Association for Endoscopic Surgery (EAES) Brussels, Belgium June 20–23, 2012, Surg Endosc 2013;27 (Suppl. 1):S168.
109. Mignon F, Brouzes S, Breitel DL, et al. Preoperative selective embolization allowing a partial splenectomy for splenic hamartoma. Ann Chir. 2003;128(2):112–6.
110. Habib NA, Spalding D, Navarra G, et al. How we do a bloodless partial splenectomy. Am J Surg. 2003;186(2):164–6.
111. Fonseca AZ, Ribeiro M Jr, Contrucci O. Torsion of a wandering spleen treated with partial splenectomy and splenopexy. J Emerg Med. 2013;44(1):e33–6.
112. Ahmed S, Oh HB, Kheng D, et al. Case report of successful partial splenectomy for a splenic abscess in a paediatric patient. Int J Surg Case Rep. 2017;38:176–9.

113. Havlik RJ, Touloukian RJ, Markowitz RI, et al. Partial splenectomy for symptomatic splenic hamartoma. J Pediatr Surg. 1990;25(12):1273–5.
114. Uranus S, Kronberger L, Kraft-Kine J. Partial splenic resection using the TA-stapler. Am J Surg. 1994;168(1):49–53.
115. Poulin EC, Thibault C, DesCoteaux JG, et al. Partial laparoscopic splenectomy for trauma: technique and case report. Surg Laparosc Endosc. 1995;5(4):306–10.
116. Balzan SM, Riedner CE, Santos LM, et al. Posttraumatic splenic cysts and partial splenectomy: report of a case. Surg Today. 2001;31(3):262–5.
117. Smith ST, Scott DJ, Burdick JS, et al. Laparoscopic marsupialization and hemisplenectomy for splenic cysts. J Laparoendosc Adv Surg Tech A. 2001;11(4):243–9.
118. Vasilescu C, Stanciulea O, Tudor S, et al. Laparoscopic subtotal splenectomy in hereditary spherocytosis: to preserve the upper or the lower pole of the spleen? Surg Endosc. 2006;20(5):748–52.
119. Diesen DL, Zimmerman SA, Thornburg CD, et al. Partial splenectomy for children with congenital hemolytic anemia and massive splenomegaly. J Pediatr Surg. 2008;43(3):466–72.
120. Mac New HG, Fowler CL. Partial splenectomy for littoral cell angioma. J Pediatr Surg. 2008;43(12):2288–90.
121. Petroianu A, Teixeira BC, de Alencar LF. Partial splenectomy for treatment of splenic hemangioma. Chirurgia (Bucur). 2009;104(4):483–6.
122. Dudi-Venkata NN, Houli N, Weinberg L, et al. Laparoscopic partial splenectomy performed by monopolar saline-cooled radiofrequency coagulation. J Laparoendosc Adv Surg Tech A. 2014;24(7):502–5.
123. Akkoyun I, Akbiyik F, Altunkeser A. Laparoscopic cystectomy for splenic hydatid cyst: a case report. J Pediatr Surg. 2011;46(11):e1–3.
124. Hodgson WJ, McElhinney AJ. Ultrasonic partial splenectomy. Surgery. 1982;91(3):346–8.
125. Chen HH, Yan NS, Hou KY, et al. Subtotal splenectomy with a microwave coagulator. Br J Surg. 1993;80(5):612–3.
126. Di Lorenzo N, Lania M, Sica GS, et al. Partial resection of the spleen for echinococcus cysts with ultrasonic surgical aspirator. Minerva Chir. 1993;48(15–16):871–3. R. Costi et al./ Journal of Pediatric Surgery 54 (2019) 1527–1538 1537
127. Breil P, Bahnini MA, Fekete F. Partial splenectomy using the TA stapler. Surg Gynecol Obstet. 1986;163(6):575–6.
128. Raschbaum G, Harnar TJ, Canizaro PC. The use of a stapler in splenic salvage as an alternative to the sutured partial splenectomy or splenorrhaphy. Surg Gynecol Obstet. 1988;166(2):179–80.
129. Bhattacharyya N, Ablin DS, Kosloske AM. Stapled partial splenectomy for splenic abscess in a child. J Pediatr Surg. 1989;24(3):316–7.
130. Bergholt T, Westphall IT, Strandberg C, et al. Partial spleen resection using a stapler. Ugeskr Laeger. 1992;154(14):938–9.
131. Yavorski CC, Greason KL, Egan MC. Splenic cysts: a new approach to partial splenectomy–case report and review of the literature. Am Surg. 1998;64(8):795–8.
132. Ho CM. Surg Endosc. 2002;16:717.
133. Wang J, Yu JC, Kang WM, et al. Laparoscopic partial splenectomy for giant hemangioma misdiagnosed as splenic cyst: a case report. Chin Med Sci J. 2010;25(3):189–92.
134. Jiao LR, Tierris I, Ayav A, et al. A new technique for spleen preservation with radiofrequency. Surgery. 2006;140(3):464–6.
135. Gumbs AA, Bouhanna P, Bar-Zakai B, et al. Laparoscopic partial splenectomy using radiofrequency ablation. J Laparoendosc Adv Surg Tech A. 2008;18(4):611–3.
136. Bong JJ, Kumar R, Spalding D. A novel technique of partial splenectomy using radiofrequency ablation. J Gastrointest Surg. 2011;15(2):371–2.
137. Liese J, Kohler S, Moench C, et al. Partial spleen resection with a radiofrequency needle device—a pilot study. Langenbecks Arch Surg. 2013;398(3):449–54.
138. 2013 Scientific session of the Society of American Gastrointestinal and Endoscopic Surgeons (SAGES) Baltimore, Maryland, USA, 17–20 April 2013, Surg Endosc 2013;27(S1):497.

139. Liu S, Lei J, Zeng Z, et al. Management of traumatic splenic rupture in adults: a single center's experience in mainland China. Hepatogastroenterology. 2014;61(132):966–71.
140. Wang WD, Lin J, Wu ZQ, et al. Partial splenectomy using a laparoscopic bipolar radiofrequency device: a case report. World J Gastroenterol. 2015;21(11):3420–4.
141. Rogulski R, Adamowicz-Salach A, Matysiak M, et al. Laparoscopic splenectomy for hereditary spherocytosis-preliminary report. Eur J Haematol. 2016;96(6):637–42.
142. Quesada R, Poves I, Iglesias M, et al. Laparoscopic partial splenectomy for giant cyst using a radiofrequency-assisted device: a case report. Surg Case Rep. 2016;2(1):82.
143. Garcia-Hernandez C, Carvajal-Figueroa L, Duenas-Ramirez JC, et al. Treatment of splenic cyst by laparoscopic partial splenectomy: case report. Cir Cir. 2010;78(1):83–5.
144. Seims AD, Breckler FD, Hardacker KD, et al. Partial versus total splenectomy in children with hereditary spherocytosis. Surgery. 2013;154(4):849–53.
145. Delarue A, Camboulives J, Bollini G, et al. Delayed cure of an omphalocele requiring abdominosternoplasty, right hepatectomy and partial splenectomy. Eur J Pediatr Surg. 2000;10(1):58–61.
146. Chazalette JP, Feigelson J, Louis D. Partial splenectomy—worth the risk. Arch Dis Child. 2003;88(7):649.
147. Burrington JD. Surgical repair of a ruptured spleen in children: report of eight cases. Arch Surg. 1977;112(4):417–9.
148. Morgenstern L, Shapiro SJ. Partial splenectomy for nonparasitic splenic cysts. Am J Surg. 1980;139(2):278–81.
149. Andreitsev AN, Kulikovskii NN, Stankevich VE, et al. Segmental resection of the spleen using a collagen sponge. Eksp Khir Anesteziol. 1976;5:43–4.
150. Thomas WE, Winfield DA. Partial splenectomy for massive splenomegaly secondary to Gaucher's disease. Postgrad Med J. 1991;67(794):1072–4.
151. Holland AJ, Ford WD, Bourne AJ. Conservative surgery for benign non-parasitic splenic cysts. Pediatr Surg Int. 1997;12(5-6):353–5.
152. Bar-Maor JA, Govrin-Yehudain J. Partial splenectomy in children with Gaucher's disease. Pediatrics. 1985;76(3):398–401.
153. el-Banna I, Fathy M, Baza O, et al. Segmental splenectomy and extraperitoneal splenic transposition with gastroesophageal devascularization in treatment of oesophageal varices. A new technique. Int Surg. 1991;76(1):6–11.
154. Li E, Zhao L, Zhu L, et al. Treating portal hypertension by subtotal splenectomy with retroperitoneal splenic transposition and devascularization: clinical study. Zhonghua Wai Ke Za Zhi. 1998;36(6):333–5.
155. Kaiwa Y, Kurokawa Y, Namiki K, et al. Laparoscopic partial splenectomies for true splenic cysts. A report of two cases. Surg Endosc. 2000;14(9):865.
156. Seshadri PA, Poulin EC, Mamazza J, et al. Technique for laparoscopic partial splenectomy. Surg Laparosc Endosc Percutan Tech. 2000;10(2):106–9.
157. Vasilescu C, Stanciulea O, Colita A, et al. Laparoscopic subtotal splenectomy in the treatment of hereditary spherocytosis. Chirurgia (Bucur). 2003;98(6):571–6.
158. Czauderna P, Vajda P, Schaarschmidt K, et al. Nonparasitic splenic cysts in children: a multicentric study. Eur J Pediatr Surg. 2006;16(6):415–9.
159. Mouttalib S, Rice HE, Snyder D, et al. Evaluation of partial and total splenectomy in children with sickle cell disease using an Internet-based registry. Pediatr Blood Cancer. 2012;59(1):100–4.
160. 20th International Congress of the European Association for Endoscopic Surgery (EAES) Brussels, Belgium, 20–23 June 2012, Surg Endosc 2013;27 (S1):43.
161. Rice HE, Englum BR, Rothman J, et al. Clinical outcomes of splenectomy in children: report of the splenectomy in congenital hemolytic anemia registry. Am J Hematol. 2015;90(3):187–92.
162. Vega EA, Yamashita S, Shin CY, et al. Laparoscopic partial splenectomy for unknown primary cancer: a stepwise approach. Ann Surg Oncol. 2017;24(4):1134.
163. Berelavichus SV, Smirnov AV, Ionkin DA, et al. Robot-assisted and laparoscopic partial splenectomy for nonparasitic cysts. Khirurgiia. 2015;7:41–8.

164. Radevic B, Jesic R, Sagic D, et al. Partial resection of the spleen and splenorenal shunt in the treatment of portal hypertension with splenomegaly and hypersplenism. Acta Chir Iugosl. 2002;49(3):93–8.
165. Radevic B, Dugalic V, Jesic R, et al. Partial resection of the spleen in the treatment of benign and traumatic lesions. Acta Chir Iugosl. 2002;49(3):67–72.
166. Gajin P, Radevic B, Nenezic D, et al. Distal splenorenal shunt with partial spleen resection. Srp Arh Celok Lek. 2007;135(5-6):293–7.
167. Chinnusamy P, Ahluwalia JS, Palanisamy S, et al. Single incision multi-trocar hepatic cyst excision with partial splenectomy. J Minim Access Surg. 2013;9(2):91–4.
168. Wei Z, Rui SG, Yuan Z, et al. Partial splenectomy and use of splenic vein as an autograft for meso-Rex bypass: a clinical observational study. Med Sci Monit. 2014;20:2235–42.
169. Chu H, Han W, Wang L, et al. Long-term efficacy of subtotal splenectomy due to portal hypertension in cirrhotic patients. BMC Surg. 2015;15:89.
170. Pincez T, Guitton C, Gauthier F, et al. Long-term follow-up of subtotal splenectomy for hereditary spherocytosis: a single-center study. Blood. 2016;127(12):1616–8.
171. Fleshner PR, Aufses AH Jr, Grabowski GA, et al. A 27-year experience with splenectomy for Gaucher's disease. Am J Surg. 1991;161(1):69–75.
172. Englum BR, Rothman J, Leonard S, et al. Hematologic outcomes after total splenectomy and partial splenectomy for congenital hemolytic anemia. J Pediatr Surg. 2016;51(1):122–7.
173. Alebouyeh M, Kheradpir MH. Partial splenectomy in homozygous beta-thalassemia. Monatsschr Kinderheilkd. 1985;133(8):549–51.
174. Rubin M, Yampolski I, Lambrozo R, et al. Partial splenectomy in Gaucher's disease. J Pediatr Surg. 1986;21(2):125–8.
175. Freud E, Cohen IJ, Neuman M, et al. Should repeated partial splenectomy be attempted in patients with hematological diseases? Technical pitfalls and causes of failure in Gaucher's disease. J Pediatr Surg. 1997;32(9):1272–6.
176. Cohen IJ, Katz K, Freud E, et al. Long-term follow-up of partial splenectomy in Gaucher's disease. Am J Surg. 1992;164(4):345–7.
177. Weinrich M, Dahmen RP, Black KJ, et al. Postoperative long-term results in highgrade traumatic ruptures of the spleen in children. Zentralbl Chir. 2014;139(6):632–7.
178. Nouri A, de Montalembert M, Revillon Y, et al. Partial splenectomy in sickle cell syndromes. Arch Dis Child. 1991;66(9):1070–2.
179. Hoekstra HJ, Tamminga RY, Timens W. Partial instead of complete splenectomy in children for the pathological staging of Hodgkin disease. Ned Tijdschr Geneeskd. 1993;137(48):2491–4.
180. Petroianu A. Subtotal splenectomy and portal variceal disconnection in the treatment of portal hypertension. Can J Surg Journal Canadien de Chirurgie. 1993;36(3):251–4.
181. Ziske CG, Müller T. Partial splenectomy. Lancet. 2002;359:9312.
182. Svarch E, Nordet I, Valdes J, et al. Partial splenectomy in children with sickle cell disease. Haematologica. 2003;88(2):222–3.
183. Gutierrez Diaz AI, Svarch E, Arencibia Nunez A, et al. Partial splenectomy in sickle cell disease. An Pediatr (Barc). 2015;82(4):228–34.
184. Petroianu A. Subtotal splenectomy preserving the inferior splenic pole for the treatment of Hodgkin's lymphoma. Int J Surg Case Rep. 2017;36:1–3.
185. de Buys Roessingh AS, de Lagausie P, Rohrlich P, et al. Follow-up of partial splenectomy in children with hereditary spherocytosis. J Pediatr Surg. 2002;37(10):1459–63.
186. Lima M, Reinberg O, Ruggeri G, et al. 3D virtual rendering before laparoscopic partial splenectomy in children. J Pediatr Surg. 2013;48(8):1784–8.
187. Feliciotti F, Sottili M, Guerrieri M, et al. Conservative ultrasound-guided laparoscopic treatment of posttraumatic splenic cysts: report of two cases. Surg Laparosc Endosc. 1996;6(4):322–5.
188. Rice HE, Crary SE, Langer JC, et al. Comparative effectiveness of different types of splenectomy for children with congenital hemolytic anemias. J Pediatr. 2012;160(4):684–689.e13.
189. Pavón Morán V, Estrada del Cueto M, Fernández Delgado N, et al. Effectiveness of partial splenectomy in the treatment of hereditary spherocytosis. Rev Invest Clin. 2000;52(3):229–33.

190. Wu HM, Kortbeek JB. Management of splenic pseudocysts following trauma: a retrospective case series. Am J Surg 2006; 191:631-4?
191. Gianom D, Wildisen A, Hotz T, et al. Open and laparoscopic treatment of nonparasitic splenic cysts. Dig Surg. 2003;20:74–8.
192. Karfis EA, Roustanis E, Tsimoyiannis EC. Surgical management of nonparasitic splenic cysts. JSLS. 2009;13:207–12.
193. Bhandarkar DS, Katara AN, Mittal G, et al. Prevention and management of complications of laparoscopic splenectomy. Indian J Surg. 2011;73:324–30.
194. Balaphas A, Buchs NC, Meyer J, et al. Partial splenectomy in the era of minimally invasive surgery: the current laparoscopic and robotic experiences. Surg Endosc. 2015;29:3618–27.
195. Esposito F, Noviello A, Moles N, et al. Partial splenectomy: a case series and systematic review of the literature. Ann Hepatobiliary Pancreat Surg. 2018;22:116–27.
196. Palmieri I, Natale E, Crafa F, et al. Epithelial splenic cysts. Anticancer Res. 2005;25:515–21.
197. Winslow ER, Brunt LM. Perioperative outcomes of laparoscopic versus open splenectomy: a meta-analysis with an emphasis on complications. Surgery. 2003;134:647–53.
198. Caronna R, Chirletti P, Fanello G. L'emostasi in chirurgia d'urgenza. In: Zanon C, editor. Emostasi in chirurgia. Torino: Ananke; 2007. p. 53–66.
199. Baiamonte M, Gargano G, Agosta G, et al. Impiego degli emostatici nelle resezioni parziali di milza. Report su di un caso clinico. Minerva Chir. 2012;67:5–14.
200. Zheng L, Deng C, Li J, et al. Treatment of hemangioma of the spleen by preoperative partial splenic embolization plus laparoscopic partial splenectomy: a case report. Medicine (Baltimore). 2018;97:e0498.
201. Rottenstreich A, Kleinstern G, Spectre G, Da'as N, Ziv E, Kalish Y. Thromboembolic events following splenectomy: risk factors, prevention, management and outcomes. World J Surg. 2018;42:675–81.
202. Ruiz-Tovar J, Priego P. Portal vein thrombosis after splenic and pancreatic surgery. Adv Exp Med Biol. 2017;906:241–51.
203. Kimmig LM, Palevsky HI. Review of the association between splenectomy and chronic thromboembolic pulmonary hypertension. Ann Am Thorac Soc. 2016;13:945–54.
204. Holdsworth RJ, Irving AD, Cuschieri A. Postsplenectomy sepsis and its mortality rate: actual versus perceived risks. Br J Surg. 1991;78:1031–8.
205. Bisharat N, Omari H, Lavi I, Raz R. Risk of infection and death among post-splenectomy patients. J Infect. 2001;43:182–6.
206. de'Angelis N, Abdalla S, Lizzi V, Esposito F, Genova P, Roy L, et al. Incidence and predictors of portal and splenic vein thrombosis after pure laparoscopic splenectomy. Surgery. 2017;162:1219–30.
207. Van Wyck DB, Witte MH, Witte CL, Thies AC. Jr Critical splenic mass for survival from experimental pneumococcemia. J Surg Res. 1980;28:14–7.
208. Liu DL, Xia S, Xu W, Ye Q, Gao Y, Qian J. Anatomy of vasculature of 850 spleen specimens and its application in partial splenectomy. Surgery. 1996;119:27–33.
209. Morgenstern L, Shapiro SJ. Partial splenectomy for nonparasitic splenic cysts. Am J Surg. 1980;139:278–81.
210. Poulin EC, Thibault C, DesCôteaux JG, Côté G. Partial laparoscopic splenectomy for trauma: technique and case report. Surg Laparosc Endosc. 1995;5:306–10.
211. Resende V, Petroianu A. Functions of the splenic remnant after subtotal splenectomy for treatment of severe splenic injuries. Am J Surg. 2003;185:311–5.
212. Wang L, Xu J, Li F, Zhan H, Liu H, Chen W, et al. Partial splenectomy is superior to total splenectomy for selected patients with hemangiomas or cysts. World J Surg. 2017;41:1281–6.
213. Petroianu A. Subtotal splenectomy for the treatment of chronic lymphocytic leukemia. Ann Hematol. 2003;82:708–9.
214. Vasilescu C, Stanciulea O, Tudor S, Stanescu D, Colita A, Stoia R, et al. Laparoscopic subtotal splenectomy in hereditary spherocytosis: to preserve the upper or the lower pole of the spleen? Surg Endosc. 2006;20:748–52.

215. Vasilescu C, Tudor S, Popa M, Tiron A, Lupescu I. Robotic partial splenectomy for hydatid cyst of the spleen. Langenbecks Arch Surg. 2010;395:1169–74.
216. Palmieri I, Natale E, Crafa F, Cavallaro A, Mingazzini PL. Epithelial splenic cysts. Anticancer Res. 2005;25:515–21.
217. Hong TH, Lee SK, You YK, Kim JG. Single-port laparoscopic partial splenectomy: a case report. Surg Laparosc Endosc Percutan Tech. 2010;20:e164–6.
218. Giulianotti PC, Buchs NC, Addeo P, Ayloo S, Bianco FM. Robot-assisted partial and total splenectomy. Int J Med Robot. 2011;7:482–8.
219. Balaphas A, Buchs NC, Meyer J, Hagen ME, Morel P. Partial splenectomy in the era of minimally invasive surgery: the current laparoscopic and robotic experiences. Surg Endosc. 2015;29:3618–27.
220. Bouchet A. New glimpses on the structure and vascularization of the greater omentum. Arch Anat Histol Embryol. 1962;45:1–31.
221. Liberati A, Altman DG, Tetzlaff J, Mulrow C, Gøtzsche PC, Ioannidis JP, et al. The PRISMA statement for reporting systematic reviews and meta-analyses of studies that evaluate health care interventions: explanation and elaboration. Ann Intern Med. 2009;151:W65–94.
222. Guyatt GH, Oxman AD, Vist GE, Kunz R, Falck-Ytter Y, Alonso-Coello P, et al. GRADE: an emerging consensus on rating quality of evidence and strength of recommendations. BMJ. 2008;336:924–6.
223. Li H, Wei Y, Peng B, Li B, Liu F. Feasibility and safety of emergency laparoscopic partial splenectomy: a retrospective analysis. Medicine (Baltimore). 2017;96:e6450.
224. Cai H, An Y, Wu D, Chen X, Zhang Y, Zhu F, et al. Laparoscopic partial splenectomy: a preferred method for select patients. J Laparoendosc Adv Surg Tech A. 2016;26:1010–4.
225. Lee SH, Lee JS, Yoon YC, Hong TH. Role of laparoscopic partial splenectomy for tumorous lesions of the spleen. J Gastrointest Surg. 2015;19:1052–8.
226. Costi R, Castro Ruiz C, Zarzavadjian L, Scerrati D, Santi C, Violi V. Spleen hydatidosis treated by hemi-splenectomy: a low-morbidity, cost-effective management by a recently improved surgical technique. Int J Surg. 2015;20:41–5.
227. Han XL, Zhao YP, Chen G, Wu WM, Dai MH. Laparoscopic partial splenectomy for splenic hemangioma: experience of a single center in six cases. Chin Med J (Engl). 2015;128:694–7.
228. de la Villeon B, Le Z, Vuarnesson H, Munoz Bongrand N, Halimi B, Sarfati E, et al. Laparoscopic partial splenectomy: a technical tip. Surg Endosc. 2015;29:94–9.
229. Wang X, Wang M, Zhang H, Peng B. Laparoscopic partial splenectomy is safe and effective in patients with focal benign splenic lesion. Surg Endosc. 2014;28:3273–8.
230. Dudi-Venkata NN, Houli N, Weinberg L, Nikfarjam M. Laparoscopic partial splenectomy performed by monopolar saline-cooled radiofrequency coagulation. J Laparoendosc Adv Surg Tech A. 2014;24:502–5.
231. Paudel GR, Agarwal R, Pathania OP, Agrawal CS. Partial splenectomy for epithelial (epidermoid) splenic cysts: report of two case. JNMA J Nepal Med Assoc. 2013;52:391–4.
232. Liese J, Kohler S, Moench C, Bechstein WO, Ulrich F. Partial spleen resection with a radiofrequency needle device—a pilot study. Langenbecks Arch Surg. 2013;398:449–54.
233. Karadayi K, Turan M, Sen M. A new technique for partial splenectomy with radiofrequency technology. Surg Laparosc Endosc Percutan Tech. 2011;21:358–61.
234. Szczepanik AB, Meissner AJ. Partial splenectomy in the management of nonparasitic splenic cysts. World J Surg. 2009;33:852–6.
235. Petroianu A, Cabezas-Andrade MA, Berindoague NR. Laparoscopic subtotal splenectomy. Surg Laparosc Endosc Percutan Tech. 2008;18:94–7.
236. Wu SC, Wang CC, Yong CC. Partial splenectomy for benign splenic cysts with the aid of a Lin clamp: technical note. World J Surg. 2007;31:2144–7.
237. Jiao LR, Tierris I, Ayav A, Milicevic M, Pellicci R, Navarra G, et al. A new technique for spleen preservation with radiofrequency. Surgery. 2006;140:464–6.
238. Balshem H, Helfand M, Schünemann HJ, Oxman AD, Kunz R, Brozek J, et al. GRADE guidelines: 3. Rating the quality of evidence. J Clin Epidemiol. 2011;64:401–6.

239. Hong TH, You YK, Lee KH. Transumbilical single-port laparoscopic cholecystectomy: scarless cholecystectomy. Surg Endosc. 2009;23:1393–7.
240. Lee YS, Kim JH, Moon EJ, Kim JJ, Lee KH, Oh SJ, et al. Comparative study on surgical outcomes and operative costs of transumbilical single-port laparoscopic appendectomy versus conventional laparoscopic appendectomy in adult patients. Surg Laparosc Endosc Percutan Tech. 2009;19:493–6.
241. Vasilescu C, Stanciulea O, Tudor S. Laparoscopic versus robotic subtotal splenectomy in hereditary spherocytosis. Potential advantages and limits of an expensive approach. Surg Endosc. 2012;26:2802–9.
242. Berelavichus SV, Smirnov AV, Ionkin DA, Kriger AG, Dugarova RS. Robot-assisted and laparoscopic partial splenectomy for nonparasitic cysts. Khirurgiia (Mosk). 2015;7:41–8.
243. Habib NA, Spalding D, Navarra G, Nicholls J. How we do a bloodless partial splenectomy. Am J Surg. 2003;186:164–6.
244. Itamoto T, Fukuda S, Tashiro H, Ohdan H, Asahara T. Radiofrequency-assisted partial splenectomy with a new and simple device. Am J Surg. 2006;192:252–4.
245. Gumbs AA, Bouhanna P, Bar-Zakai B, Briennon X, Gayet B. Laparoscopic partial splenectomy using radiofrequency ablation. J Laparoendosc Adv Surg Tech A. 2008;18:611–3.
246. Stacey MJ, Rampaul RS, Rengaragan A, Duffy JP, Macmillan RD. Use of FloSeal matrix hemostatic agent in partial splenectomy after penetrating trauma. J Trauma. 2008;64:507–8.
247. Mignon F, Brouzes S, Breitel DL, Bastie JN, Poirier H, Legendre C, et al. Preoperative selective embolization allowing a partial splenectomy for splenic hamartoma. Ann Chir. 2003;128:112–6.
248. Guan YS, Hu Y. Clinical application of partial splenic embolization. ScientificWorldJournal. 2014;2014:961345.
249. Kimura F, Ito H, Shimizu H, Togawa A, Otsuka M, Yoshidome H, et al. Partial splenic embolization for the treatment of hereditary spherocytosis. AJR Am J Roentgenol. 2003;181:1021–4.
250. Coccolini F, Montori G, Catena F, Kluger Y, Biffl W, Moore EE, et al. Splenic trauma: WSES classification and guidelines for adult and pediatric patients. World J Emerg Surg. 2017;12:40.

The Spleen and Sickle Cell Anemia

21

21.1 History and Introduction

- Sickle cell anemia (SCA) is a well-known hemoglobinopathy that was described in many countries worldwide.
- Sickle cell anemia can involve different parts of the body.
- The spleen is one of the early and commonly affected organs in the body.
- In patients with sickle cell anemia, the spleen is affected early and it enlarges during the first decade of life but then it undergoes progressive atrophy.
- This progressive atrophy develops as a result of repeated multiple small splenic infarcts. These will lead shrinkage and atrophy of the spleen. This is called auto splenectomy and sometimes the atrophied small spleen may become calcified (Figs. 21.1, 21.2, 21.3, and 21.4).
- The natural history of splenomegaly in patients with sickle cell anemia is atrophy and auto splenectomy. Sometimes splenomegaly persists in some of these patients into an older age group or sometimes to adulthood.
- This will necessitate splenectomy for a variety of complications which include:
  - Acute splenic sequestration crisis
  - Hypersplenism
  - Massive splenic infarction
  - Splenic abscess
- Patients with sickle cell anemia are prone to develop splenic complications. These are known to be associated with an increased morbidity and in some of these patients it may lead to mortality.
- Splenectomy sometimes become necessary to overcome and avoid these complications.
- Sickle cell anemia was discovered in 1904 but it existed in Africa for at least five thousand years.

A. H. Al-Salem, *The Spleen*, https://doi.org/10.1007/978-981-99-6191-7_21

INDICATIONS FOR SPLENECTOMY IN PATIENTS WITH SICKLE CELL ANEMIA

- **ACUTE SPLENIC SEQUESTRATION CRISIS**
- **HYPERSPLENISM**
- **MASSIVE SPLENIC INFARCTION**
- **SPLENIC ABSCESS**

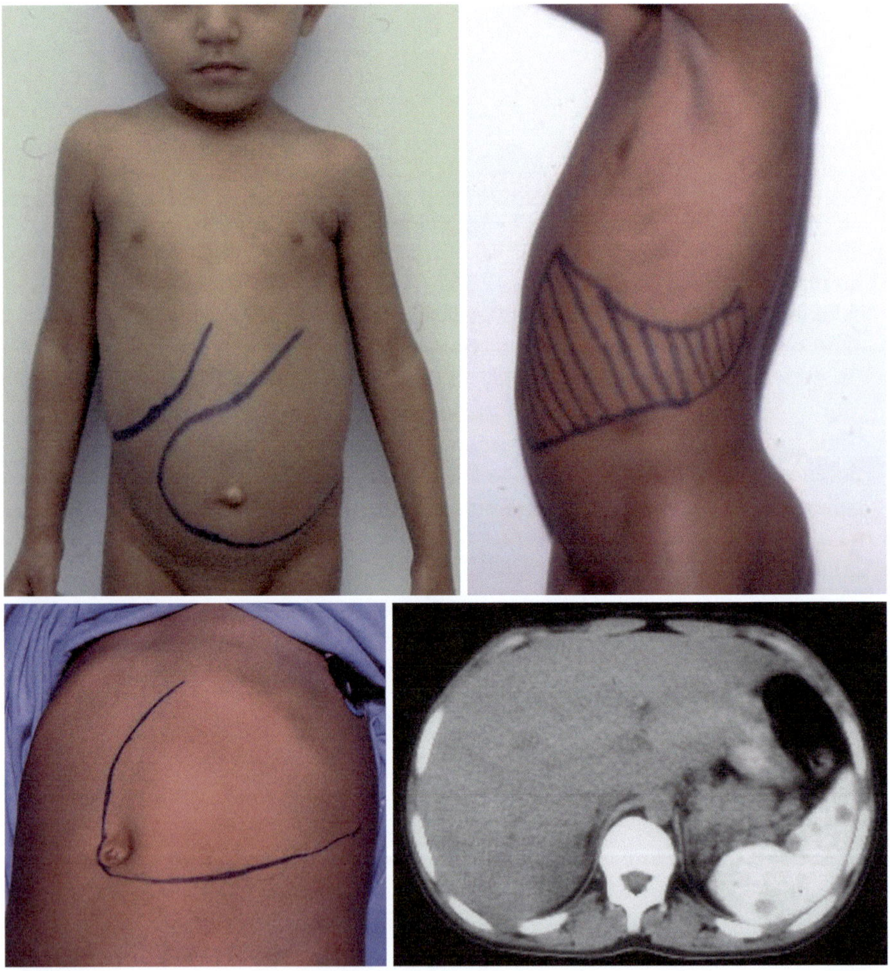

Figs. 21.1–21.4 Clincal and radilogical photographs of patients with sickle cell anemia. Note the the splenomegaly in children with sickle cell anemia which can also persist in some of these patients. Note also the atrophic small spleen which is also calcified in an older child with sickle cell anemia

- In 1904, Dr. Dresbach published the first report of the sickle red blood cells in North America. This was the observation made by one of his students who drew ellyptoid erythrocytes as he saw them under his own microscope. This was in a histology class at Ohio State University.
- In 1910, James Bryan Herrick in 1910 made the first clinical description of sickle cell anemia.
- James Bryan Herrick:
  - He was born on August 11th, 1861 in Oak Park, Illinois and died in 1954.
  - He worked as a cardiologist and professor of medicine at the Chicago Presbyterian Hospital and made a lot of contributions to the history of medical science.
  - In 1910, James Bryan Herrick (1861–1954) describe sickle cells for the first time.
  - This discovery was made by his intern Ernest Edward Irons (1877–1959).
  - Ernest Edward Irons found and described what he called as "peculiar elongated and sickle-shaped" red blood cells.
  - He found these while he was examining a peripheral blood smear of Walter Clement Noel.
  - Walter Clement Noel was a dental student who came from Grenada to become a dentist.
  - At that time, Walter Clement Noel was suffering from anemia and was admitted to the Chicago Presbyterian Hospital (Figs. 21.5, 21.6, 21.7, and 21.8).

Figs. 21.5 and 21.6 Professor James Bryan Herrick (1861–1954) who discovered sickle cell anemia in 1910 (upper picture and his Dr. Ernest Edward Irons (1877–1959) who described the abnormal and peculiar red blood cells (Wikipedia, the free encyclopedia)

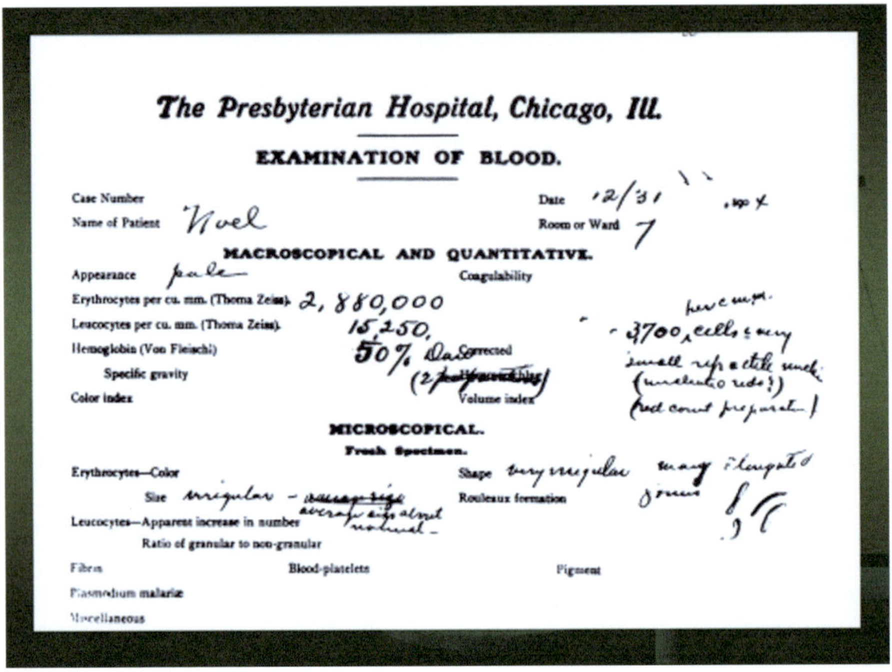

The Presbyterian Hospital, Chicago, Ill.

EXAMINATION OF BLOOD.

Case Number Date 12/31 , 190 4

Name of Patient Noel Room or Ward 7

MACROSCOPICAL AND QUANTITATIVE.

Appearance pale Coagulability

Erythrocytes per cu. mm. (Thoma Zeiss). 2,880,000

Leucocytes per cu. mm. (Thoma Zeiss). 15,250

Hemoglobin (Von Fleischl) 50% Corrected

Specific gravity

Color index Volume index

MICROSCOPICAL.

Fresh Specimen.

Erythrocytes—Color Shape very irregular many elongated

Size irregular Rouleaux formation

Leucocytes—Apparent increase in number

Ratio of granular to non-granular

Fibrin Blood-platelets Pigment

Plasmodium malariæ

Miscellaneous

Fig. 21.7 A Diagrammatic representation of the blood film which was seen by Dr. Ernest Edward Irons. This was from a blood sample of Mr. Walter Clement Noel. He described a pale patient with a total red blood cells number of 2,880,000 per cubic mm and irregular, elongated red blood cells

Fig. 21.8 A clinical photograph from a microscope showing a blood film with normal and abnormal red blood cells. Note the sickled shaped RBC pointed by the green arrow

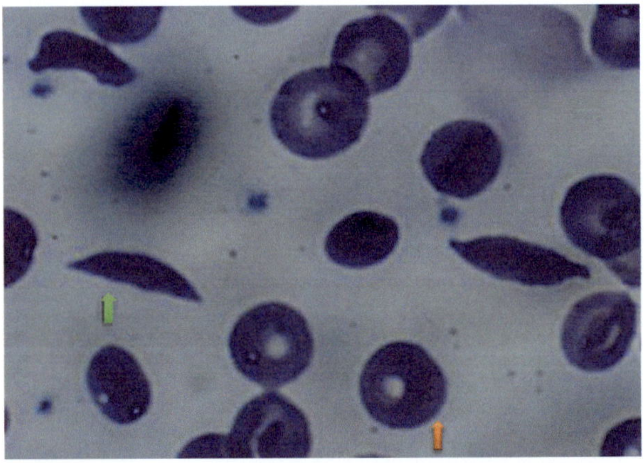

"This case is reported because of the unusual blood findings, no duplicate of which I have ever seen describe"—James B. Herrick, M.D

Fig. 21.9 Dr. Verne Rheem Mason (1889–1965). In 1922, Dr. Mason was the first to use the term sickle cell anemia to describe the disease (Wikipedia, the free encyclopedia)

- The name "sickle-cell anemia" was coined by Dr. Verne Rheem Mason in 1922 (Fig. 21.9).
- The molecule basis of sickle cell anemia was established by Dr. Pauling and his colleagues in 1949. Sickle cell anemia was shown for the first time to be caused by an abnormality in the hemoglobin molecule.
- Subsequently it was shown that sickle cell anemia abnormality is due a single change of one amino acid. The beta chain of hemoglobin is made up of 146 amino acids and sickle cell anemia develops as a result of a change of the amino acid valine instead of glutamic acid at the sixth position of the hemoglobin beta chain.
- This change of valine instead of glutamic acid leads to polymerization of the hemoglobin. This polymerization takes place when the oxygen saturation is low. As a result of this the red blood cells becomes deformed and stick to each other leading to microvascular occlusion. This microvascular occlusion is responsible for the different manifestations of sickle cell anemia (Fig. 21.10). Add to this its subsequent effects at the cellular level which include:
  - Cellular dehydration
  - The associated inflammatory response
  - Reperfusion injury

Fig. 21.10 Dr. Linus Pauling (1901–1994) was the first to establish that sickle cell anemia is caused by an abnormality in hemoglobin (Wikipedia, the free encyclopedia)

- The normal hemoglobin is made up of two components:
  - The heme part
  - The protein part
- The protein part of hemoglobin is made up of four polypeptide chains (Fig. 21.11):
  - 2 alpha chains (Each chain is made up of 141 amino acids)
  - 2 beta chains (Each chain is made up of 146 amino acids)
- There is a gene responsible for each of these chains. The alpha globin peptide chain is coded by a gene that is different from the gene that codes for the beta globin peptide chain.
- The gene responsible for the beta globin chains is the HBB gene and this is found on the short arm of chromosome 11.
- Abnormalities in beta-globin chain results from mutations in the HBB gene. These abnormalities include:
  - Hemoglobin C (HbC)
  - Hemoglobin E (HbE)
  - Hemoglobin S (HbS)
- Sickle cell anemia is caused by a mutation in the HBB gene.
- Sickle-cell anemia is inherited as an autosomal recessive and the abnormal gene is located on the short arm of chromosome 11.

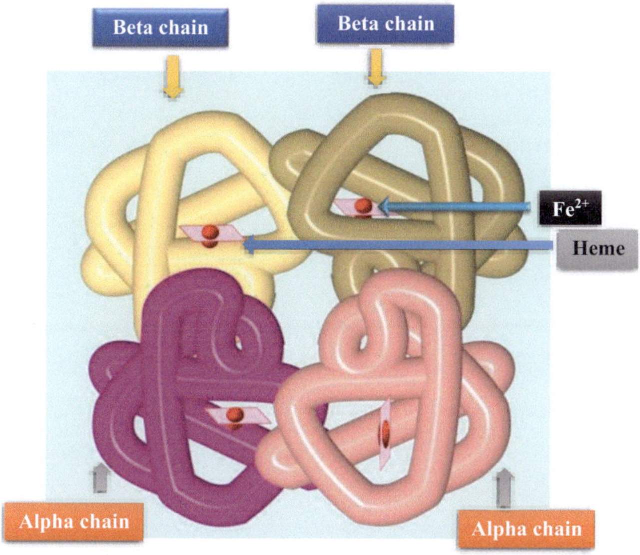

Fig. 21.11 Diagrammatic representation of the normal hemoglobin. The normal hemoglobin is made up of heme and globin. The globin part is made up of 2 alpha chains and 2 beta chains of amino acids

- In patients with sickle cell anemia, there is substitution of glutamic acid by valine. This results from alteration in the genetic code leading to a change of a single amino acid in the six position of the 146 amino acids of the beta chain of hemoglobin.
- HbS develops as a result of a single mutation. This will result in substituting thymine for adenine in the sixth codon of the beta-chain gene. GAG is changed to GTG.
  - This change will lead to coding of valine instead of glutamate in the 6 position of the Hb beta chain.
  - This alteration and relacing glutamate by valine will result in the production of HbS.
  - HbS has the physical properties of forming polymers under low oxygen conditions which will alter the shape of red blood cells.
  - HbS is also characterized by changes in solubility and molecular stability.
  - All of these changes are responsible for the clinical manifestations of sickle cell anemia.
- This single change of Valine for glutamic acid in the Hb beta chain results in hemoglobin tetramers.
- These tetramers under low oxygen tension aggregate into arrays. These aggregates lead to alteration in the shape of the red blood cell to form sickle-shaped red blood cells. The sickle shaped red blood cells are relatively inflexible and unable to negotiate their way through the capillaries leading to their obstruction (Figs. 21.12, 21.13, 21.14, 21.15, 21.16, 21.17, 21.18, and 21.19).

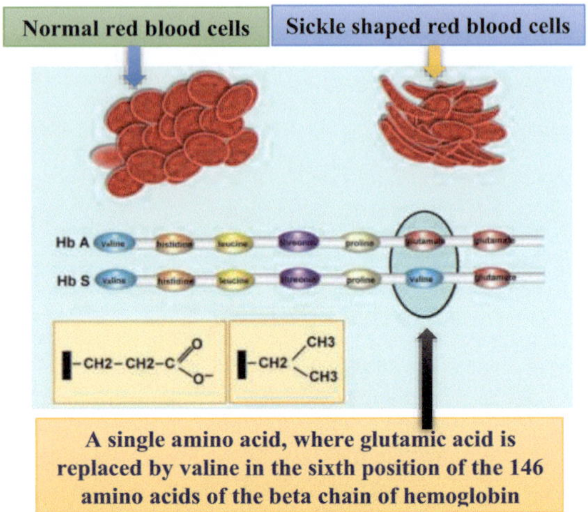

Fig. 21.12 Diagrammatic representation showing the difference between HbA and HbS where glutamic acid is replaced by valine in the sixth position of the 146 amino acids of the beta chain of hemoglobin. This single amino acid change will lead to the formation of hemoglobin S and as a result of this change, the shape of the red blood cells will change to form sickled shaped red blood cells

SEQUENCE FOR NORMAL HEMOGLOBIN

ATG	GTG	CAC	CTG	ACT	CCT	GAG	GAG	AAG	TCT	GCC
START	Val	His	Leu	Thr	Pro	Glu	Lys	Ser	Ala	Val

SEQUENCE FOR SICKLE CELL HEMOGLOBIN

ATG	GTG	CAC	CTG	ACT	CCT	GTG	GAG	AAG	TCT	GCC

Fig. 21.13 Diagrammatic representation showing the change in the codon where GAG which codes for glutamic acid normally is replaced by the GTG codon which codes for Valin amino acids in those with sickle cell anemia

- There are several changes as a result of this in the structure of hemoglobin:
  - The hemoglobin molecule will change and becomes more hydrophobic.
  - The altered hemoglobin chains when they fold into their three-dimensional shape, they will stick to each other and form long insoluble fibers of hemoglobin within the red blood cell.
  - This altered hemoglobin with result in a change in the shape of red blood cells. The red blood cells:

    Shape becomes deformed

    Become more fragile

    Rupture easily

    Adhering to each other leading to closure of the small blood vessels and capillaries (Figs. 21.20, 21.21, 21.22, and 21.23).

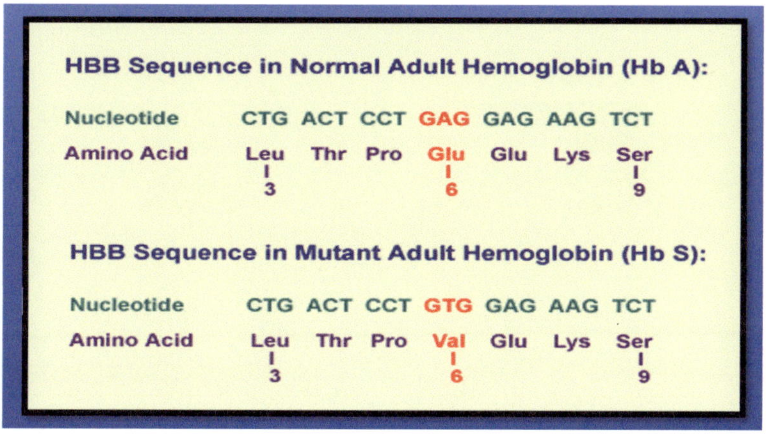

Fig. 21.14 Diagrammatic representation showing the change of Glutamic acid nucleotide (GAG) which is replaced by Valine nucleotide (GTG). In patients with sickle cell anemia, glutamic acid is replaced by Valine

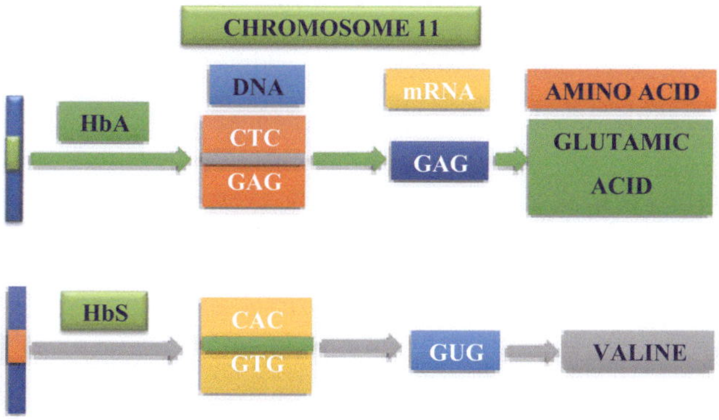

Fig. 21.15 Diagrammatic representation showing the replacement of Valine instead of Glutamic acid in patients with sickle cell anemia. These are due to changes in the nucleotide on chromosome 11 where GAG nucleotide which codes for Glutamic acid is replaced by GUG nucleotide which codes for the amino acid Valine

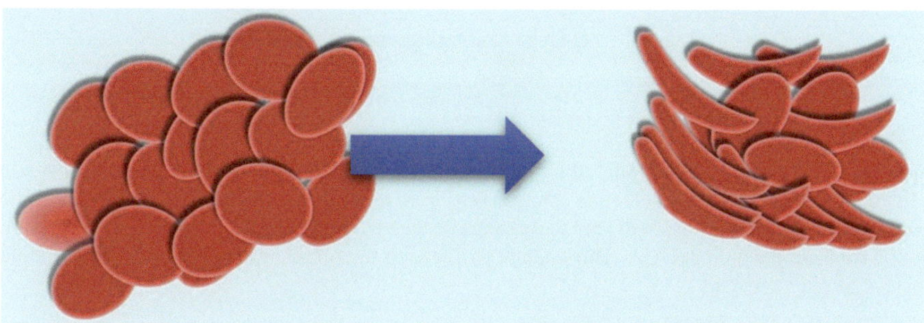

Fig. 21.16 Diagrammatic representation showing the result in the shape of red blood cells as a result of the single change of one amino acid Valine instead of Glutamic acid. The result is a change in the shape of red blood cells shape under low oxygen tension forming sickled red blood cells

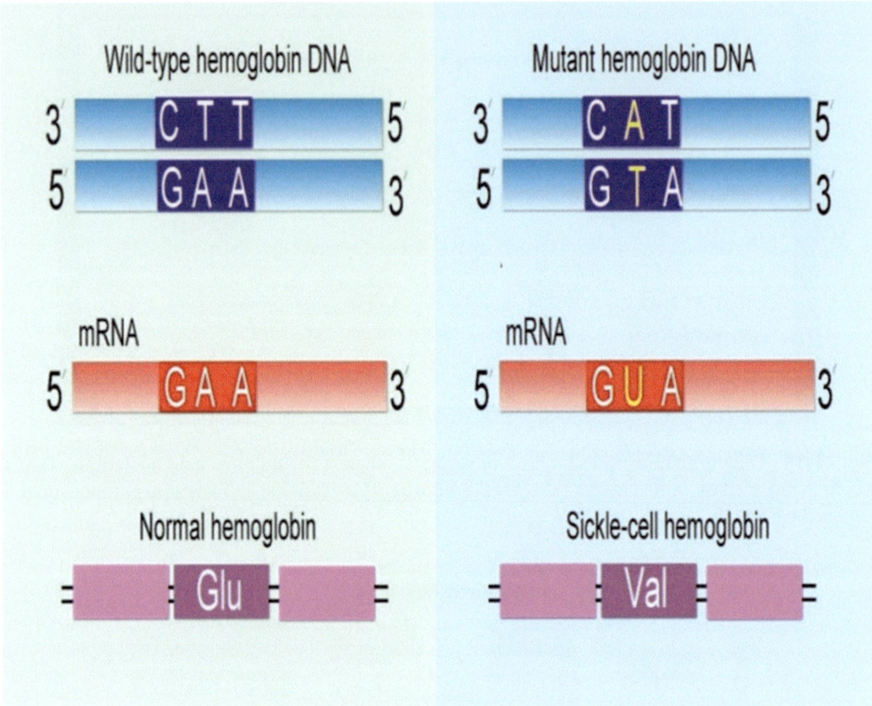

Fig. 21.17 Diagrammatic representation showing the molecular changes which leads to the formation of sickle cell anemia. This results from a single nucleotide change (A to T) in the codon for amino acid 6 of the hemoglobin beta chain. Normally Glutamic acid in the sixth position is coded by either GAA or GAG. The amino acid Valie on the other hand is coded by GUA, GUC, and GUU OR GUG. In patients with sickle cell anemia, the change mRNA converts a glutamic acid codon (GAG or GAA) to a valine codon (GUA, GUC, GUU or GUG)

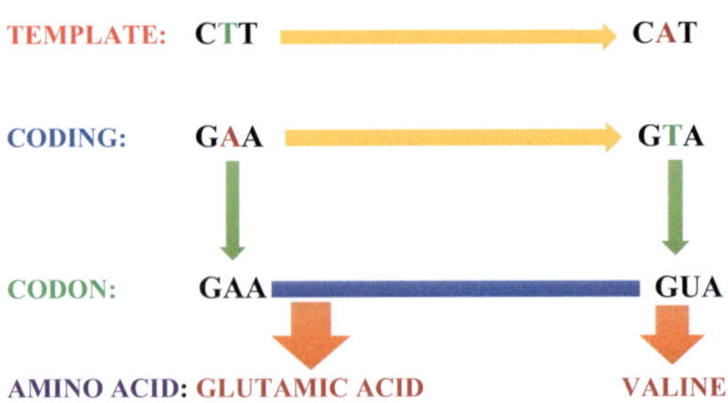

Fig. 21.18 Diagrammatic simplified representation showing the changes in patients with sickle cell anemia where a single nucleotide change (A to T) in the codon for amino acid number 6 of the hemoglobin beta chain. This will lead to a change of the codon for glutamic acid from GAA to the codon for Valine which is GUA. This change will lead to the production of the amino acid valine instead of glutamic acid

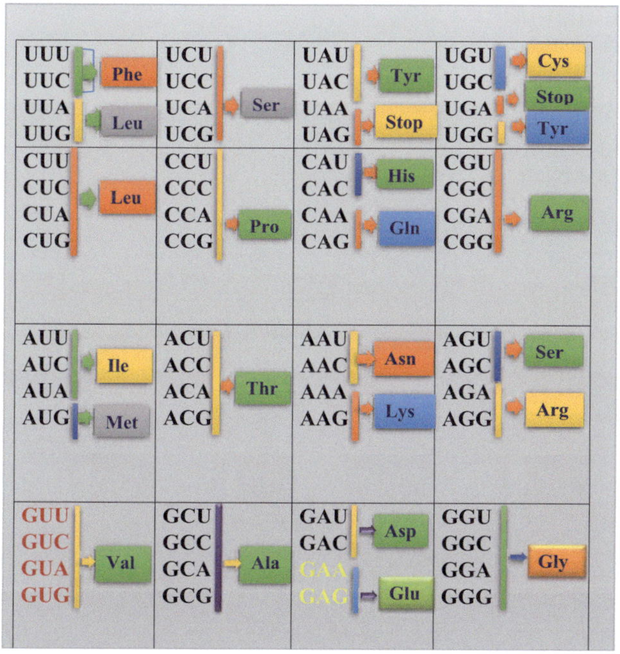

Fig. 21.19 Diagrammatic representation showing the codons for each amino acid. Glutamic acid is coded by either GAA or GAG. Valie is coded by GUA, GUC, and GUU OR GUG. In patients with sickle cell anemia, glutamic acid at the sixth position among the 146 amino acids of the hemoglobin beta change is replaced by Valine. This change will lead to alteration of the hemoglobin molecule. As a result of this change, the hemoglobin tends to stick to each other and form long insoluble fibers within the red blood cell. The red blood cells shape becomes altered and they also become more fragile, adhering to each other and rupture easily

- Sickle cell anemia is inherited as an autosomal recessive disorder.
- For people to be affected by sickle cell anemia they must have two copies of the sickle cell gene (HbS gene).
- They must have homozygous genotype = SS.
- People who have only one copy of the sickle cell gene they are only carriers of the sickle cell gene and referred to as sickle cell trait.
- They are heterozygous with the genotype = AS (Figs. 21.24, 21.25, 21.26, 21.27, and 21.28).

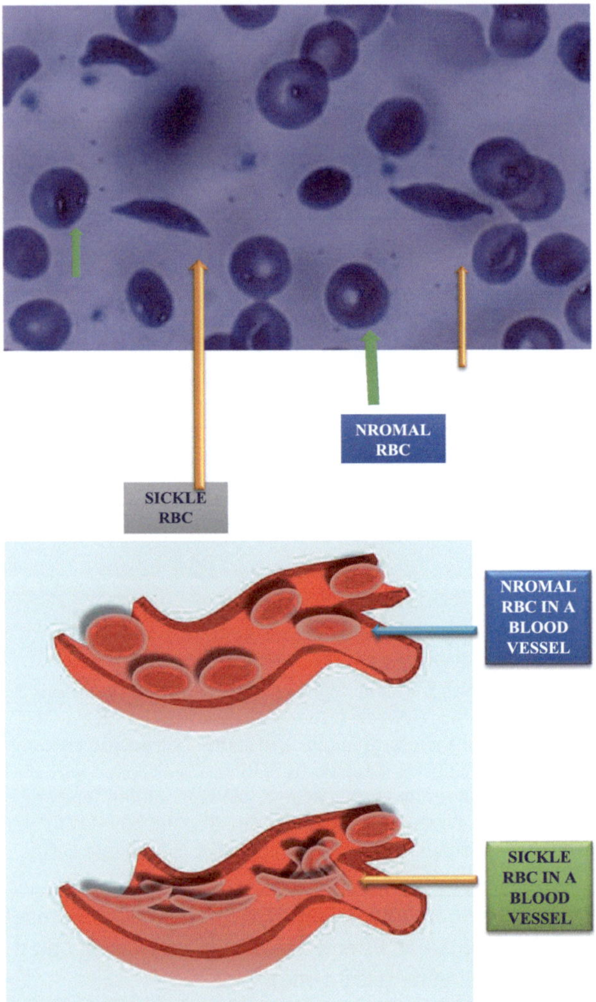

Figs. 21.20 and 21.21 A blood film showing abnormal sickled red blood cells and a normal looking red blood cells. Note the abnormal sickled shape red blood cells. Note also a diagrammatic representation showing normal red blood cells within a blood vessel moving easily and the abnormal sickled red blood cells sticking to each other and causing occlusion of small blood vessels and capallaries

PRESENTATIONS OF PATIENTS WITH SICKLE CELL TRAIT

- **They are usually asymptomatic**
- **They will be discovered during laboratory testing**
- **Patients with sickle cell trait can however develop complications under low oxygen tension such as high altitude and travelling in an unpressurized aircraft.**

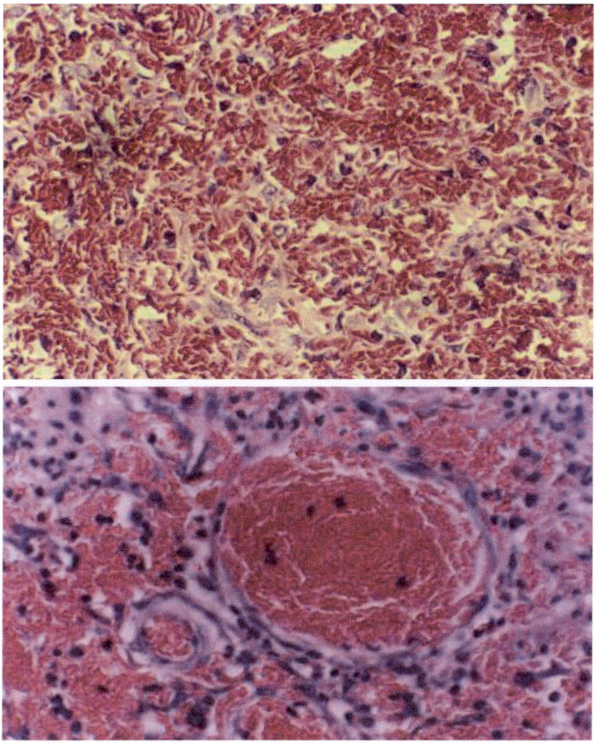

Figs. 21.22 and 21.23 A microscopic picture showing sickling of red blood cells within the small blood vessels. They are deformed and stick to each other causing blockage of small blood vessels. Note also the presence of hemosiderosis

- **These complications include:**
  - **Vaso-occlusive attacks**
  - **Renal medullary carcinoma: This is a very rare complication seen in patients with sickle cell trait.**
  - **Asymptomatic hematuria**
  - **Renal papillary necrosis**
  - **Hyposthenuria: This is inability to concentrate urine or excretion of urine of low specific gravity.**
  - **Splenic infarcts: These can develop in patients with sickle cell trait when they are at high altitude or travelling in unpressurized aircraft.**
  - **Urinary tract infection**
  - **Rarely, sudden death during physical exertion**

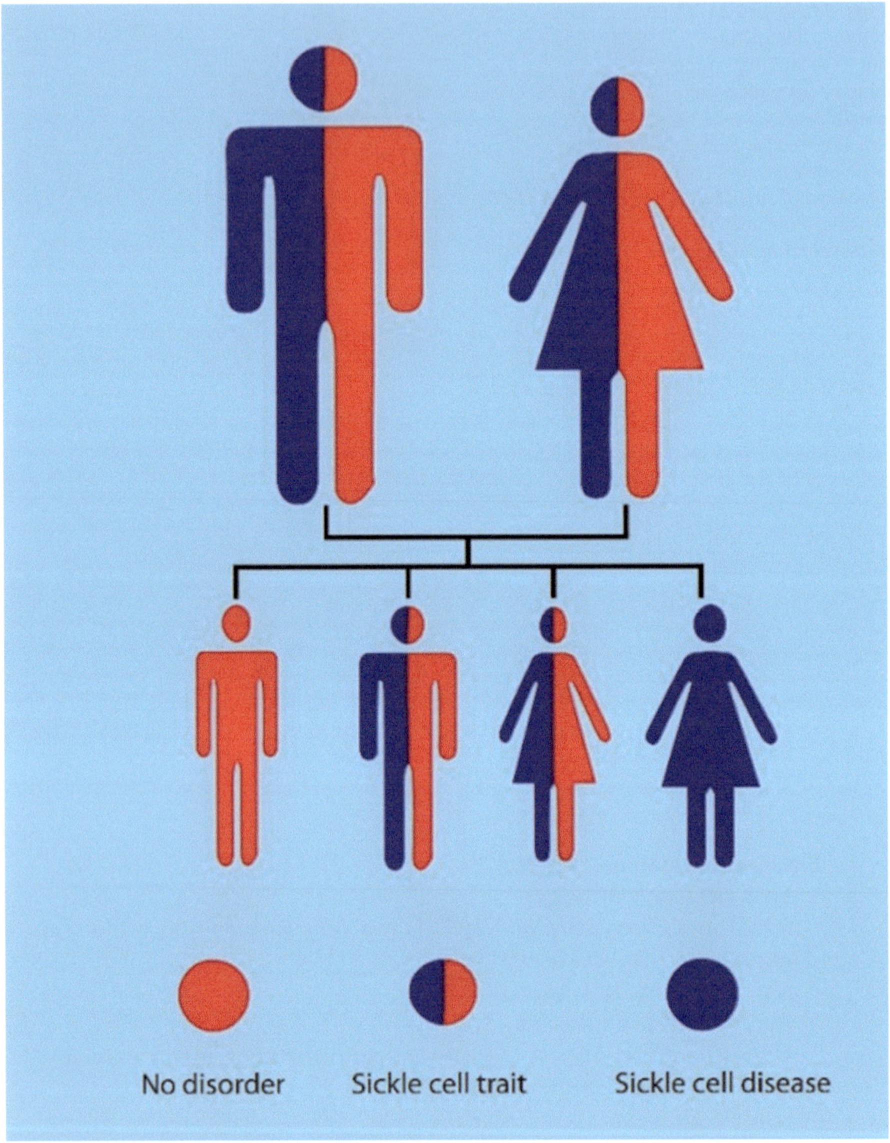

Fig. 21.24 Sickle cell anemia is inherited as an autosomal recessive and this is a diagrammatic representation of inheritance if both parents have sickle cell trait. 25% of children will be affected by sickle cell anemia (Homozygous), 25% of children will be healthy and 50% will have sickle cell trait (Heterozygous)

- Sickle cell anemia (SCA) is inherited as an autosomal recessive disorder.
  - Sickle cell anemia develops as a result of an abnormal gene and this abnormal gene is located on the short arm of chromosome 11.
  - This abnormal gene leads to the production of an abnormal form of hemoglobin.

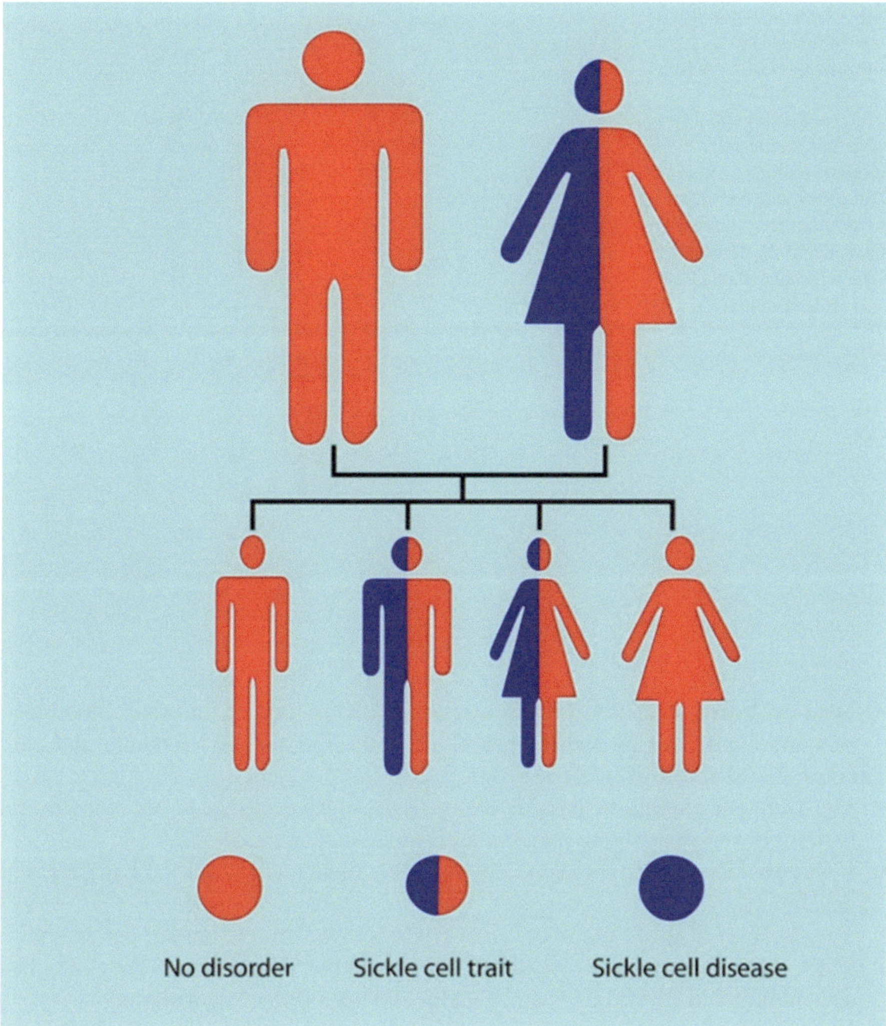

Fig. 21.25 Diagrammatic representation of inheritance if one parent has sickle cell trait. If only one parent has sickle cell trait then 50% of children will be healthy and the other 50% will have sickle cell trait (Heterozygous)

  - This abnormal hemoglobin is called hemoglobin S (HbS).
  - Hemoglobin S is characterized by a single amino acid change where Glutamic acid is replaced by Valine at the sixth position of the 146 amino acids of hemoglobin beta chain.
  - A single change of one amino acid in these patients can cause significant morbidity and in some of them mortality (Fig. 21.29).
- Patients affected with sickle cell anemia (SCA) usually present early during childhood with sickle cell anemia related complications.
- The most common clinical complication of sickle cell anemia vaso-occlusive crisis (Painful crisis).

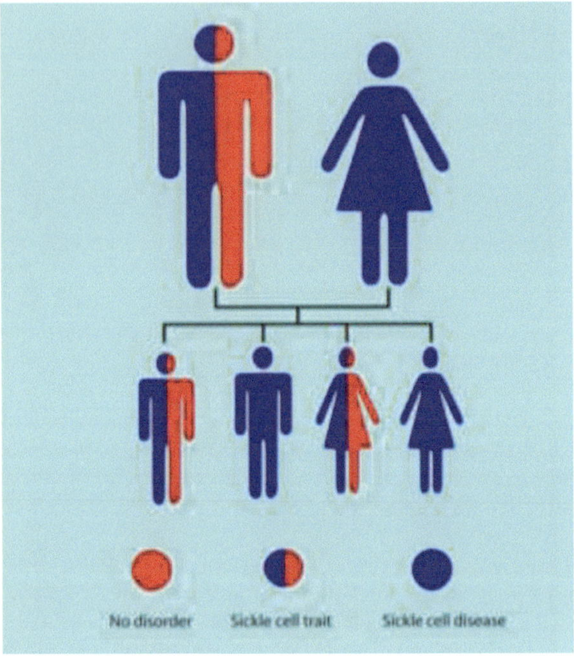

Fig. 21.26 Diagrammatic representation of inheritance if one parent has sickle cell trait (Heterozygous) and the other has sickle cell anemia (Homozygous). The result will be 50% will be affected by sickle cell anemia and the remaining 50% will have sickle cell trait (Heterozygous)

- Vaso-occlusive crisis develops as a result of obstruction of the small blood vessels and capillaries by sickled red blood cells. This causes ischemia and pain (Figs. 21.30 and 21.31).
- The frequency of these painful crisis (vasooclusive crisis) is not known and extremely variable among patients with sickle cell anemia.
- It was estimated that approximately half the patients with sickle cell anemia will develop vaso-occlusive crisis.
- The duration of these vaso-occlusive crisis is also variable among patients with sickle cell anemia and during different crisis in the same patient. They may last several hours to several days and usually end up suddenly as it started.
- In patients with sickle cell anemia, the spleen at birth is morphologically and functionally normal. Commonly, these patients lose the spleen function as a result of repeated and progressive injury to the spleen.
- This develops as a result of vaso-occlusion within the spleen followed by ischemia, leading to progressive fibrosis and atrophy, resulting in 'auto splenectomy'.
- In healthy adults the Hb is made up as follows:
  - 95% of the Hb is Hb A ($\alpha_2\beta_2$)
  - A small amount (<3.5%) of Hb A_2 ($\alpha_2\delta_2$)
  - A small amount of Hb F ($\alpha_2\gamma_2$) present.

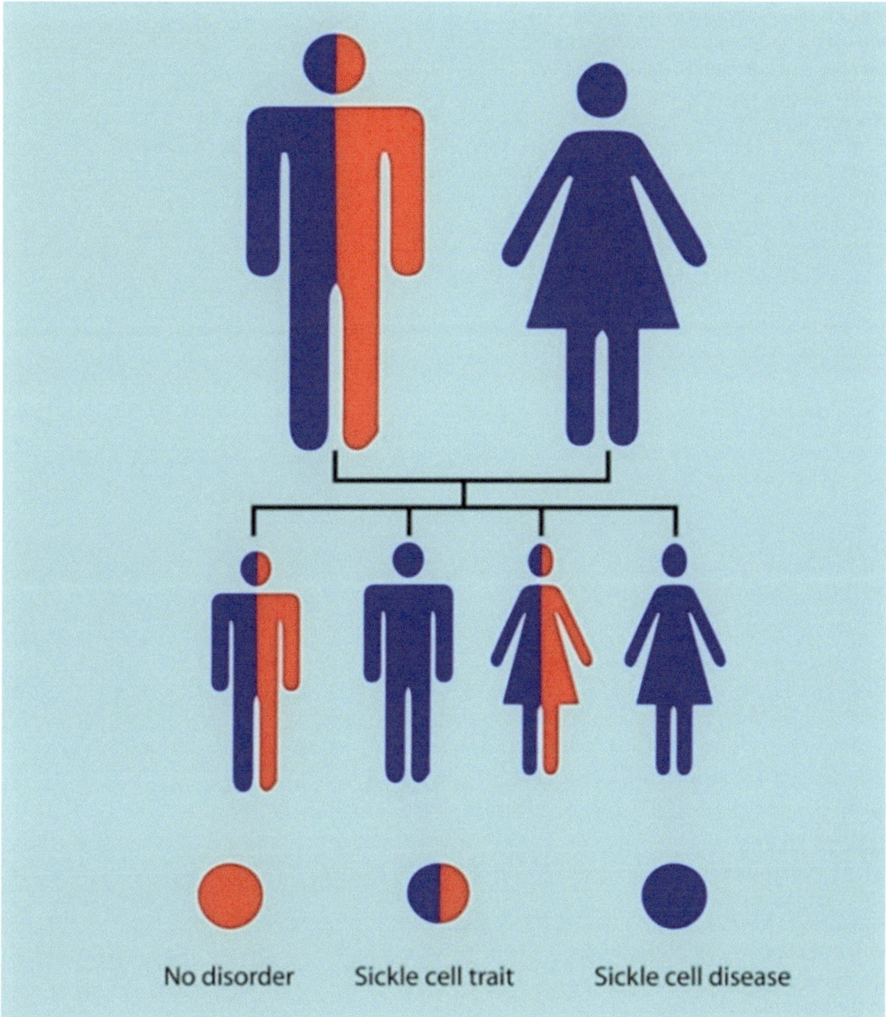

Fig. 21.27 Diagrammatic representation of inheritance if one parent has sickle cell anemia and the other is normal. In this case all children will have sickle cell trait (Heterozygous)

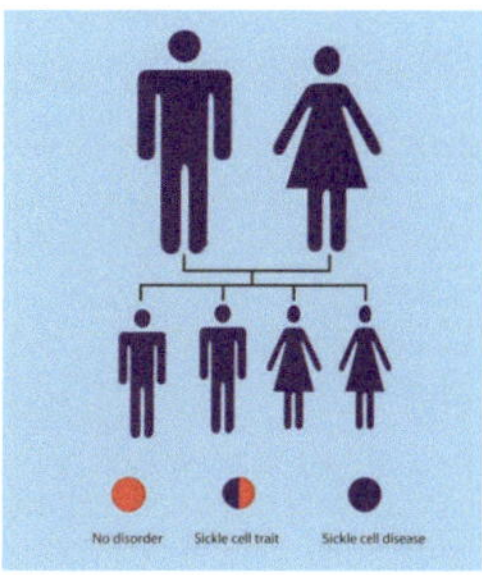

Fig. 21.28 Diagrammatic representation of inheritance if both parents have sickle cell anemia. If both parents have sickle cell anemia (Homozygous) then all children will have sickle cell anemia (Homozygous)

Fig. 21.29 A clinical photograph showing a child with severe sickle cell anemia. Note the significant morbidity in this patient as a result of sickle cell anemia

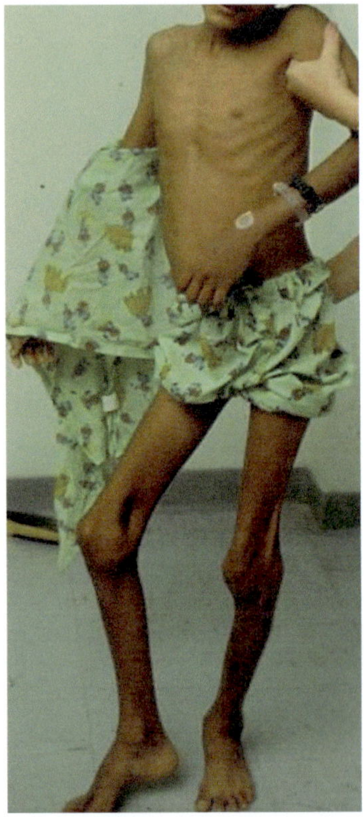

Figs. 21.30 and 21.31 Microscopic photographs showing normal red blood cells and sickle red blood cells. Note the shape of rec blood cells in the lower picture. They are sickled RBC

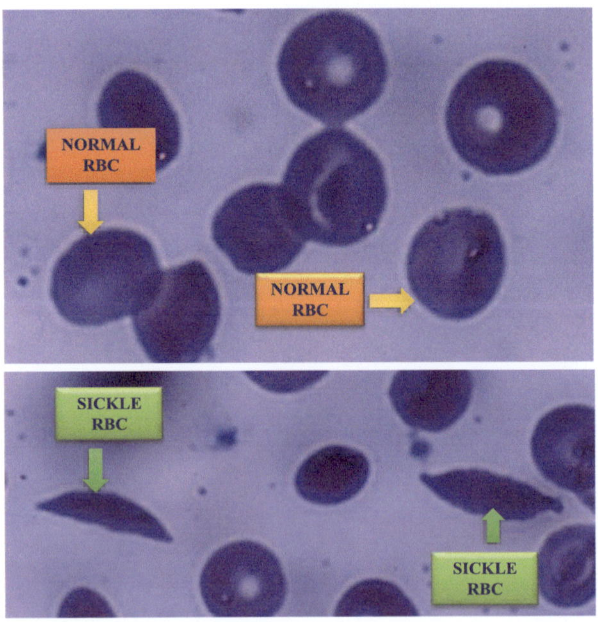

HEMOGLOBIN	ADULTS	FETAL
Hb A (2alpha and 2 beta)	96-98%	15-20%
Hb A2 (2 alpha and 2 gamma)	Undetectable	2.5%-3.5%
Hb F (2 alpha and 2 lambda)	< 1%	80-85%

NORMAL HEMOGLOBINS
- **Hemoglobin A: Two alpha chains and two beta chains (a_2b_2)**
- **Hemoglobin A2: Two alpha chains and two delta chains (a_2d_2)**
- **Hemoglobin F: Two alpha chains and two gamma chains (a_2g_2)**

NORMAL HEMOGLOBINS IN ADULTS
- **Hb A: 95–98%**
- **Hb A2: 2–3%**
- **Hb F: 0.8–2%**
- **Hb S: 0%**
- **Hb C: 0%**

HOMOZYGOUS HbS
HbA: ABSENT ULESS TANSFUSED
HbS: OVER 50%
HbF: UP TO 10%. HIGHER FOR INFANTS
HbA2: 2–4%

SICKLE CELL TRAIT
HbA: TYPICALLY, 50–65%
HbS: TYPICALLY, 35–45%
HbA2: NORMAL (1.5–4%)
HbF: HIGHER FOR INFANTS

The Sickle-Beta Thalassemia Syndromes

- **Sβ^+-thal: HbS and beta plus-thalassemia**
  - **HbS: always dominates, typically 60–80%**
  - **HbA: 15–30%**
  - **HbA2: elevated to 3–8%**
  - **HbF: 2–20%, higher for age in infants**
- **Sβ^o-thal: HbS and beta null thalassemia**
  - **HbS: 75–90%**
  - **HbA: absent unless transfused**
  - **HbA2: 5–8%**
  - **HbF: 5–20% in adults, higher for age in infants**

- In 1935, Diggs was the first to document the histological description of the spleen in patients with sickle cell anemia.
  - The 'Splenic cords become filled with sickled red blood cells. They appear as masses of elongated, pointed, curved and bizarre shaped red blood cells.
  - Gamma–Gandy bodies:

    These are a characteristic finding in the spleen in patients with sickle cell anemia.

    They appear as granuloma-like nodules.

    They develop as a result of periarteriolar hemorrhage followed by fibrosis and impregnation of iron pigments.
  - The spleen is often affected early in patients with sickle cell anemia with repeated attacks of vaso-occlusion with subsequent ischemia and multiple splenic infarcts. These will lead to progressive fibrosis and atrophy of the spleen and ultimately ending in a small fibrotic spleen (Auto splenectomy).
  - Auto splenectomy is common in patients with sickle cell anemia and it is thought to be complete by 3–5 years of age.
- Functional asplenia:
  - Pearson et al was the first to describe functional asplenia in 1969.
  - Functional asplenia is defined as a palpable enlarged spleen without demonstrable scintigraphic uptake.
- The spleen is affected very early in patients with sickle cell anemia and this is clearly seen with a sharp rise in PIT counts which start after 6 months of age.
- 86% of patients with sickle cell anemia will have loss of splenic function before 12 months of age. This is as assessed by ^{99}mTc Sulphur-colloid liver spleen scintigraphy.
- The following findings may coexist in a small percentage of children with sickle cell anemia:
  - Splenomegaly
  - Hypersplenism
  - Functional hyposplenism (Figs. 21.32, 21.33, and 21.34).

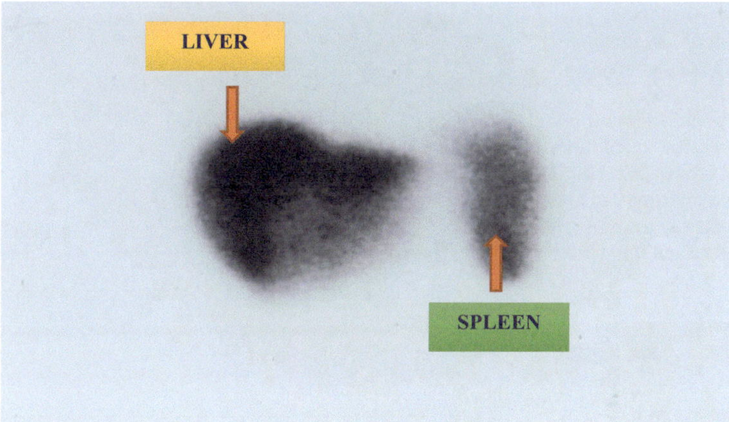

Fig. 21.32 A technetium sulfur colloid scan showing a normally functioning spleen in a patient with sickle cell anemia

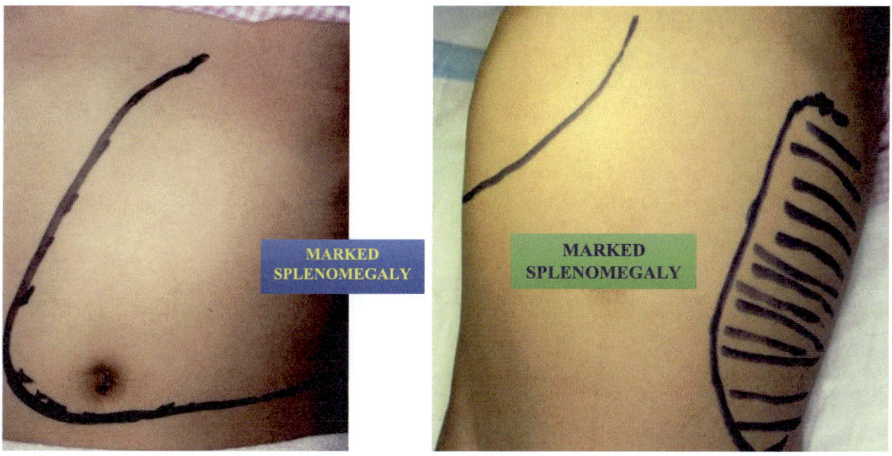

Figs. 21.33 and 21.34 Clinical photographs of two children with sickle cell anemia. Note the persistent splenomegaly. The spleen may reach a large size in some of these patients

- The manifestations of sickle cell anemia start early usually after the first 6 months of life.
- Infants below 6 months of age are relatively protected by the elevated levels of HbF in their blood. Once the level of HbF level decreases, the manifestations of sickle cell anemia become evident.

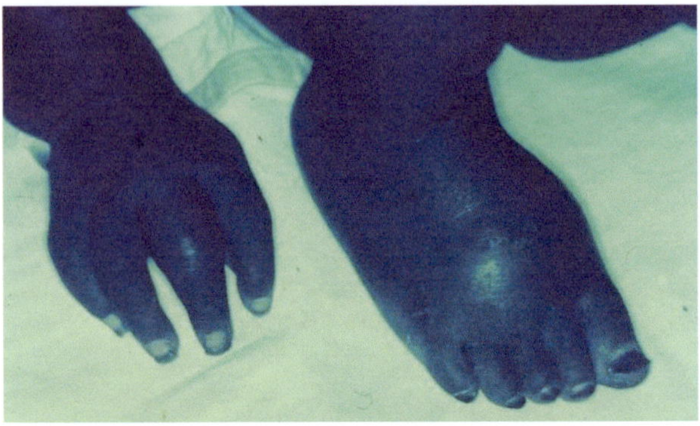

Fig. 21.35 A clinical photograph of an infant with sickle cell anemia showing dactylitis affecting both hands and feet. This is also called the hand-foot syndrome. The hand-foot syndrome can affect the hands, feet or both together and it is very painful

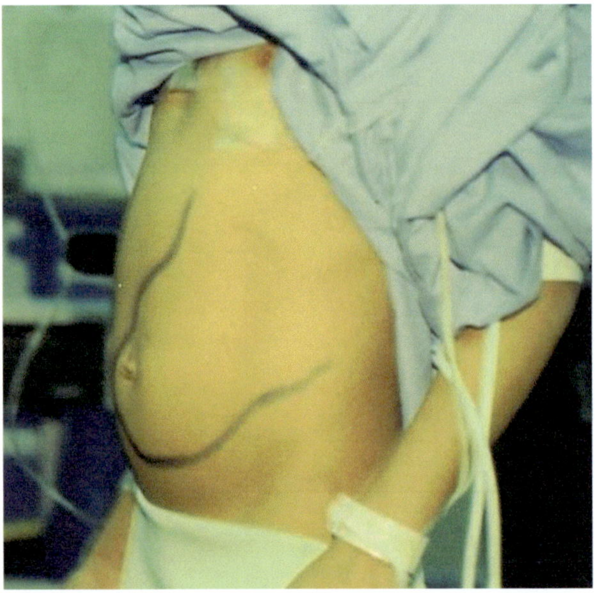

Fig. 21.36 A clinical photograph of a child with sickle cell anemia showing persistent splenomegaly. Note the markedly enlarged spleen reaching to the lower part of the abdomen

- The manifestations of sickle cell anemia during infancy include:
  - Repeated attacks of infections
  - Failure to thrive
  - Dactylitis:

    This is one of the early manifestations of sickle cell anemia during infancy.

    It is called dactylitis

    It is also called the hand-foot syndrome.

    It is characterized by sudden onset of pain and swelling of the dorsum of the hands and or feet (Fig. 21.35).
  - Anemia
  - Splenomegaly (Fig. 21.36)

- Sickle cell anemia is known to affect almost any organ in the body and this will manifest in several complications.
- The clinical severity of sickle cell anemia is variable. Patients with sickle cell anemia may develop some or all of the known complications.

MANIFESTATIONS OF SICKLE CELL ANEMIA DURING INFANCY

- **ANEMIA**
- **REPEATED INFECTIONS**
- **FAILURE TO THRIVE**
- **DACTYLITIS (HAND-FOOT SYNDROME)**
- **SPLENOMEGALY**

- The spleen:
  - The spleen is an important and valuable organ of the body and functionally it is considered part of the lymphatic system.
  - The spleen is the largest lymphoid tissue of the human body. It accounts for 25% of total body lymphocytes.
  - Splenectomy should be avoided unless it becomes pathological.
  - In sickle cell anemia, the spleen is one of the commonly and early organs to be affected.
  - Commonly the spleen enlarges early and young children with sickle cell anemia usually have splenomegaly but then and as result of repeated attacks of vasoocclusive crisis the spleen undergoes progressive atrophy leading to auto splenectomy.
  - The end result is a small shrunken and nonfunctional spleen.
  - The spleen may disappear completely or become like a small fibrotic nodule.
  - This small fibrotic spleen, ssometimes becomes calcified and takes different shapes.
  - This however is not the case always and sometimes splenomegaly persists in some of these patients to an older age group and sometimes to adults (Figs. 21.37 and 21.38).
  - The reason for persistent splenomegaly is not exactly known.
  - An elevated level of Hb F which is found in some of these patients is known to contribute to persistence of splenomegaly in these patients.
  - Normally, the spleen is enlarged during the first decade of life but done and as a result of repeated splenic infarction undergoes progressive atrophy leading to auto splenectomy (Fig. 21.39 and 21.40).
  - Persistence of splenomegaly in some patients with sickle cell anemia is known to predispose them to serious complications necessitating splenectomy.
  - These splenic complications include:

    Persistent splenomegaly with a non-functioning spleen
    Minor acute splenic sequestration crisis
    Major acute splenic sequestration crisis
    Splenic abscess
    Hypersplenism
    Massive splenic infarction

SPLENIC COMPLICATIONS OF SICKLE CELL ANEMIA

- **PERSISTENT SPLENOMEALY**
- **ACUTE SPLENIC SEQUESTRATION CRISIS (MINOR AND MAJOR ATTACKS)**
- **SPLENIC ABSCESS**
- **HYPERSPLENISM**
- **MASSIVE SPLENIC INFARCTION**

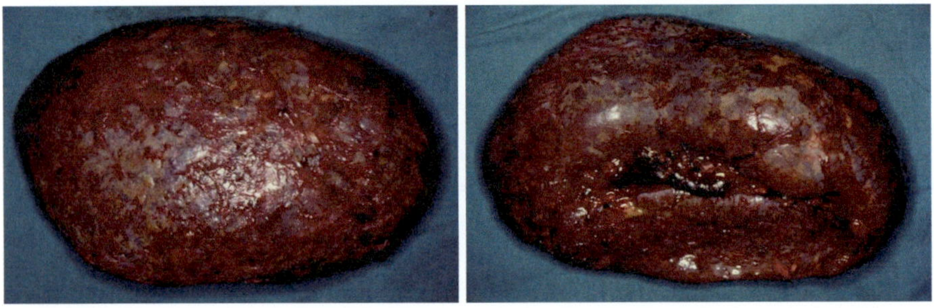

Figs. 21.37 and 21.38 Clinical photographs showing an enlarged spleen from a child with sickle cell anemia who underwent splenectomy because of repeated splenic sequestration crisis. Note also the pathological changes and multiple small infarcts of the spleen

- Sickle cell anemia is a multisystem disease and can affect any part of the body. The clinical features and complications are also variable and these can cause diagnostic and therapeutic dilemmas to the treating physicians.
- In patients with sickle cell anemia, the spleen is one of the most common and early organs to be affected and the splenic complications are known to cause morbidity and mortality (Figs. 21.41, 21.42, 21.43, 21.44, 21.45, and 21.46).
- Patients with sickle cell anemia are known to develop ssplenic complications which are known to be associated with an increased morbidity and in some it may lead to mortality.
- Splenectomy and in order to avoid these complications becomes an essential part of their management. Splenectomy is not curative for sickle cell anemia and should be avoided as much as possible but sometimes it is valuable, and in certain patients may be lifesaving.

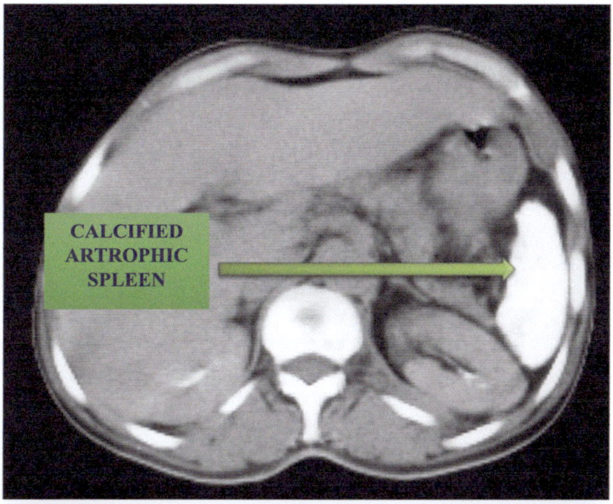

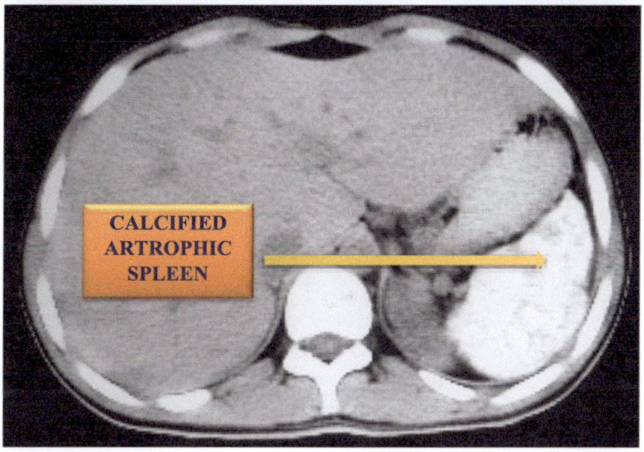

Figs. 21.39 and 21.40 Abdominal CT-scan showing of two children with sickle cell anemia showing small atrophied and calcified spleens. Note the different shapes of the atrophied and calcified spleens. A large number of patients with sickle cell anemia will end up with autosplenectomy

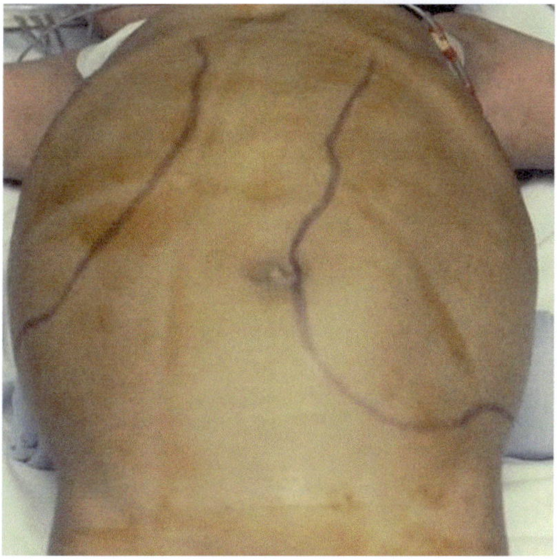

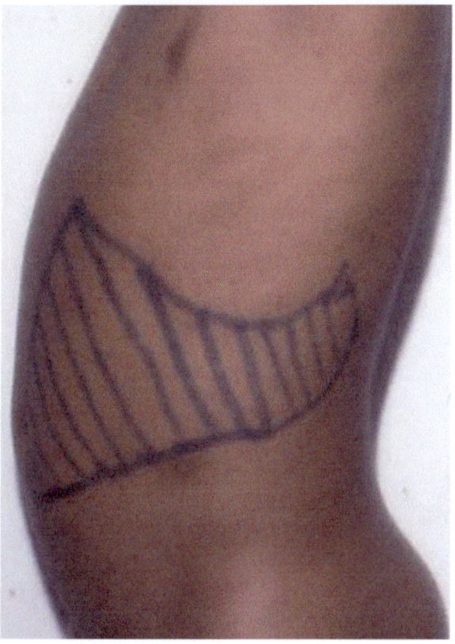

Figs. 21.41 and 21.42 Clinical photographs of two children with sickle cell anemia. Note the marked and persistent splenomegaly

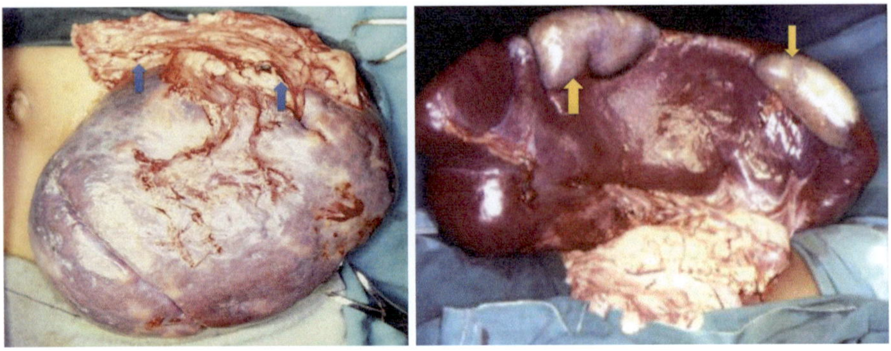

Figs. 21.43 and 21.44 Intraoperative phorographs of two children with sickle cell anemia showing markedly enlarged spleens. Note the splenic infarcts and also the omentum adherent to the splenic infarcts

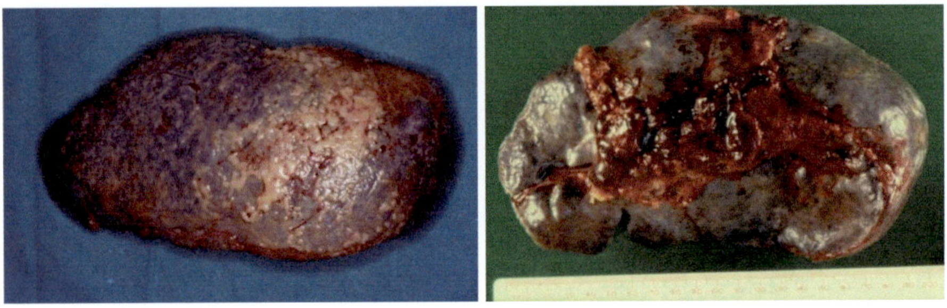

Figs. 21.45 and 21.46 Two clinical photographs showing removed spleens in two patients with sickle cell anemia. Note the enlarged spleens and also the multiple infarcts in both of them

21.2 Functions of the Spleen

- The spleen is made up of two main parts:
  - The red pulp
  - The white pulp
- The Red pulp:
  - The red pulp is made up of thin-walled venous sinusoids.
  - These sinusoids are lined by special type of endothelial cells with a discontinuous wall.
  - This discontinuity in sinusoids wall allows the passage of red blood cells freely between the sinusoids and splenic cords.
  - The red pulp sinuses are separated by splenic cords.
  - These are called cords of Billroth and they contain macrophages.

  - The red pulp has several important functions which include:
    Filtration: The red pulp filters red blood cells and ingests old RBC, damaged RBCs such as sickle cells or spherocytes, and also antibody coated RBCs.
    The red pulp also removes Heinz bodies and other RBC inclusions such as Howell-Jolly bodies.
- The White pulp:
  - The White pulp is made up of sheaths of lymphoid cells.
  - These sheaths of lymphoid cells are located around arteries.
  - They are called periarteriolar lymphatic sheath.
  - These lymphatic sheaths are made up of T cells and lymphoid follicles (B cells).
  - Their main function is to trap antigens from the blood stream for processing.
  - The white pulp is important immunologically for the body and has the following functions:
    It is responsible for the active immune response through both humoral and cell-mediated pathways.
- There are other important functions of the spleen which include:
  - Production of the followings:
    Opsonins
    Properdin
    Tuftsin
  - Production of natural antibodies:
    These are produced by IgM memory B cells in the marginal zone of the spleen.
    They are important and required for phagocytosis.
    These are immunoglobulins that facilitate phagocytosis either directly or by complement deposition on the capsule.
  - Production of red blood cells:
    Normally, the bone marrow is the primary site of hematopoiesis.
    The spleen has important hematopoietic functions up until the fifth month of gestation.
    After birth, the erythropoietic functions of the spleen cease, except in certain conditions and for some hematologic disorders.
    The spleen is a major lymphoid organ in the body and part of the reticulo-endothelial system.
    The spleen continues to produce lymphocytes and, as such, remains a hematopoietic organ.
  - Storage:
    The spleen act as a storage of red blood cells, lymphocytes and other formed elements.
    The spleen stores red blood cells and these can be released into the blood stream when needed.

It has been estimated that in humans, up to a cup (236.5 mL) of red blood cells can be stored in the spleen and released in cases of hypovolemic shock.

The spleen can also store platelets in case of an emergency.

The spleen also stores lymphocytes and it was estimated that up to a quarter of lymphocytes can be stored in the spleen at any one time.

– The spleen function:

  The spleen function can be measured by the number of Howell-jolly bodies and less commonly Heinz bodies in blood smears.

  Their increase in number is an indication of decreased or absence of splenic function.

  An isotope scan (technetium sulfur colloid scan) is more important to evaluate the function of the liver and spleen (Figs. 21.47 and 21.48).

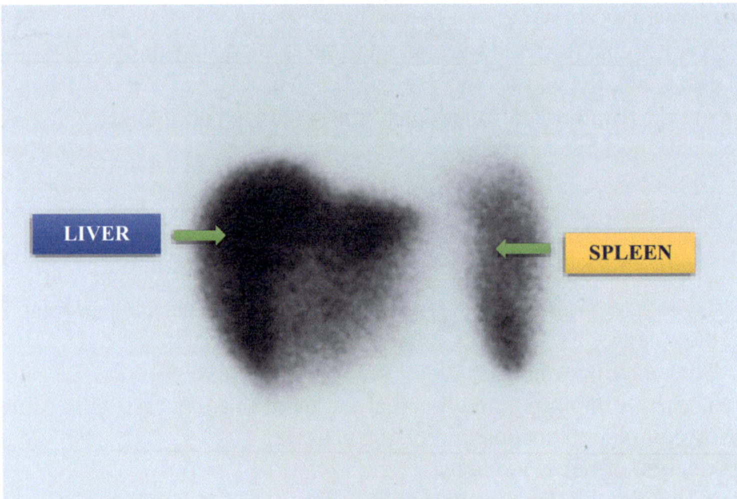

Fig. 21.47 An isotope scan of a child with sickle cell anemia showing a normal functioning spleen

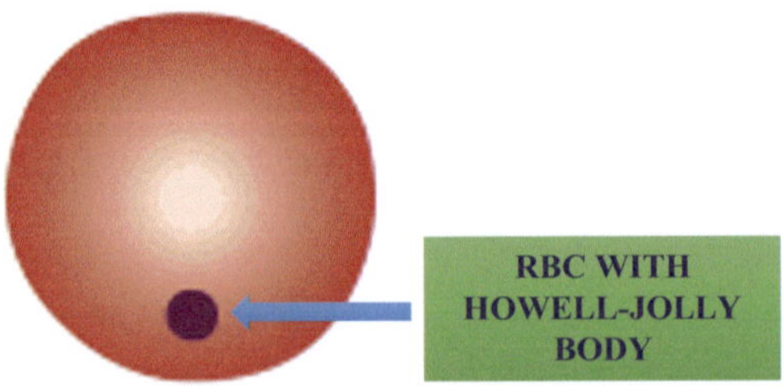

Fig. 21.48 RBC with Howell-Jolly body. The increased number of Howell-Jolly bodies and less commonly Heinz bodies in the blood smears is an indication of decreased or absence of splenic function

21.3 Splenic Sequestration Crisis

- Acute splenic sequestration crisis (ASSC) results from the rapid sequestration of red blood cells in the spleen.
- In patients with sickle cell anemia, acute splenic sequestration crisis develops as a result of obstruction of the venous outflow from the spleen by sickled red blood cells. In acute splenic sequestration crisis, there is good inflow of blood into the spleen via the splenic artery but the blood outflow from the spleen via the splenic vein is obstructed by sickled RBCs.
- This will lead to pooling of blood into the spleen sinusoids leading to splenomegaly, anemia and sometimes hypovolemic shock.
- Acute splenic sequestration crisis is a serious complication of sickle cell anemia.
- In patients with sickle cell anemia, acute splenic sequestration crisis is considered the second leading cause of death after infection in the first decade of life.
- Acute splenic sequestration crisis is usually seen in infants and young children commonly between 5 months and 2 years of age.
- Acute splenic sequestration crisis is commonly seen during the first 5 years of life in children with sickle cell anemia.

CAUSES OF ANEMIA IN PATIENTS WITH SICKLE CELL ANEMIA

- **HEMOLYTIC ANEMIA (ACUTE AND CHRONIC)**
- **APLASTIC CRISIS**
- **ACUTE SPLENIC SEQUESTRATION CRISIS**
- **HYPERHEMOLYSIS**
- **DELAYED TRANSFUSION REACTIONS FROM ALLOANTIBODIES**
- **HEPATIC SEQUESTRATION CRISIS**
- **SEVERE INFECTION AND MULTIORGAN DAMAGE**

- Splenic sequestration crisis although commonly seen during the first five years of age, it can also occur at an older age group.
- This is seen in those with persistent splenomegaly.
  - The exact reason for this is not known.
  - One contributing factor is persistence of splenomegaly into an older age group.
  - Persistent splenomegaly is partly attributed to the high hemoglobin F level in these patients which has a protective effect.
- There are two types of acute splenic sequestration crisis.
- This classification depends on the severity and degree of anemia
- Acute splenic sequestration crisis is divided into:
  - Major splenic sequestration crisis
  - Minor splenic sequestration crisis

- In minor splenic sequestration crisis, there is a moderate increase in the size of the spleen with a sudden drop of hemoglobin of 2–3 g/dL.
- In major splenic sequestration crisis, there is a significant sudden increase in spleen size, a greater drop of hemoglobin sometimes decreasing to reach as low as 2–3 g/dL and hypovolemia.

- Acute splenic sequestration crisis is characterized by:
  - The sudden onset of anemia which can be life-threatening in those with major splenic sequestration crisis where the hemoglobin level can drop to a very low level as low as 2 g/dL.
  - Sudden appearance of splenomegaly or a sudden increase in the size of an already enlarged spleen.
  - Supportive evidence of increased activity of the bone marrow as shown by an increase in the reticulocyte count.
  - Regression and decrease in the size of the spleen after giving blood transfusion.
  - Spleen regression after blood transfusion differentiates acute splenic sequestration crisis from hypersplenism where the spleen size is chronically enlarged and does not regress after blood transfusion.
- Acute splenic sequestration crisis and depending on the severity is divided into two types:
  - Minor attack
  - Major attack
- Minor acute splenic sequestration crisis is characterized by:
  - A moderate and sudden increase in the size of the spleen
  - A sudden drop of hemoglobin of 2–3 g/dL.
- Major acute splenic sequestration crisis is characterized by:
  - A significant sudden increase is the size of the spleen
  - A greater drop of hemoglobin sometimes decreasing to as low as 2–3 g/dL
  - Hypovolemia

TYPES OF ACUTE SPLENIC SEQUESTRATION CRISIS

- **ASSC is divided depending on the severity of the attack into:**
  - **Minor attack**
  - **Major attack**
- **Minor attacks are characterized by:**
  - **A moderate increase in the size of the spleen**
  - **A sudden drop of hemoglobin of 2–3 g/dL**
- **Major attacks are characterized by:**
  - **A significant sudden increase is spleen size**
  - **A greater drop of hemoglobin sometimes decreasing to as low as 2–3 g/dL**
  - **Hypovolemia**

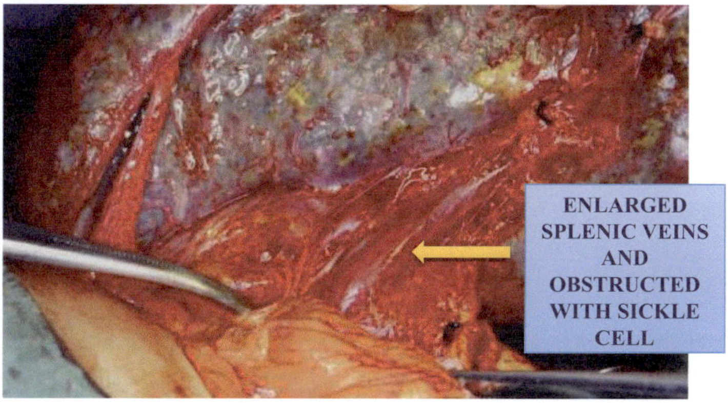

Fig. 21.49 A clinical intraoperative photograph of a child with sickle cell anemia showing a very large spleen following an attack of acute splenic sequestration crisis. Note the multiple small splenic infarcts. Note also the enlarged congested splenic vein. This is caused by obstruction of the splenic vein by sickled red blood cells

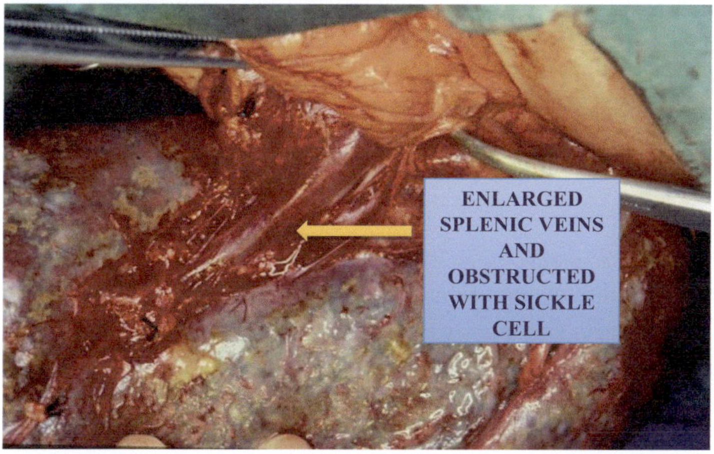

Fig. 21.50 A clinical intraoperative photograph of a child with sickle cell anemia who underwent splenectomy for acute splenic sequestration crisis. Note the enlarged spleen with multiple small infarcts. Note also the enlarged congested splenic vein which is obstructed by sickled red blood cells

- Acute splenic sequestration crisis is considered a medical emergency.
- Acute splenic sequestration crisis is considered the second leading cause of death following infection in children younger than 5 years of age (Figs. 21.49, 21.50, 21.51, and 21.52).
- In order to diagnose acute splenic sequestration crisis early and avoid the associated morbidity and mortality, it is very important to teach the mothers how to palpate the spleen and recognize a sudden enlargement. They also should report to the hospital if there is sudden enlargement of the spleen.

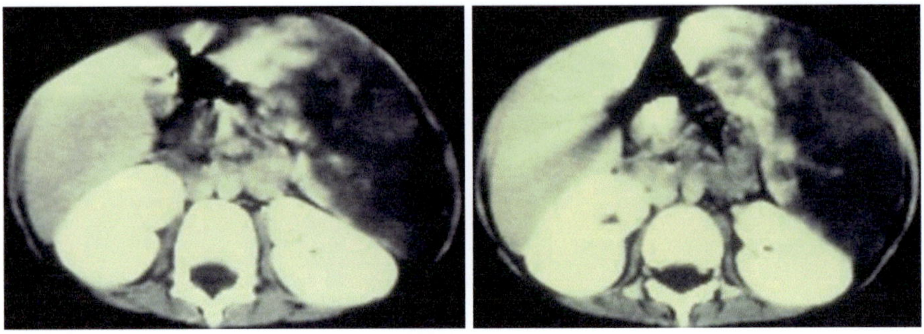

Figs. 21.51 and 21.52 CT-scan of a child with acute splenic sequestration crisis. Note the enlarged spleen which is filled with sequested blood as a result of obstruction of splenic vein by sickled red blood cells

- The cause of acute splenic sequestration crisis is not exactly known.
- There is an association between acute splenic sequestration crisis with upper respiratory tract infection suggesting a precipitating viral cause.
- Infection with human Para virus B19 causing aplastic crisis is another precipitating factor of major acute splenic sequestration crisis.
- Treatment of acute splenic sequestration crisis depends on:
  - Early recognition and proper diagnosis
  - Close and careful monitoring
  - Appropriate blood transfusion
- Acute splenic sequestration crisis tends to be recurrent and to overcome this, some physicians advocate long-term blood transfusion.
- Others advocate splenectomy if the child develops one major attack or two minor attacks of splenic sequestration crises.

CHARACTERISTICS OF ACUTE SPLENIC SEQUESTRATION CRISIS

- **SUDDEN ONSET OF LIFE-THREATENING ANEMIA**
- **SUDDEN ENLARGMENT OF THE SPLEEN**
- **HIGH RETICULOCYTE COUNT**
- **REGRESSION OF THE SPLEEN SIZE AFTER BLOOD TRANSFUSIONS**

- Diagnosis of acute splenic sequestration crisis:
  - The diagnosis of ASSC is based on the followings:

    Sudden onset of acute anemia

    Sudden enlargement of the spleen

    Regression of spleen size after blood transfusion

    In these patients, it is important to keep a record of the spleen size in each visit to the clinic or hospital. This will help to recognize a sudden enlargement of the spleen when the present to the emergency room. It is also important to teach the mothers how to palpate the spleen and how to recognize sudden enlargement.

  - CBC will show anemia.
  - The severity of anemia in these patients depends on the type of acute splenic sequestration crisis.
  - CBC will also show reticulocytosis as an evidence of active bone marrow.
  - The WBC and platelets may also decrease.
- Treatment of acute splenic sequestration crisis:
- The treatment of acute splenic sequestration crisis is initially supportive and includes the followings:
  - Clinical evaluation and close monitoring
  - Patients with major attacks should be admitted to ICU for close observations
  - Adequate hydration with intravenous fluids
  - Appropriate blood transfusion with packed red blood cells

    It is important to avoid over transfusing these patients as this will lead to an increase in blood viscosity with its associated risks.

    It is also important to remember that in acute splenic sequestration crisis blood is sequestered and stored in the spleen and once blood transfusion is given some of this stored blood will be released into the circulation. This can also lead to an increase in blood viscosity.

    The aim should be to give blood transfusion to increase the hemoglobin level to around 10–11 g/dL.
  - It is important to teach the parents how to palpate the spleen of their affected children and recognize sudden enlargement. They should also report to the hospital early if they recognize sudden enlargement of the spleen (Figs. 21.53, 21.54, and 21.55).

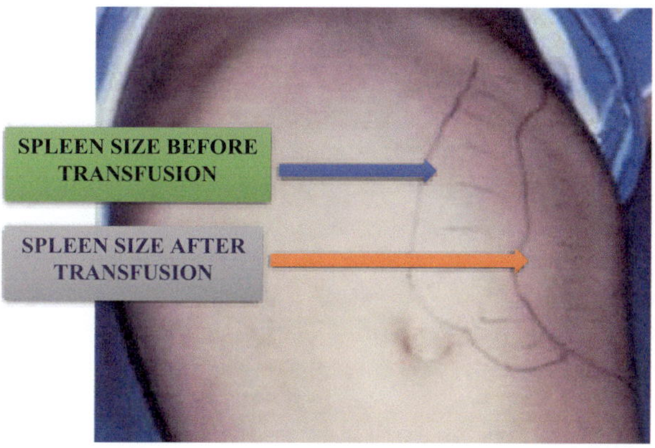

Fig. 21.53 A clinical photograph of a child with sickle cell anemia who developed acute splenic sequestration crisis. Note the original spleen size following the attack of acute splenic sequestration crisis and the regression of the size of the spleen following blood transfusion. This is an important distinguishing criterion between acute splenic sequestration crisis and hypersplenism. In hypersplenism the spleen size does not regress after blood transfusion

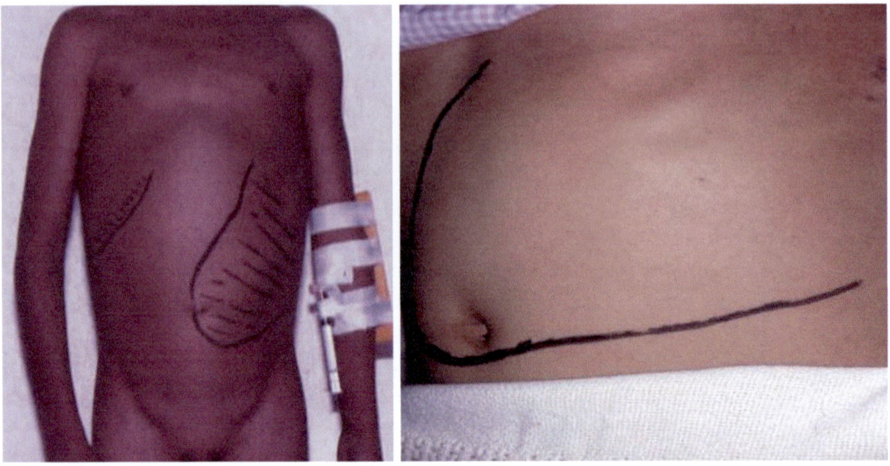

Figs. 21.54 and 21.55 Clinical photographs of two children with sickle cell anemia showing persistent splenomegaly. Note the markedly enlarged spleens

- These patients should be immunized with pneumococcal, meningococcal and H. influenza vaccines.
- Indication for splenectomy in major acute splenic sequestration crisis:
  The indication for splenectomy in major attacks is well established.
  Splenectomy is advocated if the child develops one major attack.
  Major acute splenic sequestration crisis is a serious complication of sickle cell anemia and it is known to be associated with a high mortality.
  Some of these patients after a major acute splenic sequestration crisis may be brought to the hospital comatosed with subsequent neurological deficit as a result of hypovolemia, shock and hypoxic encephalopathy.
  The possibility of a hidden mortality in these patients as a result of deaths prior to arrival to hospital cannot be excluded.

- The role of splenectomy in minor attacks is still controversial.
  In patients with sickle cell anemia, minor splenic sequestration crisis tends to be recurrent.
  Some physicians advocate chronic blood transfusion as a therapy to prevent recurrent attacks.
  Chronic blood transfusion program is effective on the short term but it is not without risks including:
  Allosensitization
  Hepatitis
  Iron overload
  Other blood born infections
  The duration of chronic blood transfusion is also not known.
  It was also found that a significant number of sickle cell anemia patients on chronic blood transfusion therapy developed acute splenic sequestration crisis when attempts were made to stop blood transfusion or shortly after completing the blood transfusion program.

There are other factors that make makes chronic blood transfusion unsuitable form of therapy for patients with sickle cell anemia.
These factors include:
Blood is not readily available
Blood is not without risks and complications
The poor compliance of parents
It is advocated that splenectomy should be performed if the child develops two minor attacks of acute splenic sequestration crisis.
Splenectomy with avoid a recurrence risk of 40%–50% and a mortality rate as high as 20% in these patients.

21.4 Post-Splenectomy Complications

- An increase in white blood cells (Leucocytosis)
- An increase in platelets (Thrombocytosis)
  - It is well known that post-splenectomy the platelet count will increase and sometimes the platelet count may rise to abnormally high levels (thrombocytosis).
  - Thrombocytosis is associated with an increased risk of potentially fatal clot formation or portal vein thrombosis.
- An increase in the number of Howell-Jolly bodies in the blood smears.
- Less commonly an increase in the number of Heinz bodies in the blood smears.
- Heinz bodies are usually found in patients suffering from G6PD (Glucose-6-Phosphate Dehydrogenase) deficiency and chronic liver disease.
- Post splenectomy there is a diminished responsiveness to some vaccines. It is advisable to immunize children at least two weeks prior to splenectomy.
- Post splenectomy sepsis:
  - Splenectomy is known to be associated with an increased susceptibility to infections.
  - Post splenectomy sepsis is commonly by bacteria and protozoa.
  - There is an increased risk of post splenectomy sepsis from polysaccharide encapsulated bacteria such as pneumococci, meningococci and H influenza.

21.5 Immunizations and Splenectomy (Table 21.1)

- Splenectomy is known to be associated with an increased risk of sepsis.
- Post splenectomy sepsis is commonly caused by encapsulated organisms namely:
  - *S. pneumoniae*
  - *Haemophilus influenzae*
  - *Meningococci*
- These patients especially children should receive immunization at least two weeks prior to splenectomy for elective cases and vaccinations should be given the day prior to discharge for those who had emergency splenectomy.

Table 21.1 Recommended vaccinations for patients undergoing splenectomy

	Initial vaccines	Vaccines to be given at 8 weeks	Long-term vaccines
Pneumococcal 13- valent conjugate (PCV13 – Prevnar 13) 0.5 mL IM	YES		
Pneumococcal polysaccharide (PPSV23 – Pneumovax 23) 0.5 mL IM		YES	Repeat every 5 years to equal 2 doses below the age of 65 plus one dose after age 65
Meningococcal serogroups A, C, Y, and W (MedQuadfi) 0.5 mL IM	YES	YES	Repeat every 5- years for life.
Meningococcal serogroup B (Bexsero) 0.5 mL IM	YES	YES	Repeat at 1-year and every 3-years thereafter for life
Haemophilus influenza type B (Hib - ActHIB) 0.5 mL IM	YES		

- These vaccines include:
  - Pneumococcal vaccine
  - Meningococcal vaccines
  - Haemophilus influenza type b vaccine
- These vaccines should be given at least two weeks prior to splenectomy and followed subsequently by poster doses. Giving these vaccines prior to splenectomy to insure a better response to these vaccines in the presence of the spleen.
- The recommended vaccines for these patients should include:
  - Pneumococcal vaccine, PCV13 (Prevnar-13) – 0.5 mL IM
  - Haemophilus influenzae vaccine (HiB) – 0.5 mL IM
  - Meningococcal vaccine
    Serogroups MenACWY (Menveo) – 0.5 mL IM
    Serogroup B MenB-4C (Bexsero) – 0.5 mL IM
- If splenectomy is done for an emergency cause it is recommended to administer these vaccines postoperatively prior to discharge from the hospital.

INITIAL VACCINES TO BE ADMINISTERED

- **13 VALENT PNUMOCCOCAL CONJUGATE VACCINE (13v PCV)**
- **QUADRIVALENT (ACWY) MENINGOCOCCAL CONJUGATE VACCINE (4vMenCV)**
- **MENINGOCOCCAL B RECOMBINANT VACCINE (4CMenB)**
- **HAEMOPHILUS INFLUENA TYPE B CONJUGATE VACCINE (Hib)**

FOLLOW-UP VACCINES TO BE ADMINISTERED

- **23 VALENT PNEUMOCOCCAL POLYSACCHARIDE VACCINE (23vPPV)**
- **QUADRIVALENT (ACWY) MENINGOCOCCAL CONJUGATE VACCINE (4vMenCV)**
- **MENINGOCOCCAL B RECOMBINANT VACCINE (4CMenB)**

- These patients should also receive follow-up vaccines to boost their immunity. These vaccines should be given at least 8 weeks after the initial dose of vaccination.
- It is recommended that these patients should receive annual influenza vaccination as the prevention of influenza may also reduce the risk of secondary bacterial infection.
- These patients should receive a second dose of Pneumovax 23 (23vPPV) after 5 years from the initial vaccination dose.
- It is also recommended that these patients should receive a third dose of Pneumovax 23 (23vPPV) at a minimum of 5 years after the second dose.

RECOMMENDATIONS FOR ANTIBIOTICS PROPHYLAXIS AGAINST INFECTION IN ASPLENIC PATIENTS

- Quadrivalent (ACWY) meningococcal conjugate vaccine (4vMenCV) should be repeated at 5 yearly intervals.
- Children who undergo splenectomy should also receive prophylactic antibiotics.
- The duration of prophylactic antibiotics is still controversial.
- It is recommended to give prophylactic antibiotics for a minimum of 2 years post splenectomy depending on the patient age at the time of splenectomy (Table 21.2).
- This is because overwhelming post-splenectomy sepsis is commonly seen within the first two years post splenectomy.
- The recommended prophylactic antibiotic is penicillin either orally or intramuscularly. Oral penicillin is preferable.

Table 21.2 Antibiotics recommendations for asplenic patients

	Under 2 years	2–5 years	More than 5 years
Antibiotic prophylaxis	Amoxycillin (20 mg/kg daily) essential until 2 years age then Amoxycillin (250 mg daily) recommended until 5 years of age but optional thereafter	Amoxycillin (250 mg daily) recommended until 5 years of age but optional thereafter	Amoxycillin (250 mg daily) optional

21.6 Hypersplenism

- The exact definition of hypersplenism is not known.
- Hypersplenism is defined as follows:
  - The presence of splenomegaly with:

    Anemia: This is defined hen the blood transfusion requirements exceed 250 mL/kg of packed red blood cells per year.

    Thrombocytopenia: This is defined when the platelets count is less than $100{,}000/mm^3$.

    Neutropenia: This is defined when the white blood cell count is less than $4000/mm^3$.
  - The presence of these can be either singly or in combination
- Splenomegaly:
  - The presence of splenomegaly in hypersplenism is persistent
  - The size of splenomegaly does not change after blood transfusion
  - This is to differentiate hypersplenism from acute splenic sequestration crisis.
  - In acute splenic sequestration crisis, the size of the spleen will regress after giving blood transfusion.
  - In hypersplenism the spleen may reach a large size and, in some cases, it may reach the level of massive splenomegaly.
  - In hypersplenism, these hematological changes should return to normal levels following splenectomy.
- In patients with sickle cell anemia, acute splenic sequestration crisis is much more common than hypersplenism.
- The incidence of hypersplenism is relatively low in patients with sickle cell anemia.
- The reason for the low incidence of hypersplenism in patients with sickle cell anemia is not known.
- Early auto splenectomy in patients with sickle cell anemia is a contributing factor.
- In the past minor attacks of acute splenic sequestration crises were confused with hypersplenism, but when strict definition criteria are applied, hypersplenism is relatively rare in patients with sickle cell anemia.
- Hypersplenism is more commonly seen in patients with Sickle-B^+-Thalassemia.
- Patients with Sickle-B^+-Thalassemia tend to have clinical features more like thalassemia with splenomegaly and features of hypersplenism.

- Splenectomy is known to be associated with several postoperative complications. One of these serious complications is overwhelming post splenectomy infection (OPSI).
- In order to avoid this risk, there are several treatment modalities other than total splenectomy to treat patients with hypersplenism.
- These treatment options include:
  - Chronic blood transfusions therapy
  - Partial splenectomy
  - Percutaneous intraluminal occlusion of the splenic artery
  - Embolic therapy
- Chronic blood transfusion therapy is known to be associated with several complications including:
  - Alloimmunization:

    Alloimmunization was reported to develop in up to 20% of multiply transfused patients.

    This subsequently makes it very difficult to find matching blood.
  - Transmission of blood – borne infections including hepatitis B and C and acquired immune deficiency.
  - Iron overload
  - The poor compliance of parents and the fact that blood is not readily available makes this form of therapy unsuitable for some of these patients.

HYPERSPLENISM

Splenomegaly associated with:

- **Anemia: Transfusion requirements exceeds 250 mL/kg of packed red blood cells per year**
- **Thrombocytopenia: <100,000/mm^3**
- **Neutropenia: <4000/mm^3**

Either singly or in combination

- Partial splenic embolization:
  - Partial splenic embolization has been shown to be safe and reliable method of treatment of hypersplenism.
  - Partial splenic embolization is known to be associated with complications which makes it unsuitable form of therapy for patients with sickle cell anemia.
- Splenectomy:
  - Splenectomy when performed for sickle cell anemia patients with hypersplenism was shown to be safe and beneficial for the following reasons:

    Splenectomy decreases their blood transfusion requirements.

    Splenectomy eliminates the complications associated with chronic blood transfusion.

    Splenectomy eliminates the discomfort from mechanical pressure of the enlarged spleen and improve their appetite.

  - Post-splenectomy, there is a significant increase in the hemoglobin level, platelets, white blood cells and a significant decrease in the levels of reticulocytes.
  - Splenectomy however is not without complications and known to be associated with an increased risk of post splenectomy sepsis.
- Partial splenectomy:
  - Partial splenectomy is a valuable alternative to total splenectomy.
  - The aim of partial splenectomy is to preserves part of the functioning spleen that will help protect these patients from overwhelming post splenectomy sepsis.
  - This benefit is however is temporary as the spleen remnant tends to grow with time and lead to recurrence of hypersplenism.
  - Add to this the fact the spleen in a large number of patients with sickle cell anemia is either non-functional or hypofunctional. Preserving a nonfunctional spleen in these patients is not beneficial in these patients .
  - Partial splenectomy may be beneficial in children who are younger than 2 years old and require splenectomy.
  - These children are at a much higher risk of post splenectomy sepsis and partial splenectomy will be protective for them and buy them sometime. Total splenectomy if required can be then subsequently when they are older.

21.7 Splenic Abscess and Sickle Cell Anemia

- Splenic abscess is relatively rare. It accounts for 0.14% to 0.7% of necropsy specimens.
- Recently, a noticeable change was observed in the spectrum of splenic abscess.
- There are reports of unusual organisms causing splenic abscess and more cases of splenic abscesses being reported in immunocompromised patients.
- There are several factors that can predispose to splenic abscess, but sepsis is the commonest predisposing factor.
- Infective endocarditis is the commonest cause of splenic abscess.
- Splenic abscess is very rare in patients with sickle cell anemia.
- Patients with sickle cell anemia tend to have early auto splenectomy and this one reason why splenic abscess is considered very rare in patients with sickle cell anemia.
- Splenic abscess in patients with sickle cell anemia:
  - Persistence of splenomegaly in some patients with sickle cell anemia predisposes them to the development of splenic infarction.
  - Splenic infarcts are relatively common in patients with sickle cell anemia.
  - These infarcts are usually small and repetitive leading to fibrosis and ultimately auto splenectomy.
  - Sometimes these infarcts are large in size.
  - Patients with sickle cell anemia are also predisposed to the early development of functional asplenia.

- The functional asplenia makes them liable to develop systemic infections.
- The development of systemic infection and in the presence of splenic infarction predisposes them to the development of splenic abscess.
- Splenic abscess in patients with sickle cell anemia is very rare and remains a diagnostic challenge and the diagnosis can be delayed for the following reasons.

  This is specially so in children with sickle cell anemia.

  These children frequently present with fever due to other infective causes.

  They also present with abdominal pain which is commonly attributed to vasoocclusive crisis.

  Add to this the rarity of splenic abscess.
- Physicians caring for patients with sickle cell anemia should be aware of such a complication and the possibility of splenic abscess should be considered in differential diagnosis of children with sickle cell anemia who present with fever and abdominal pain specially if they have a tender enlarged spleen.

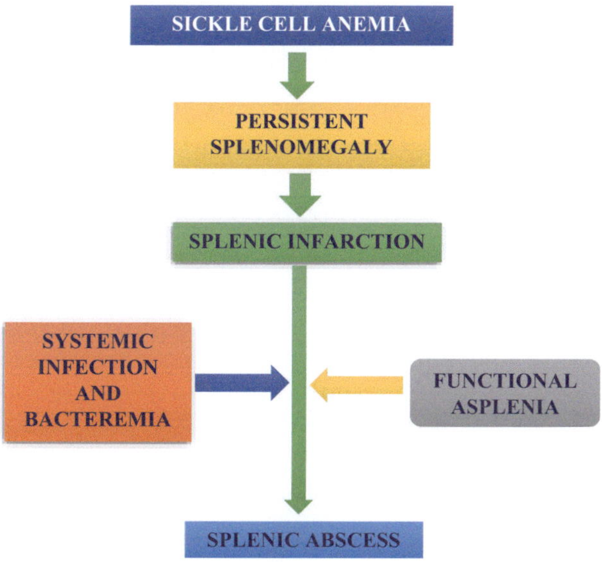

- Presentation:
  - The presentation of splenic abscess is variable and to avoid any delay in diagnosis, a high index of suspicion is important for early diagnosis and treatment.
  - Patients with splenic abscess usually present with:

    Fever

    And left upper quadrant abdominal pain

    Most of these patients are already receiving antibiotics for some other reason and this may mask their fever.

    Clinically, these patients usually have a tender enlarged spleen.

- Diagnosis:
  - Chest and abdominal x-rays:

    In patient with splenic abscess, chest and abdominal x-rays may reveal a soft tissue density in the left upper quadrant and elevated left hemidiaphragm with or without left pleural effusion (Figs. 21.56, 21.57, 21.58, 21.59, and 21.60).
  - Abdominal ultrasound and CT-scan:

    Abdominal ultrasound is a simple noninvasive and easily repeatable investigation. It should be used frequently in these patients.

    The diagnosis of splenic abscess can be confirmed by abdominal CT-scan (Figs. 21.61 and 21.62).

    Although abdominal ultrasound is diagnostic but CT-scan is more valuable. Abdominal CT-scan allows more accurate anatomical localization of site and size of the splenic abscess (Figs. 21.63, 21.64, 21.65, 21.66, 21.67, and 21.68).
- Splenectomy is known to be associated with an increased risk of post splenectomy sepsis and in an order to avoid this risk, splenic preservation procedures were advocated even in patients with splenic abscess.

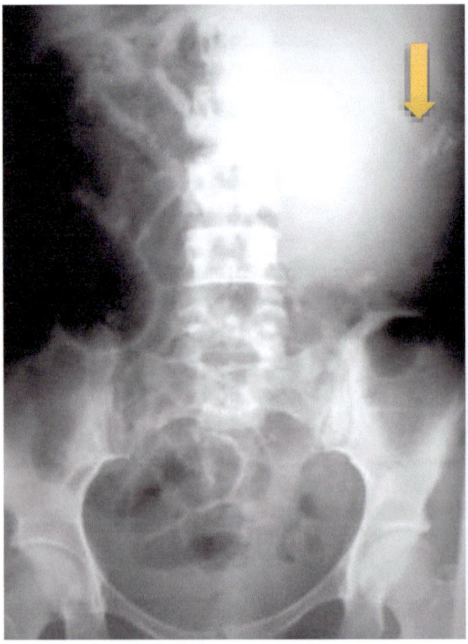
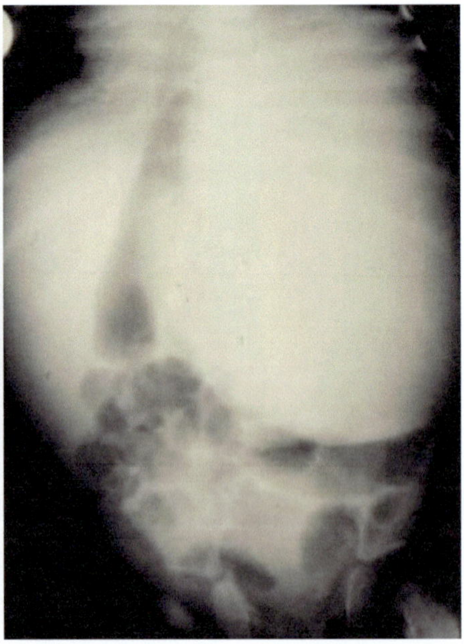

Figs. 21.56 and 21.57 Abdominal radiographs of two patients with sickle cell anemia and splenic abscess. Note the left upper quadrant soft tissue density caused by splenomegaly. Note also the calcification in the first photograph. These represent calcifications in previous splenic infarcts

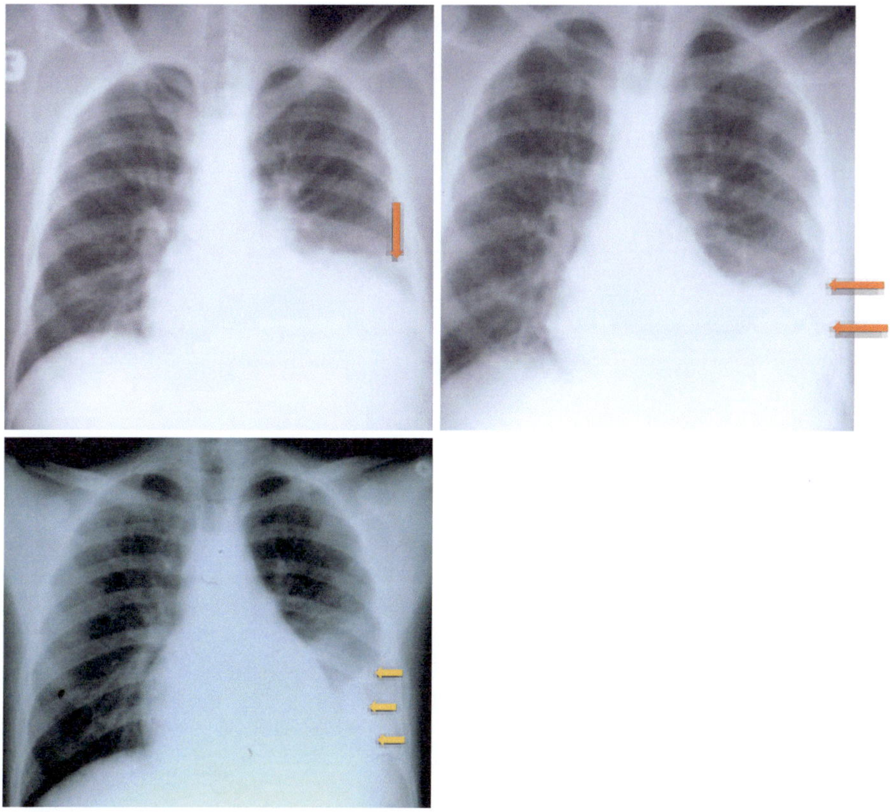

Figs. 21.58–21.60 Chest x-rays in three patients with sickle cell anemia who developed splenic abscesses. Note the elevated left hemidiaphragm and mild left pleural effusion

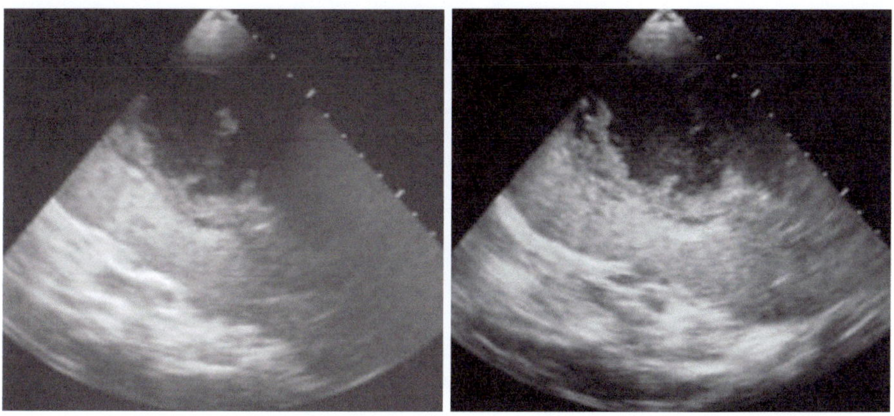

Figs. 21.61 and 21.62 Abdominal ultrasounds of two patients with sickle cell anemia showing large splenic abscess

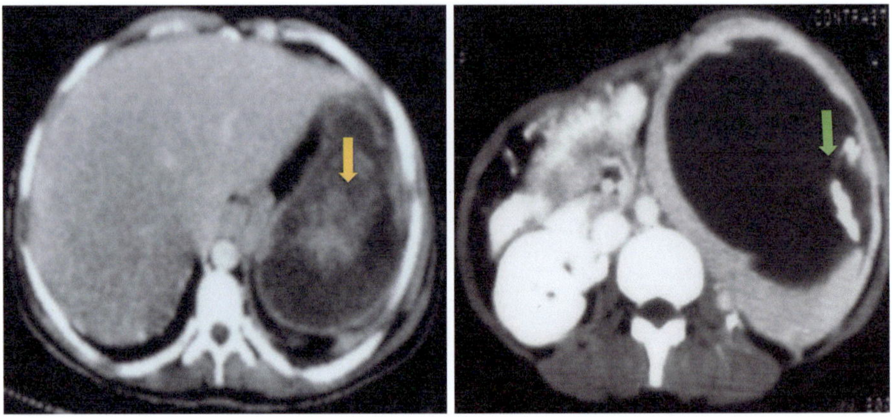

Figs. 21.63 and 21.64 Abdominal CT-scans of two patients with sickle cell anemia showing large splenic abscess almost replacing the whole spleen. Note the splenic sequasterum in the upper photographs and calcifications in the lower photograph

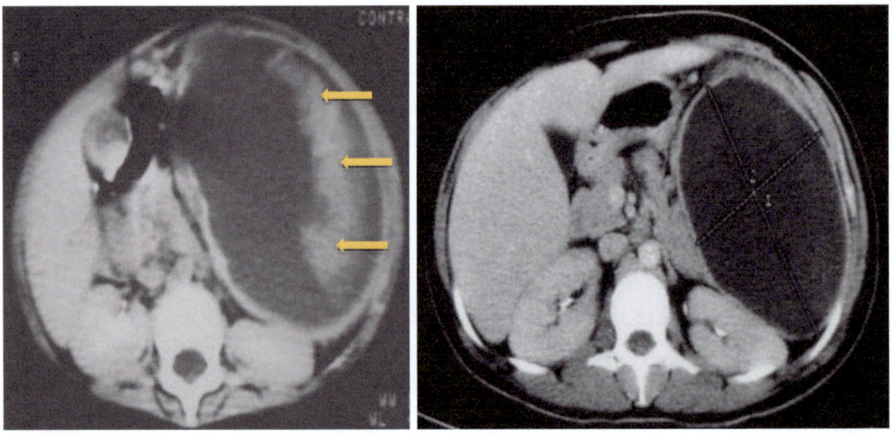

Figs. 21.65 and 21.66 Abdominal CT-scans showing splenic abscess in two patients with sickle cell anemia. Note the splenic sequasterum in the upper photographs and the large size of the splenic abscess in the lower photograph. The splenic abscess is involving all of the spleen which is replaced by the abscess cavity

- There are several procedures advocated to preserve the spleen including:
  - Partial splenectomy
  - CT-guided percutaneous catheter drainage of splenic abscess
  - Noninterventional treatment of splenic abscess with the use of antibiotics.
  - In children with sickle cell anemia who present with splenic abscess, splenectomy and antibiotic coverage is the treatment of choice.
  - In children with sickle cell anemia and splenic abscess there is no point in preserving the spleen which is nonfunctioning in the majority of these patients.

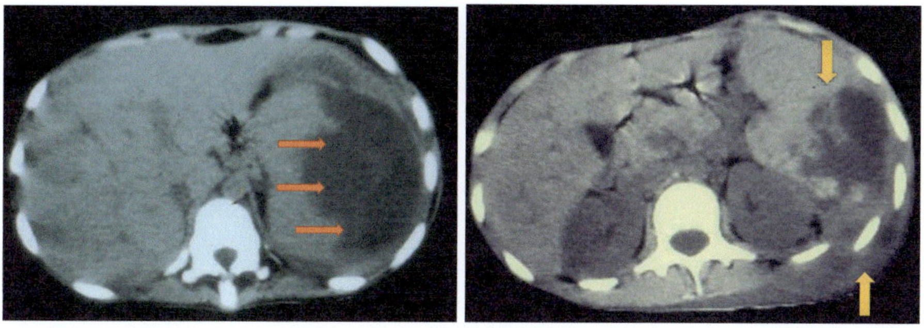

Figs. 21.67 and 21.68 Abdominal CT-scan showing splenic abscess in two patients with sickle cell anemia. Note the size of the splenic abscess which is involving a large part of the spleen in the upper photograpgh. Note also the soft tissue swelling in the second photograph. This is a result of ruptur of splenic abscess

Figs. 21.69 Abdominal CT-Scan showing large localized splenic abscess in a patient with sickle cell anemia

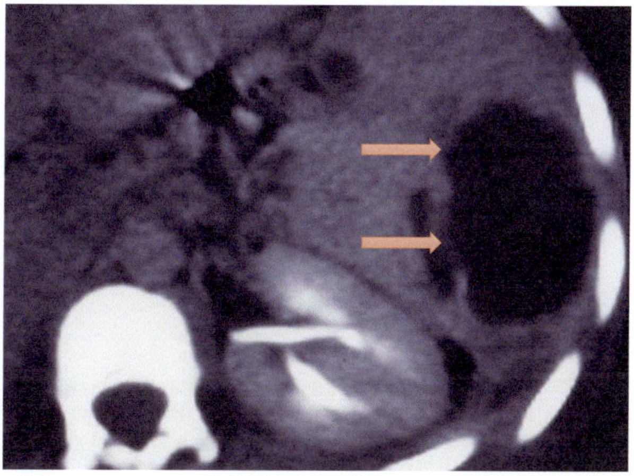

- Percutaneous catheter drainage of splenic abscess can be used as a temporary measure in patients who are sick to undergo immediate laparotomy.
- Percutaneous catheter drainage of splenic abscess however requires the availability of personal experienced in this technique.

- The size of the splenic abscess is variable.
- Sometimes the whole spleen is affected and replaced totally by an abscess (Fig. 21.69).
- Occasionally, part of splenic tissue is found in the middle of the abscess cavity. This is a sequasterum of splenic tissue within the abscess cavity (Figs. 21.70 and 21.71).
- There are several organisms which can cause splenic abscess including (Figs. 21.72, 21.73, 21.74, 21.75, 21.76, 21.77, and 21.78):
  - Staphylococci
  - Streptococci

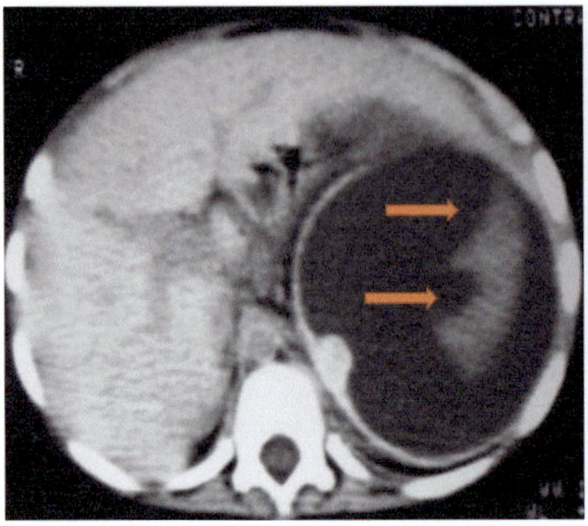
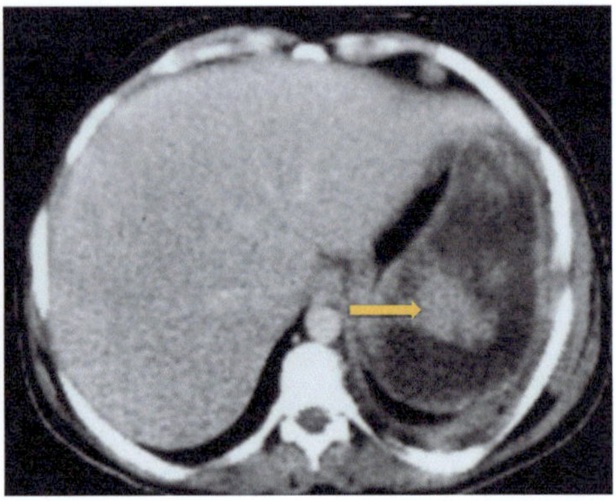

Figs. 21.70 and 21.71 Abdominal CT-Scan of two patients with sickle cell anemia showing a very large splenic abscess. Note also the sequestrum of splenic tissue in the middle of the abscess cavity

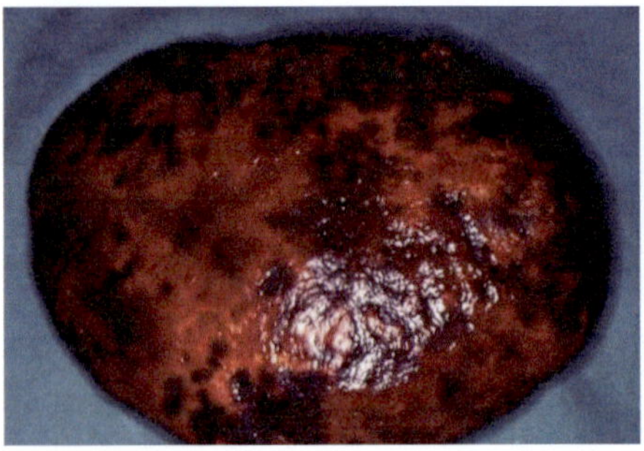
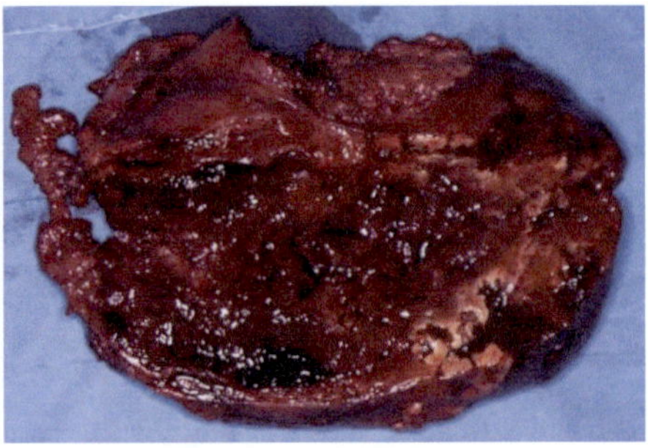

Figs. 21.72 and 21.73 Clinical operative photographs showing a removed spleen in two patients with sickle cell anemia. Note also the splenic infarcts. The First photograph shows an intact spleen with a large splenic abscess while the lower photograph shows a ruptured large splenic abscess

Fig. 21.74 A cut-section of the spleen in a patient with sickle cell anemia showing splenic abscess. Note the abscess cavity

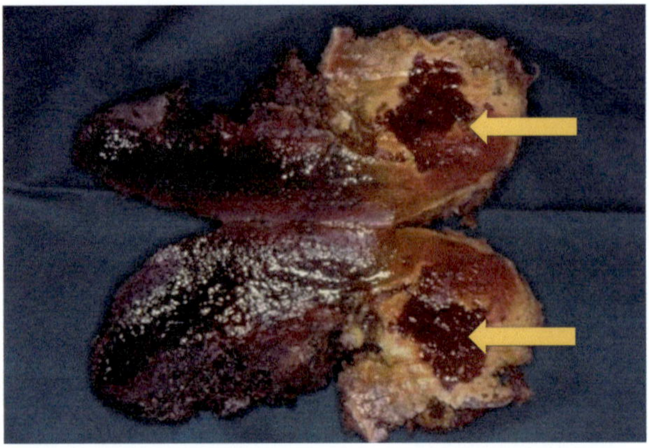

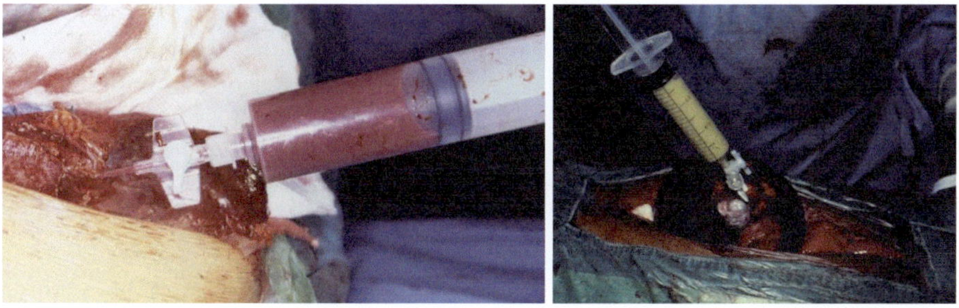

Fig. 21.75 and 21.76 Clinical intraoperative photographs in two patients with sickle cell anemia showing aspiration of a large splenic abscess

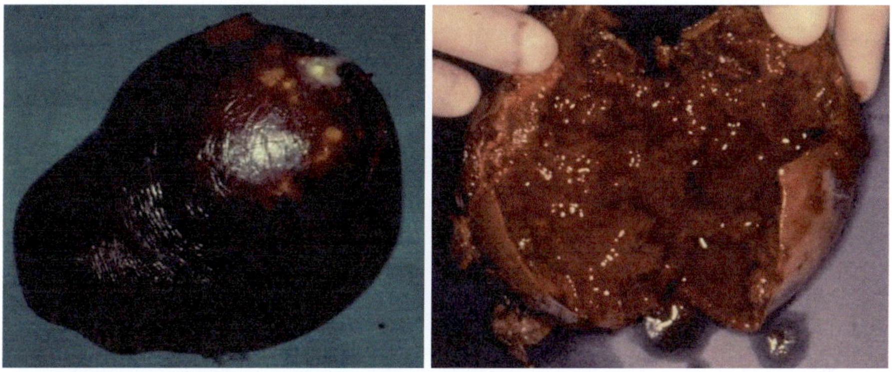

Figs. 21.77 and 21.78 Clinical photographs of two patients with sickle cell anemia showing an excised spleen for splenic abscess. Note the large abscess cavity as shown in the open spleen in the lower. The abscess cavity destroyed and replaced the whole splenic tissue. This will require splenectomy as a definite treatment

- Gram negative bacilli
- Salmonella: An interesting finding was the isolation of Salmonella as a causative organism of splenic abscess in patients with sickle cell anemia.
- Salmonella is a known causative organism for a variety of infections in patients with SCA including:
  Septic arthritis
  Osteomyelitis
  Septicemia.
  Splenic abscess
- This is important to remember when considering to start antibiotic coverage as empirical antibiotic therapy should include antibiotics against Salmonella till the culture and sensitivity results become available.

- Treatment of splenic abscess:
  - In the past, total splenectomy was the treatment of choice for splenic abscess.
  - Splenectomy however is associated with an increased risk of post splenectomy sepsis.
  - To avoid the risk of overwhelming post-splenectomy Infection, a variety of techniques were developed to preserve the spleen even in those with splenic abscess.
  - A variety of procedures have been proposed to treat splenic abscess.
  - These procedures include:

    Partial splenectomy

    CT-guided percutaneous catheter drainage

    Non-interventional treatment of splenic abscess with antibiotics.
  - Percutaneous abscess drainage can be used as a temporary measure in patients who are sick and unfit to undergo emergency laparotomy.
  - This a valuable option in patients who are sick and septicemic. The drainage of the abscess will help control the infection and gain time till the patient is fit to undergo splenectomy.
  - Percutaneous abscess drainage however requires the availability of personal experienced in this technique.
  - Any delay in the diagnoses and treatment of splenic abscess can lead to rupture of splenic abscess and peritonitis.
  - This is a serious complication which is known to be associated with high morbidity and mortality. This is specially so in children with sickle cell anemia who are already sick and immunocompromised (Figs. 21.79, 21.80, 21.81, and 21.82).
  - Splenectomy remains the treatment of choice for patients with sickle cell disease as there is no point to preserve a non-functioning spleen which is present in the majority of patients with splenic abscess.
  - Antibiotics and percutaneous drainage are valuable alternatives to treat patients who are sick and unfit to undergo laparotomy and splenectomy.

Splenic Abscess: Causative Organisms

- **Most frequent organisms:**
  - **Staphylococci**
  - **Streptococci**
  - **Gram –ve bacilli**
- **Other organisms include:**
  - **Pseudomonas**
  - **Enterobacter**
  - **Salmonella**
  - **Serratia**
  - **Fungi**
- **In patients with sickle cell anemia, Salmonella is the most common causative organism of splenic abscess**

Splenic Abscess: Treatment Options

- **Non-interventional treatment with intravenous antibiotics**
- **CT-guided percutaneous catheter drainage**
- **Ultrasound-guided percutaneous drainage**
- **Partial splenectomy**
- **Total splenectomy**

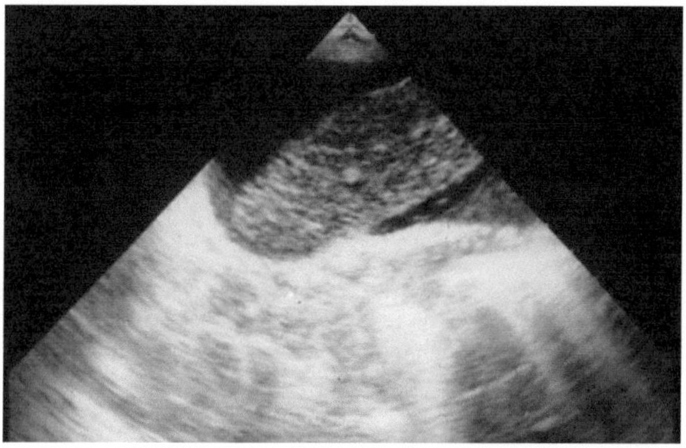

Fig. 21.79 Abdominal ultrasound of a child with sickle cell anemia who presented with ruptured splenic abscess. Note the fluid seen around the spleen

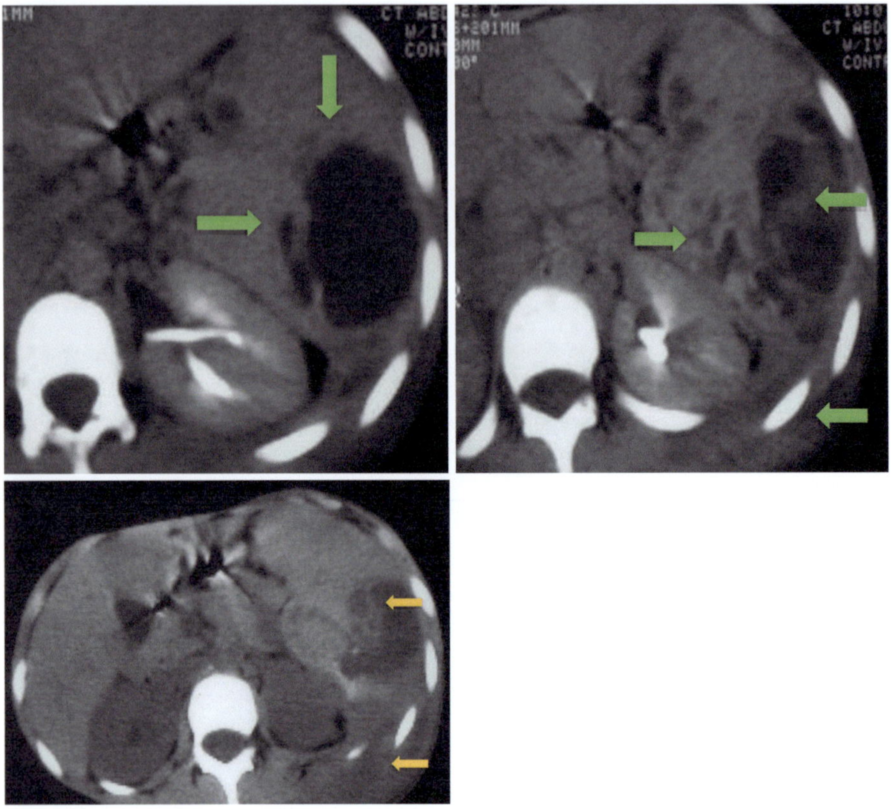

Figs. 21.80–21.82 Abdominal CT-scans of children with sickle cell anemia who presented with ruptured splenic abscess. Note the ruptured splenic abscess in these patients and the edema and swelling of the surrounding tissues

21.8 Massive Splenic Infarction

- The spleen is one of the early and commonly affected organs in patients with sickle cell anemia.
- In patients with sickle cell anemia, the spleen is commonly enlarged during the first few years of life but then undergoes progressive atrophy as a result of repeated multiple splenic infarcts.
- In most cases, these infarcts are followed by splenic scarring and atrophy leading to a small fibrotic and sometimes calcified splenic nodule (auto splenectomy).
- Sometimes the enlarged spleen becomes outlined by calcifications. These most likely represent calcified small splenic infarcts (Figs. 21.83 and 21.84).
- Splenic infarcts in patients with sickle cell anemia are common and usually small, repetitive. They usually pass unnoticed by the patients leading to auto splenectomy and a small splenic nodule that sometimes can be calcified spleen (Fig. 21.85).

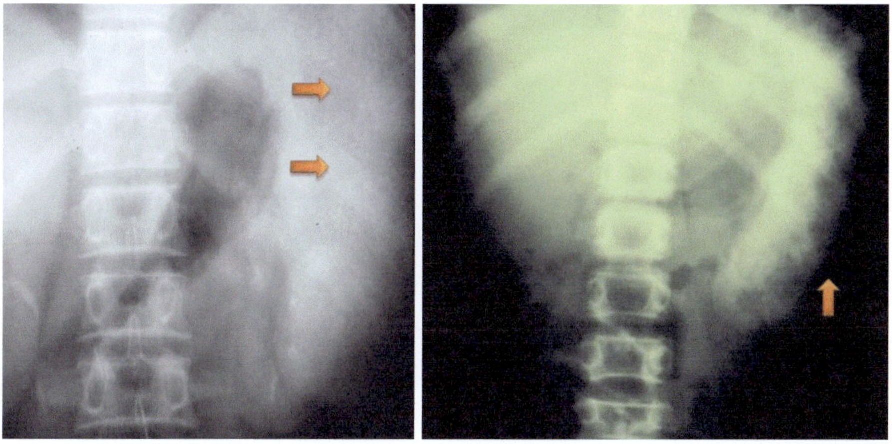

Figs. 21.83 and 21.84 Abdominal x-rays showing enlarged spleen in patients with sickle cell anemia. Note the calcifications of the spleen

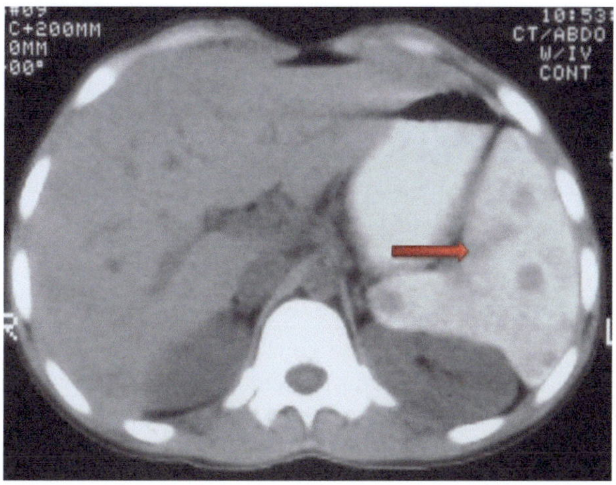

Fig. 21.85 Abdominal CT-scan showing a small atrophied and calcified spleen in a patient with sickle cell anemia

- This however is not the case always and sometimes splenomegaly persists into an older age group or sometimes into adulthood.
- The reason for persistent splenomegaly in some of these patients is not known.
- It has been suggested that a high Hb F level in some of these patients may play a role in persistence of splenomegaly in these patients.
- Hemoglobin F exhibits its favorable effect through inhibition of hemoglobin S polymerization and thus a longer survival of hemoglobin F containing cells and less sickling.
- It was reported that splenomegaly persists in about 10% of adults with sickle-cell anemia in the West Indies and such cases are usually characterized by high levels of hemoglobin F in their blood.
- Persistence of splenomegaly in these patients makes them liable to develop splenic complications including massive splenic infarctions (Figs. 21.86 and 21.87).

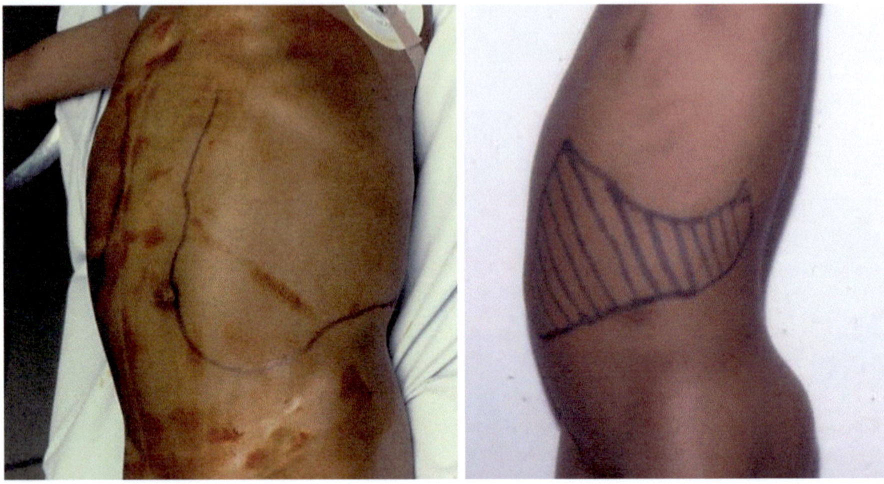

Figs. 21.86 and 21.87 Clinical photographs showing enlarged spleens in patients with sickle cell disease. Persistent splenomegaly in some of these patients can be complicated by the development of massive splenic infarction

- Massive splenic infarction (MSI) is a very rare condition.
- Massive splenic infarction is arbitrarily defined as infarction involving more than 50% of the spleen size.
- There several causes of massive splenic infarction including hematological and non-hematological causes.
- These include:
  - Evans syndrome
  - Endocarditis
  - Polycythemia Vera
  - Paroxysmal nocturnal hemoglobinuria
  - Sickle cell anemia
  - Sickle cell hemoglobin C disease
  - Sickle cell-beta-thalassemia
  - Sickle cell trait
  - Hemoglobin SE
- Massive splenic infarction is very rare among adults with sickle cell anemia.
- There are however reports of massive splenic infarcts in patients with:
  - Sickle cell trait
  - Hemoglobin SE disease
  - Hemoglobin SC disease
- Massive splenic infarctions in these patients are seen particularly in association with stress or following air travel especially in an unpressurized aircraft.
- In patients with sickle cell trait, large splenic infarcts in the normal sized spleen were also reported following:
  - Exposure to hypoxia during high altitude

- Flights in an unpressurized air plane
- Mountain climbing
- Sleep apnea with its associated hypoxia was also implicated as a possible cause of splenic infarcts in patients with sickle cell trait without apparent precipitating factor.

- Massive splenic infarcts had also been reported in other sickle cell variants where splenomegaly is common, mainly in:
  - Hb SC disease
  - $S\beta^+$ Thalassemia
  - Hb SE disease
- Massive splenic infarction is considered vey rare and unique complication of sickle cell anemia.
- Patients with sickle cell anemia complicated by splenic infarction commonly present with:
  - Left upper quadrant abdominal pain
  - A clinically tender enlarged spleen
  - This should hint to the possibility of splenic infarction and physicians caring for these patients must keep this in mind.
  - The availability of ultrasound examination has made it possible to diagnose splenic infarction early and to follow the progress of this complication by repeated ultrasounds.
- Clinical features:
  - Massive splenic infarction in patients with sickle cell anemia (SCA) is unique and extremely rare (Figs. 21.88, 21.89, 21.90, 21.91, 21.92, 21.93, 21.94, 21.95, 21.96, 21.97, 21.98 and 21.99).
  - The precipitating factor for massive splenic infarction in patients with sickle cell anemia is usually hypoxia induced by high-altitude.
  - This is typically seen in:

    Patients travelling in unpressurized airplanes

    Patients doing mountain climbing

    Massive splenic infarction has been reported to occur at mountain altitudes of 5000–7000 ft. above sea level.

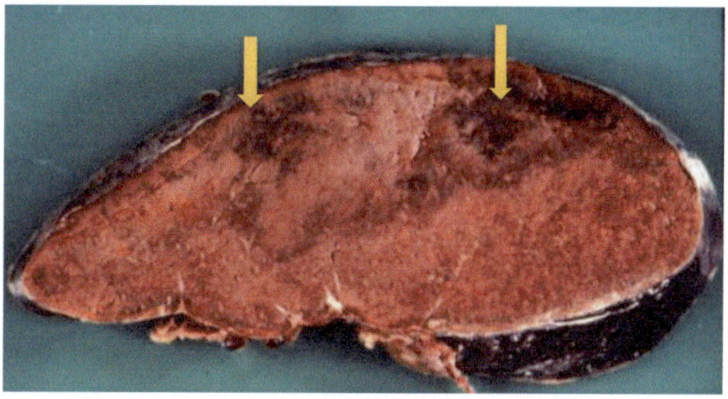

Fig. 21.88 A clinical photograph showing a cut section of a spleen from a patient with sickle cell anemia who developed splenic infarcts. Note the splenic infarcts which are usually small and multiple but can attain a large size also

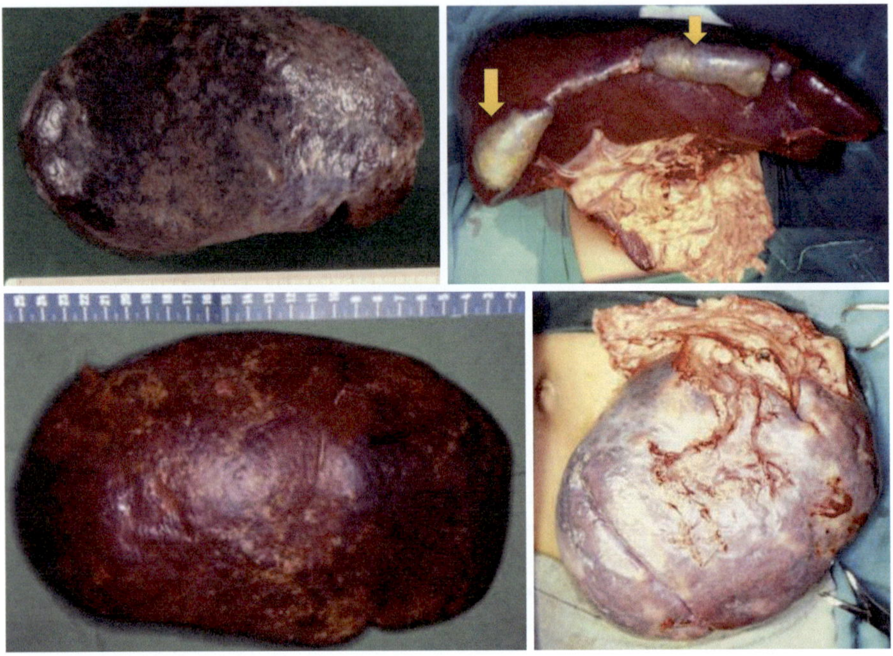

Figs. 21.89–21.92 Clinical photographs showing excised spleens of patients with sickle cell anemia. Note the size of the spleens and also the presence of small and multiple infarcts. Note also the variable extent and size of these infarcts

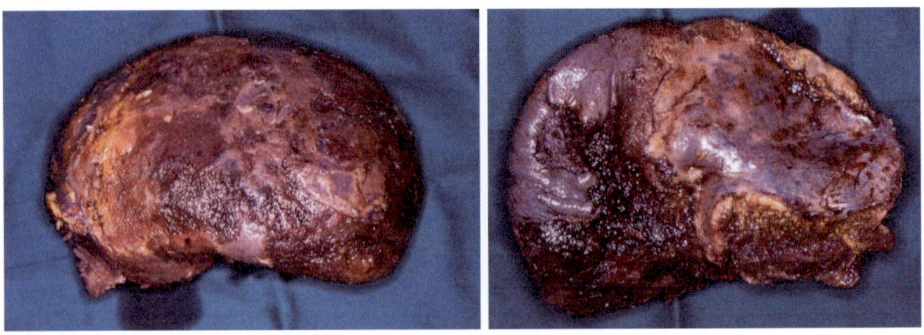

Figs. 21.93 and 21.94 Clinical photographs of excised spleens showing massive splenic infarction in a patient with sickle cell anemia. Note the massive infarction which is involving almost the whole spleen

- These factors promote sickling by metabolic effects such as acidosis, hypoxia and dehydration.
- Other precipitating factors for massive splenic infarction include:
  - Septicemia
  - Severe vasoocclusive crisis
  - Acute chest syndrome

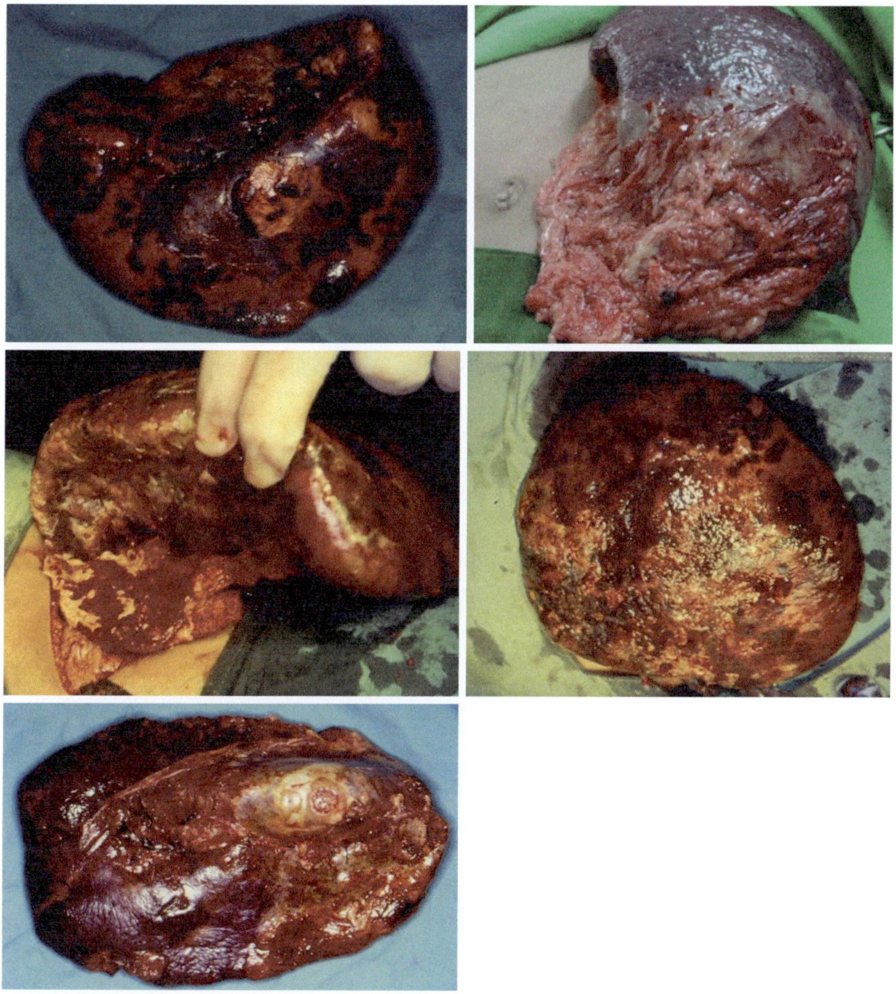

Figs. 21.95–21.99 Clinical photographs showing three spleens from patients with sickle cell anemia who presented with massive splenic infarction. Note the variation in the size of the infarction involving almost the whole spleen in the upper two pictures and more than half the size of the spleen in the third picture

The additional hypoxemia caused by acute chest syndrome is another contributing factor to the development of massive splenic infarction in these patients.

– This may explain the occurrence of massive splenic infarction in patients with sickle cell anemia admitted to the hospital with other medical complications such as:

  Acute chest syndrome
  Severe vasoocclusive crisis,
  Septicemia

- The presentation of massive splenic infarction is variable but the majority of patients present with:
  Sudden onset of severe left upper quadrant abdominal pain.
  This may be associated with nausea, vomiting, fever, and chills.
- These symptoms are non-specific and may be confused with other causes of acute abdominal pain.
- Physicians caring for these patients should be aware of this.
- Clinically, these patients will have a tender enlarged spleen.
- A history of recent travel to a high-altitude place, mountain climbing or travelling in unpressurized aircraft is important.
- This information should hint to the possibility of splenic infarction.
- The availability of ultrasound examination made it possible to diagnose splenic infarction early.

• Diagnosis of massive splenic infarction:
  - Abdominal ultrasound is the main non-invasive investigation.
    Abdominal ultrasound is the first line of investigations.
    It is cheap, non-invasive and can be repeated easily.
  - Abdominal CT-scan however is more valuable.
    It accurately outlines the site and size of splenic infarction.
    It helps in differentiating this from other conditions affecting the spleen namely splenic abscess (Figs. 21.100, 21.101, 21.102, 21.103, and 21.104).

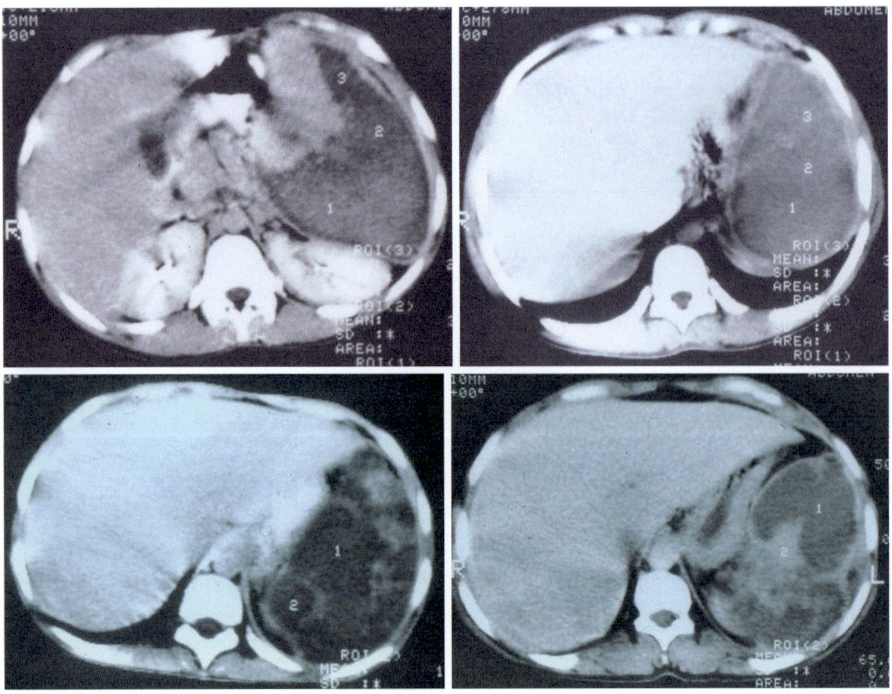

Figs. 21.100–21.103 Abdominal CT-scans showing massive splenic infarction in four patients with sickle cell anemia. Note the variable extent of the infarction which involves more than half the spleen size in some and almost involving the whole spleen in others

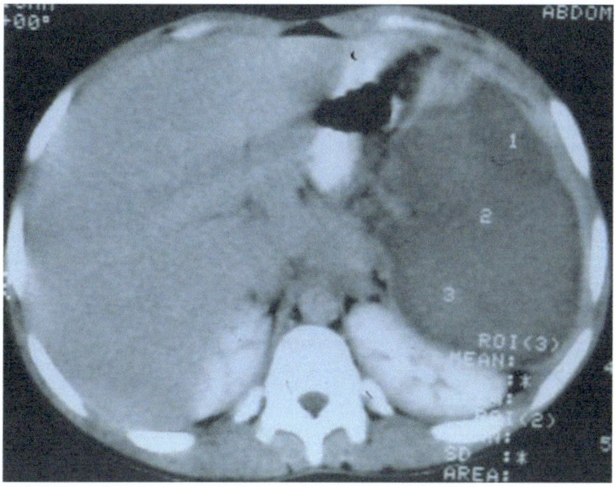

Fig. 21.104 Abdominal CT-scan showing massive splenic infarction in a patient with sickle cell anemia. Note the extent of splenic infarction which affect almost the whole spleen

Computerized tomography (CT) of the abdomen is more valuable in outlining the extent of the infarction.

- The clinical presentation and radiological features of massive splenic infarction can resemble those with splenic abscess.
- It is important to differentiate between the two as massive splenic infarction can be treated conservatively ultimately leading to auto splenectomy.
- The management of splenic abscess is different.
- Management of massive splenic infarction:
  - The treatment of patients with massive splenic infarction is conservative and should be supportive and directed toward correcting any predisposing conditions.
  - Adequate hydration with I.V fluids
  - Adequate oxygenation
  - Blood transfusion where necessary
  - Analgesia
  - It is important to closely follow-up of these patients as they are liable to develop complications including:
    - Splenic rupture
    - Splenic hemorrhage
    - Development of splenic abscess
    - Pseudocyst formation
  - The management of such complications is challenging both to the physicians and the surgeons, and a team work approach is most fruitful in these patients.
- The majority of patients with massive splenic infarction are febrile at the time of presentation, and in the presence of massive splenic infarct, the possibility of splenic abscess although rare must be excluded.
- These two conditions can be confused with each other and in doubtful cases they can also be differentiated by percutaneous aspiration under ultrasound guidance.

MASSIVE SPLENIC INFARCTION

- **MASSIVE SPLENIC INFARCTION HAS BEEN REPORTED IN PATIENTS WITH:**
  - **Sickle cell trait**
  - **Evans syndrome**
  - **Endocarditis**
  - **Polycythemia Vera**
  - **Paroxysmal nocturnal hemoglobinuria**
  - **Sickle cell hemoglobin C disease**
  - **Sickle cell-beta-thalassemia**
  - **Sickle cell anemia**
  - **Hemoglobin SE**
- **MASSIVE SPLENIC INFARCTION IN THESE WAS ASSOCIATED WITH STRESS**
- **MASSIVE SPLENIC INFARCTION IS RARE IN PATIENTS WITH SICKLE CELL ANEMIA**

- These patients should be covered with antibiotics after obtaining the necessary cultures.
- The role of splenectomy in patients with massive splenic infarction remains unclear.
- Most cases of massive splenic infarction are treated conservatively and they tend to be self-limiting ultimately leading to atrophy of the spleen
- Splenectomy for massive splenic infarction is indicated for:
  Cases where there is doubt about the possibility of splenic abscess
  Those with persistent left upper quadrant abdominal pain.
  Those who develop complications such as hemorrhage, rupture, abscess, or persistent pseudocyst.
- Splenectomy is also indicated if there is difficulty differentiating splenic infarction from splenic abscess (Figs. 21.105, 21.106, and 21.107).

Massive Splenic Infarction: Treatment

- **Conservative**
- **Splenectomy**
- **Suspicion of superadded infection**
- **Persistent left upper abdominal pain**
- **Development of complications**

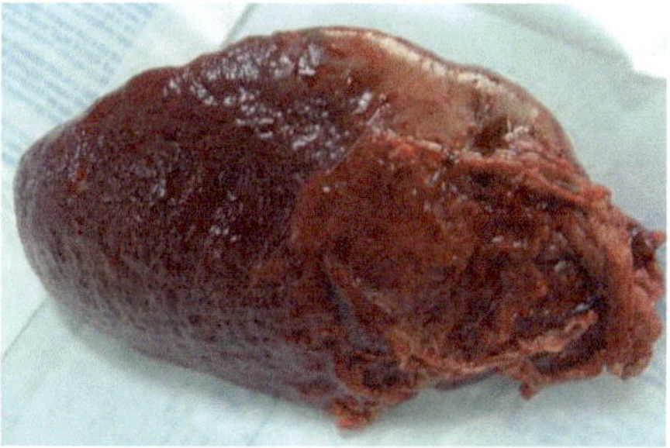

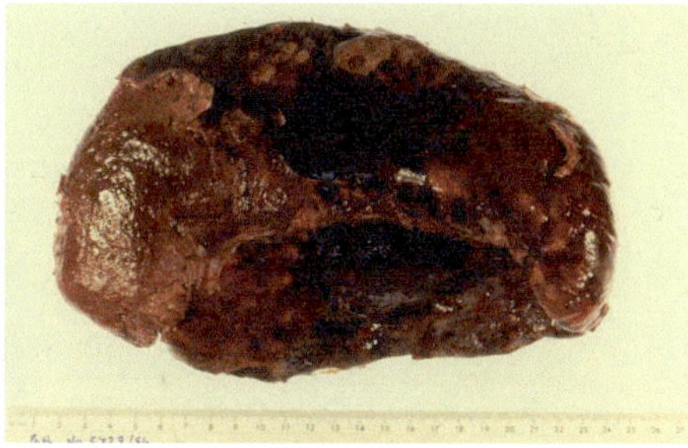

Figs. 21.105 and 21.106 Clinical photographs showing removed spleens from two patients with sickle cell anemia with massive splenic infarcts. Note the extent of the infarct which is involving about 50% of the size of the spleen in the upper one and almost involving the whole spleen in the second one

Fig. 21.107 A clinical photograph showing a cut section of a spleen from a child with sickle cell anemia showing massive splenic infarction

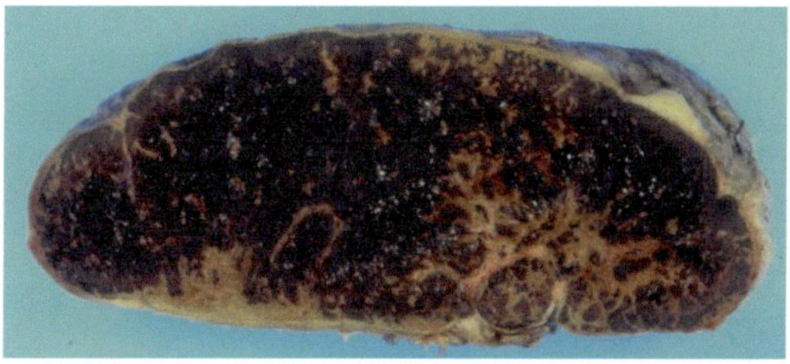

21.9 Partial Splenectomy

- Splenectomy is usually done through an open incision (upper midline, left subcostal or left upper transverse).
- This has changed over the years and currently and with the recent advances in minimal invasive surgery, laparoscopic splenectomy is the preferred procedure.
- Laparoscopic splenectomy is currently the preferred procedure when splenectomy is done electively and in cases where the spleen is not too large in size.
- Total splenectomy is known to be associated with subsequent complications specially post splenectomy sepsis.
- Total splenectomy should however be avoided whenever possible.
- This is to avoid the increased risk of post-splenectomy sepsis.
- Vaccination and antibiotics provide a reasonable protection against the risks of post splenectomy sepsis.
  - This however is not 100% effective
  - Vaccines are not always available especially in poorer countries
  - There is poor compliance when it comes to log term use of antibiotics specially children.
- To overcome this, partial splenectomy or partial splenic arterial embolization has been advocated (Figs. 21.108, 21.109, 21.110, 21.111, 21.112, and 21.113).

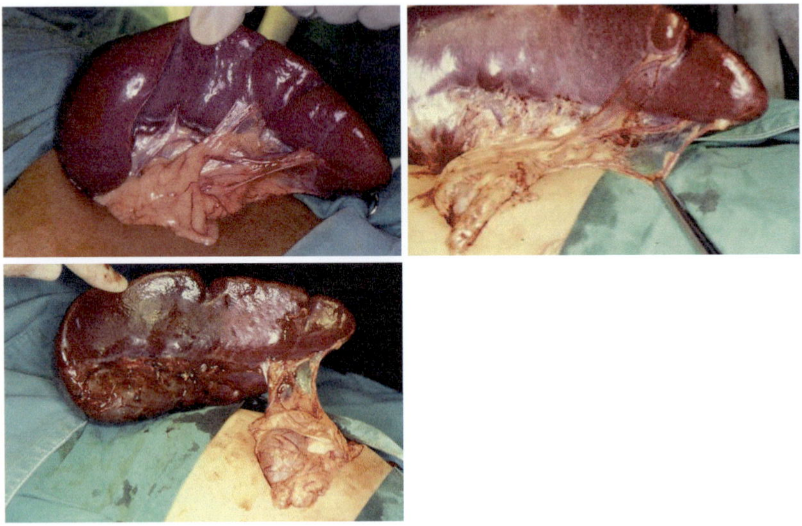

Figs. 21.108–21.110 Clinical photographs showing enlarged spleens. Note also the vascular supply and distribution of vessels at the hilum of the spleen. Note also preservation of the lower polar vessels for partial splenectomy. The lower polar vessels or short gastric vessels can also be preserved when the upper pole of the spleen is to be preserved

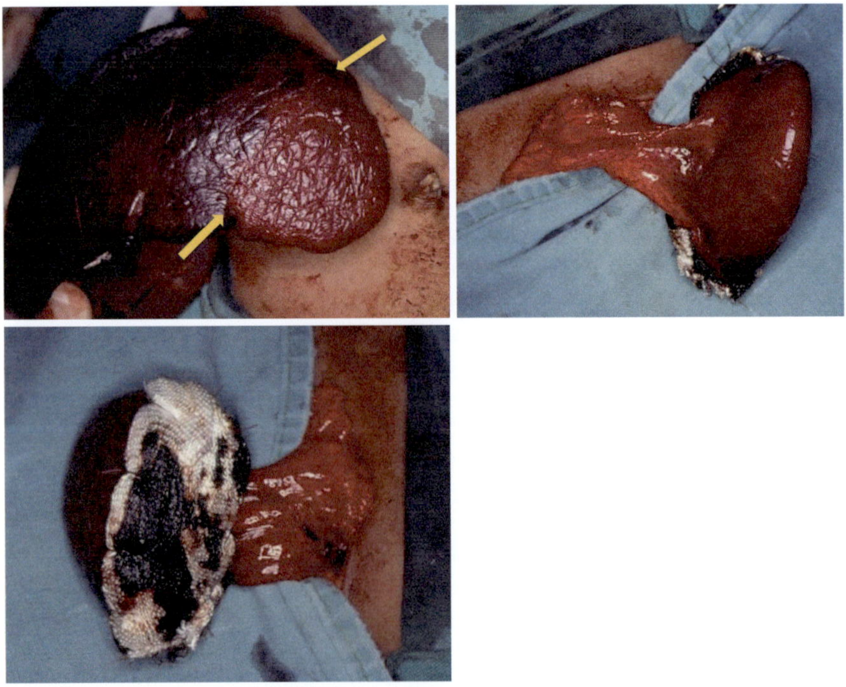

Figs. 21.111–21.113 Clinical photographs showing partial splenectomy. Note the preservation of the lower pole of the spleen and the line of demarcation. Note also the surgicele which is used for hemostasis of the cut surface of the spleen

- It was shown that the spleen's protective functions can be maintained if a small part of the spleen can be preserved and left behind.
- It was estimated that up to 1/3 of the size of the normal spleen is sufficient to provide the spleen's protective function.
- Currently and where clinically appropriate, every attempt should be made to preserve the spleen and perform either surgical subtotal (partial) splenectomy, or partial splenic embolization.
- Partial splenectomy is usually performed through an open technique but recently, laparoscopic partial splenectomy was shown to be feasible and safe.
- As it may take some time for the preserved splenic tissue to provide full immunological protection, it is advisable to immunize these patients preoperatively and cover them with prophylactic antibiotics.

21.10 Laparoscopic Splenectomy

- In the past, open splenectomy was the only option to treat splenic complications of children with sickle cell anemia.
- Open splenectomy is done through an upper transverse, a left subcostal incision or an upper midline incision (Figs. 21.114 and 21.115).
- Large abdominal wounds are ugly and known to be associated with more pain which may interfere with respiration. This can lead to immobility and increases the risk of developing acute chest syndrome in these patients (Figs. 21.116, 21.117, and 21.118).

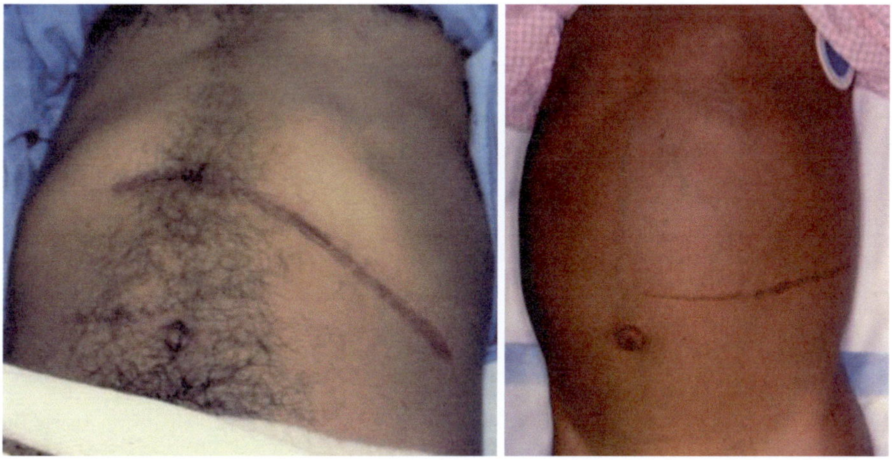

Figs. 21.114 and 21.115 A clinical photograph showing a large wound following open splenectomy for two patients with sickle cell anemia. In the upper one this was done through a left subcostal incision while in the lower one it was done through a left upper transverse incision

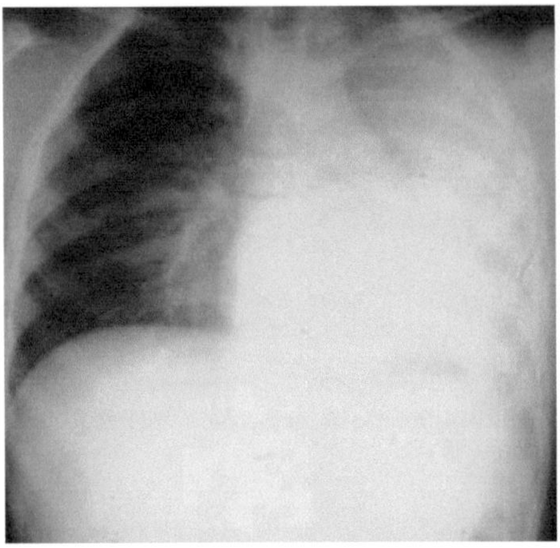

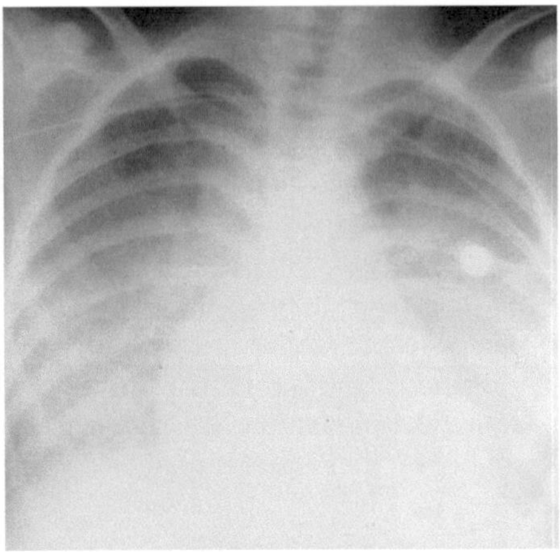

Figs. 21.116 and 21.117 Chest x-rays of two patients with sickle cell anemia showing acute chest syndrome involving the left lung in the upper x-ray and involving both lungs in the second one

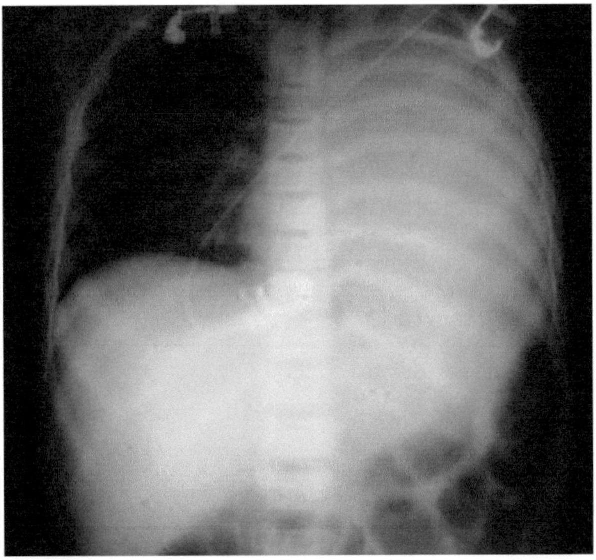

Fig. 21.118 Chest x-ray in a patient with sickle cell anemia showing severe acute chest syndrome involving the whole of the left lung

- In 1991, Delaitre performed the first laparoscopic splenectomy and since then it has become the procedure of choice to treat patients with hematological diseases requiring splenectomy.
- Laparoscopic splenectomy was reported to be feasible and safe both in adults and children.
- Laparoscopic splenectomy should be the procedure of choice to treat patients with sickle cell anemia related splenic complications (Figs. 21.119, 21.120, 21.121, 21.122, 21.123, and 21.124).

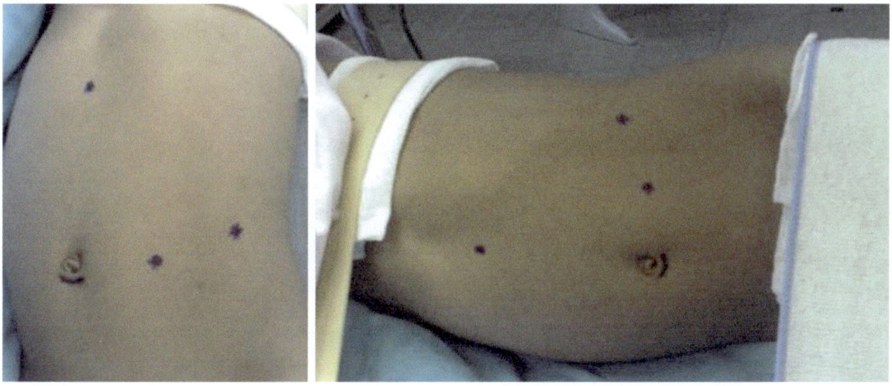

Figs. 21.119 and 21.120 Clinical photographs showing the sites of trocars used for laparoscopic splenectomy. These sites can be modified according to the size of the spleen and whether the position of the patient is supine or a lateral position

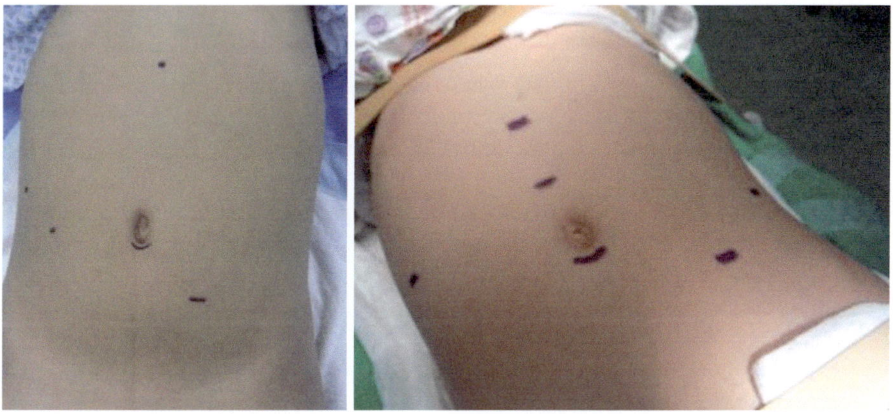

Figs. 21.121 and 21.122 Clinical photographs showing the sites of trocars used for laparoscopic splenectomy and cholecystectomy whether a supine or a lateral position is used. The number of ports can also be modified depending on the size of the patient and spleen

- Another common complication of sickle cell anemia is development of cholelithiasis.
- The frequency of cholelithiasis in patients with sickle cell anemia is however variable but its frequency is known to increase with increasing age.
- It was estimated that overall, 70% of patients with sickle cell anemia develop gallstones.
- The treatment of symptomatic gallstones in patients with sickle cell anemia is well established and currently laparoscopic cholecystectomy is the treatment of choice.
- The treatment of asymptomatic gallstones in patients with sickle cell anemia is still controversial.

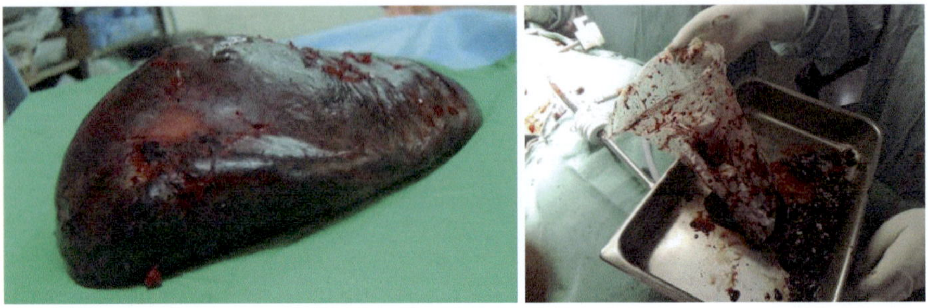

Figs. 21.123 and 21.124 Clinical intraoperative photographs showing an enlarged spleen that was removed intact via a large laparotomy wound and a large spleen that was remove laparoscopically, put into a retrieval bag and removed in pieces via small trocar wounds

- With good perioperative management, laparoscopic splenectomy is feasible, and safe in children with sickle cell anemia.
- Splenectomy can also be done concomitantly with cholecystectomy for patients with sickle cell anemia.
- This was shown to be feasible and safe.
- There is a definite learning curve where laparoscopic splenectomy and cholecystectomy takes more operative time than the open technique. This is specially so for children with sickle cell anemia who are known to have large spleens with severe adhesions.
- Laparoscopic splenectomy was shown to be superior to open splenectomy with regard to hospital stay, postoperative recovery, complications, and cosmetic appearance.
- Laparoscopic partial splenectomy was also shown to be feasible and safe in experienced hands.
- In the past, large spleens are removed through large laparotomy wounds but currently these are removed laparoscopically through few small wounds. This may prove beneficial in reducing the frequency of post-operative acute chest syndrome (Fig. 21.123).

21.11 Perioperative Management of Children with SCA

- In the past surgery for patients with sickle cell anemia was associated with a perioperative mortality as high as 10% and a postoperative morbidity up to 50%.
- For these reasons, surgery was not advocated in patients with sickle cell anemia except in symptomatic patients.
- This however is not the case nowadays and several factors have contributed to decrease the perioperative complications of sickle cell anemia.
- These factors include:
  - A better understanding of sickle cell anemia
  - Improved perioperative care

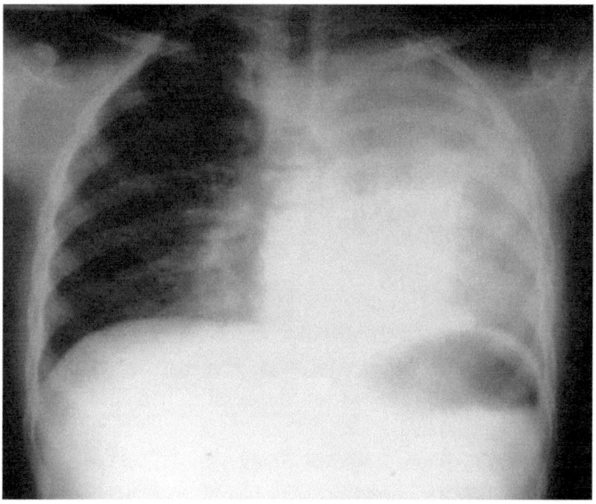

Fig. 21.125 A chest x-ray showing acute chest syndrome in a patient who had splenectomy. The acute chest syndrome involves almost the whole left lung

- One of the common post-splenectomy complications in patients with sickle cell anemia is acute chest syndrome. This can be severe and life threatening (Fig. 21.125).
- To overcome this, a perioperative management protocol for all patients with sickle cell anemia undergoing surgery should be adopted.
- The perioperative management protocol includes:
  - Preoperative clinical evaluation.
  - Preoperative adequate hydration with intravenous fluids at a rate of 1.5 their maintenance. This should start the night before surgery and continues postoperatively till the patient have adequate oral intake.
  - All patients should be immunized with pneumococcal, meningococcal and H. influenza vaccines at least 10–14 days preoperatively. These vaccines are to be given postoperatively for those who require an emergency splenectomy.
  - Preoperative blood transfusion is given preoperatively to increase their hemoglobin level to 10 g/dL with packed red blood cells and using this formula (weight in Kg X (hemoglobin desired – actual hemoglobin) × 3) is advocated.
  - At the time of surgery, hypoxia, acidosis, hypothermia, and hypercarbia should be avoided.
  - They should receive adequate analgesia postoperatively.
  - Early mobilization
  - Chest physiotherapy using incentive spirometry.

21.12 Overwhelming Post-splenectomy Infection (OPSI)

- One of the serious and life-threatening complication following splenectomy is overwhelming post splenectomy infection.
- The true incidence of OPSI however, remains unknown.

- OPSI is commonly seen in children in whom splenectomy was performed in the first few years of life.
- OPSI is seen commonly within the first two years following splenectomy.
- Streptococcus pneumonia is the commonest organism causing OPSI, followed by Hemophilus influenza and Neisseria meningitides.
- To avoid OPSI, children undergoing splenectomy should be immunized with the following vaccines:
  - Pneumococcal
  - H. influenza
  - Meningococcal
- These children should also be covered with penicillin prophylaxis.
- Penicillin prophylaxis can be given either orally or intramuscularly for 2–3 years following splenectomy depending on the age of the child at the time of splenectomy.
- The effectiveness of this prophylaxis in an era of increasing microbial resistance to antibiotics needs however to be re-evaluated.
- The parents of these children should be educated about OPSI and to report to the hospital at the earliest sign of any infection.
- A point of concern in children with sickle cell anemia is their increased susceptibility to Salmonella infections. This must be kept in mind and should be evaluated further in term of long-term antibiotic prophylaxis as well as the feasibility of vaccination against Salmonella.

Further Reading

1. Al-Salem AH. Indications and complications of splenectomy for children with sickle cell disease. J Pediatr Surg. 2006;41:1909–15.
2. Haricharan RN, Roberts JM, Morgan TL, et al. Splenectomy reduces packed red cell transfusion requirements in children with sickle cell disease. J Pediatr Surg. 2008;43(6):1052–6.
3. Al Jam'a A, Al Salem AH, Al Dabous I. Massive splenic infarction in Saudi patients with sickle cell anemia: A unique manifestation. Am J Haematol. 2002;69:205–9.
4. Al-Salem AH, Qaisaruddin S, Al Jam'a A, Al-Kalaf J, Al-Bashier AM. Splenic abscess and sickle cell disease. Am J Haematol. 1998;58:100–4.
5. Al-Awamy BH, Niazi GA, El-Mouzan MI, et al. Relationship of haemoglobin F and alpha-thalassemia to severity of sickle cell anemia in the eastern province of Saudi Arabia. Ann Trop Pediatr. 1986;6:261–5.
6. Elbashier AM, Al-Salem AH, Aljama A. Salmonella as a causative organism of various infections in patients with sickle cell disease. Ann Saudi Med. 2003;23:358–62.
7. Lerner RM, Spataro RF. Splenic abscess: percutaneous drainage. Radiology. 1984;153:643–5.
8. Fernandes ET, Tavares PB, Gaarcette CBM. Conservative management of splenic abscess in children. J Pediatr Surg. 1992;27:1578–9.
9. Bhattacharyya N, Ablin DS, Kosloske AM. Stapled partial splenectomy for splenic abscess in a child. J Pediatr Surg. 1989;24:316–7.
10. Platt O, Brambilla DJ, Roses WF, et al. Mortality in sickle cell disease: life expectancy and risk factors for early death. N Engl J Med. 1994;330:1639–44.
11. Pugliese R, Sansonna F, Scandroglio I, et al. Laparoscopic splenectomy: a retrospective review of 75 cases. Int Surg. 2006;91:82–6.

12. Al-Wabari A, Parida L, Al-Salem AH. Laparoscopic splenectomy and/or cholecystectomy for children with sickle cell disease. Pediatr Surg Int. 2009;25(5):417–21.
13. Waghorn DJ. Overwhelming infection in asplenic patients: current best practice preventive measures are not being followed. J Clin Pathol. 2001;54(3):214–8.
14. Corcione F, Pirozzi F, Aragiusto G, Galante F, Sciuto A. Laparoscopic splenectomy: experience of a single center in a series of 300 cases. Surg Endosc. 2012;26(10):2870–6.
15. Al-Salem AH. Splenic complications of sickle cell anemia and the role of splenectomy. ISRN Hematol. 2011;2011:864257.
16. Al-Salem AH. Massive splenic infarction in children with sickle cell anemia and the role of splenectomy. Pediatr Surg Int. 2013;29(3):281–5.
17. Al-Wabari A, Parida L, Al-Salem AH. Laparoscopic splenectomy for children with hematological disorders: our initial experience with 45 cases. J Pediatr Surg Specialt. 2008;2(3):18–20.
18. Airede AI. Acute splenic sequestration in a five-week-old infant with sickle cell disease. J Pediatr. 1992;120:160.
19. Alkindi S, Al-Maini M, Pathare A. Clinical and laboratory characteristics of patients with sickle-cell and autoimmune/connective tissue diseases. Rheumatol Int. 2012;32:373–8.
20. Al-Salem AH, Al-Aithan S, Bhamidipati P, Al-Jam'a A, Al Dabbous I. Sonographic assessment of spleen size in Saudi patients with sickle cell disease. Ann Saudi Med. 1998;18:217–20.
21. Ammann AJ, Addiego J, Wara DW, Lubin B, Smith WB, Mentzer WC. Polyvalent pneumococcal-polysaccharide immunization of patients with sickle-cell anemia and patients with splenectomy. N Engl J Med. 1977;297:897–900.
22. Aneni EC, Hamer DH, Gill CJ. Systematic review of current and emerging strategies for reducing morbidity from malaria in sickle cell disease. Tropical Med Int Health. 2013;18:313–27.
23. Aquino VM, Norvell JM, Buchanan GR. Acute splenic complications in children with sickle cell-hemoglobin C disease. J Pediatr. 1997;130:961–5.
24. Baethge BA, Bordelon TR, Mills GM, Bowen LM, Wolf RE, Bairnsfather L. Antinuclear antibodies in sickle cell disease. Acta Haematol. 1990;84:186–9.
25. Barrett-Connor E. Bacterial infection and sickle cell anemia. An analysis of 250 infections in 166 patients and a review of the literature. Medicine. 1971;50:97–112.
26. Barrios NJ, Livaudais F, McNeil J, Humbert JR, Corrigan J. Reversible splenic hypofunction in hypertransfused children with homozygous sickle cell disease. J Natl Med Assoc. 1993;85:677–80.
27. Brousse V, Elie C, Benkerrou M, Odièvre MH, Lesprit E, Bernaudin F, Grimaud M, Guitton C, Quinet B, Dangiolo S, de Montalembert M. Acute splenic sequestration crisis: cohort of 190 paediatric patients. Br J Haematol. 2012;156:643–8.
28. Brown AK, Sleeper LA, Miller ST, Pegelow CH, Gill FM, Waclawiw MA. Reference values and hematologic changes from birth to 5 years in patients with sickle cell disease. Cooperative Study of Sickle Cell Disease. Arch Pediatr Adolesc Med. 1994;148:796–804.
29. Buffet PA, Milon G, Brousse V, Correas JM, Dousset B, Couvelard A, Kianmanesh R, Farges O, Sauvanet A, Paye F, Ungeheuer MN, Ottone C, Khun H, Fiette L, Guigon G, Huerre M, Mercereau-Puijalon O, David PH. Ex vivo perfusion of human spleens maintains clearing and processing functions. Blood. 2006;107:3745–52.
30. Cameron PU, Jones P, Gorniak M, Dunster K, Paul E, Lewin S, Woolley I, Spelman D. Splenectomy associated changes in IgM memory B cells in an adult spleen registry cohort. PLoS One. 2011;6:e23164.
31. Campbell PJ, Olatunji PO, Ryan KE, Davies SC. Splenic regrowth in sickle cell anaemia following hypertransfusion. Br J Haematol. 1997;96:77–9.
32. Chen LT, Weiss L. Electron microscopy of the red pulp of human spleen. Am J Anat. 1972;134:425–57.
33. Crary SE, Buchanan GR. Vascular complications after splenectomy for hematologic disorders. Blood. 2009;114:2861–8.
34. Deplaine G, Safeukui I, Jeddi F, Lacoste F, Brousse V, Perrot S, Biligui S, Guillotte M, Guitton C, Dokmak S, Aussilhou B, Sauvanet A, Cazals Hatem D, Paye F, Thellier M, Mazier D, Milon

G, Mohandas N, Mercereau-Puijalon O, David PH, Buffet PA. The sensing of poorly deformable red blood cells by the human spleen can be mimicked in vitro. Blood. 2011;117:e88–95.
35. Di Sabatino A, Carsetti R, Corazza GR. Post-splenectomy and hyposplenic states. Lancet. 2011;378:86–97.
36. Diggs LW. Siderofibrosis of the spleen in sickle cell anemia. JAMA. 1935;104:538–41.
37. Donald DE, Aarhus LL. Active and passive release of blood from canine spleen and small intestine. Am J Physiol. 1974;227:1166–72.
38. El Nemer W, Gane P, Colin Y, Bony V, Rahuel C, Galacteros F, Cartron JP, Le Van Kim C. The Lutheran blood group glycoproteins, the erythroid receptors for laminin, are adhesion molecules. J Biol Chem. 1998;273:16686–93.
39. Emond AM, Collis R, Darvill D, Higgs DR, Maude GH, Serjeant GR. Acute splenic sequestration in homozygous sickle cell disease: natural history and management. J Pediatr. 1985;107:201–6.
40. Ferster A, Bujan W, Corazza F, Devalck C, Fondu P, Toppet M, Verhas M, Sariban E. Bone marrow transplantation corrects the splenic reticuloendothelial dysfunction in sickle cell anemia. Blood. 1993;81:1102–5.
41. Fontana V, Jy W, Ahn ER, Dudkiewicz P, Horstman LL, Duncan R, Ahn YS. Increased procoagulant cell-derived microparticles (C-MP) in splenectomized patients with ITP. Thromb Res. 2008;122:599–603.
42. Gaston MH, Verter JI, Woods G, Pegelow C, Kelleher J, Presbury G, Zarkowsky H, Vichinsky E, Iyer R, Lobel JS, Diamond S, Holbrook CT, Gill FM, Richey K, Faletta JM, for the Prophylactic Penicillin Study Group. Prophylaxis with oral penicillin in children with sickle cell anemia. A randomized trial. N Engl J Med. 1986;314:1593–9.
43. Gladwin MT, Sachdev V, Jison ML, Shizukuda Y, Plehn JF, Minter K, Brown B, Coles WA, Nichols JS, Ernst I, Hunter LA, Blackwelder WC, Schechter AN, Rodgers GP, Castro O, Ognibene FP. Pulmonary hypertension as a risk factor for death in patients with sickle cell disease. N Engl J Med. 2004;350:886–95.
44. Gotthardt M, Broker S, Schlieck A, Bauhofer A, Herbst B, Behe M, Corstens FH, Behr TM, Gorg C. Scintigraphy with 99mTc-labeled heat-altered erythrocytes in diagnosing hyposplenia: prospective comparison to 99mTc-labeled colloids and colour-coded duplex ultrasonography. Nuklearmedizin. 2007;46:135–40.
45. den Haan JM, Kraal G. Innate immune functions of macrophage subpopulations in the spleen. J Innate Immun. 2012;4:437–45.
46. Harrod VL, Howard TA, Zimmerman SA, Dertinger SD, Ware RE. Quantitative analysis of Howell-Jolly bodies in children with sickle cell disease. Exp Hematol. 2007;35:179–83.
47. Higgs DR, Aldridge BE, Lamb J, Clegg JB, Weatherall DJ, Hayes RJ, Grandison Y, Lowrie Y, Mason KP, Serjeant BE, Serjeant GR. The interaction of alpha-thalassemia and homozygous sickle-cell disease. N Engl J Med. 1982;306:1441–6.
48. Hirst C, Owusu-Ofori S. Prophylactic antibiotics for preventing pneumococcal infection in children with sickle cell disease. Cochrane Database Syst Rev. 2012;9:CD003427.
49. Holroyde CP, Oski FA, Gardner FH. The "pocked" erythrocyte. Red-cell surface alterations in reticuloendothelial immaturity of the neonate. N Engl J Med. 1969;281:516–20.
50. Imbert P, Rapp C, Buffet PA. Pathological rupture of the spleen in malaria: analysis of 55 cases (1958–2008). Travel Med Infect Dis. 2009;7:147–59.
51. Kalpatthi R, Kane ID, Shatat IF, Rackoff B, Disco D, Jackson SM. Clinical events after surgical splenectomy in children with sickle cell anemia. Pediatr Surg Int. 2010;26:495–500.
52. Kato GJ, Hebbel RP, Steinberg MH, Gladwin MT. Vasculopathy in sickle cell disease: biology, pathophysiology, genetics, translational medicine, and new research directions. Am J Hematol. 2009;84:618–25.
53. Kinney TR, Ware RE, Schultz WH, Filston HC. Long-term management of splenic sequestration in children with sickle cell disease. J Pediatr. 1990;117:194–9.
54. Koduri PR, Nathan S. Acute splenic sequestration crisis in adults with hemoglobin S-C disease: a report of nine cases. Ann Hematol. 2006;85:239–43.

55. Komba AN, Makani J, Sadarangani M, Ajala-Agbo T, Berkley JA, Newton CR, Marsh K, Williams TN. Malaria as a cause of morbidity and mortality in children with homozygous sickle cell disease on the coast of Kenya. Clin Infect Dis. 2009;49:216–22.
56. Lane PA, O'Connell JL, Lear JL, Rogers ZR, Woods GM, Hassell KL, Wethers DL, Luckey DW, Buchanan GR. Functional asplenia in hemoglobin SC disease. Blood. 1995;85:2238–44.
57. Lee A, Thomas P, Cupidore L, Serjeant B, Serjeant G. Improved survival in homozygous sickle cell disease: lessons from a cohort study. BMJ. 1995;311:1600–2.
58. Lesher AP, Kalpatthi R, Glenn JB, Jackson SM, Hebra A. Outcome of splenectomy in children younger than 4 years with sickle cell disease. J Pediatr Surg. 2009;44:1134–8. discussion 1138
59. Makani J, Komba AN, Cox SE, Oruo J, Mwamtemi K, Kitundu J, Magesa P, Rwezaula S, Meda E, Mgaya J, Pallangyo K, Okiro E, Muturi D, Newton CR, Fegan G, Marsh K, Williams TN. Malaria in patients with sickle cell anemia: burden, risk factors, and outcome at the outpatient clinic and during hospitalization. Blood. 2010;115:215–20.
60. McAuley CF, Webb C, Makani J, Macharia A, Uyoga S, Opi DH, Ndila C, Ngatia A, Scott JA, Marsh K, Williams TN. High mortality from Plasmodium falciparum malaria in children living with sickle cell anemia on the coast of Kenya. Blood. 2010;116:1663–8.
61. McCarville MB, Luo Z, Huang X, Rees RC, Rogers ZR, Miller ST, Thompson B, Kalpatthi R, Wang WC. Abdominal ultrasound with scintigraphic and clinical correlates in infants with sickle cell anemia: baseline data from the BABY HUG trial. AJR Am J Roentgenol. 2011;196:1399–404.
62. McCavit TL, Quinn CT, Techasaensiri C, Rogers ZR. Increase in invasive Streptococcus pneumoniae infections in children with sickle cell disease since pneumococcal conjugate vaccine licensure. J Pediatr. 2011;158:505–7.
63. Naik RP, Streiff MB, Haywood C Jr, Nelson JA, Lanzkron S. Venous thromboembolism in adults with sickle cell disease: a serious and under-recognized complication. Am J Med. 2013;126:443–9.
64. Ohene-Frempong K, Weiner SJ, Sleeper LA, Miller ST, Embury S, Moohr JW, Wethers DL, Pegelow CH, Gill FM. Cerebrovascular accidents in sickle cell disease: rates and risk factors. Blood. 1998;91:288–94.
65. Owusu-Ofori S, Hirst C. Splenectomy versus conservative management for acute sequestration crises in people with sickle cell disease. Cochrane Database Syst Rev. 2013;5:CD003425.
66. Parent F, Bachir D, Inamo J, Lionnet F, Driss F, Loko G, Habibi A, Bennani S, Savale L, Adnot S, Maitre B, Yaici A, Hajji L, O'Callaghan DS, Clerson P, Girot R, Galacteros F, Simonneau G. A hemodynamic study of pulmonary hypertension in sickle cell disease. N Engl J Med. 2011;365:44–53.
67. Payne AB, Link-Gelles R, Azonobi I, Hooper WC, Beall BW, Jorgensen JH, Juni B, Moore M. Invasive pneumococcal disease among children with and without sickle cell disease in the United States, 1998–2009. Pediatr Infect Dis J. 2013;32:1308–12.
68. Pearson HA, Spencer RP, Cornelius EA. Functional asplenia in sickle-cell anemia. N Engl J Med. 1969;281:923–6.
69. Pearson HA, Cornelius EA, Schwartz AD, Zelson JH, Wolfson SL, Spencer RP. Transfusion-reversible functional asplenia in young children with sickle-cell anemia. N Engl J Med. 1970;283:334–7.
70. Pearson HA, Gallagher D, Chilcote R, Sullivan E, Wilimas J, Espeland M, Ritchey AK. Developmental pattern of splenic dysfunction in sickle cell disorders. Pediatrics. 1985;76:392–7.
71. Piccin A, Rizkalla H, Smith O, McMahon C, Furlan C, Murphy C, Negri G, Mc Dermott M. Composition and significance of splenic Gamna-Gandy bodies in sickle cell anemia. Hum Pathol. 2012;43:1028–36.
72. van der Plas EM, van den Tweel XW, Geskus RB, Heijboer H, Biemond BJ, Peters M, Fijnvandraat K. Mortality and causes of death in children with sickle cell disease in the Netherlands, before the introduction of neonatal screening. Br J Haematol. 2011;155:106–10.
73. Powell RW, Levine GL, Yang YM, Mankad VN. Acute splenic sequestration crisis in sickle cell disease: early detection and treatment. J Pediatr Surg. 1992;27:215–8; discussion 218–219

74. Quinn CT, Rogers ZR, McCavit TL, Buchanan GR. Improved survival of children and adolescents with sickle cell disease. Blood. 2010;115:3447–52.
75. Ram S, Lewis LA, Rice PA. Infections of people with complement deficiencies and patients who have undergone splenectomy. Clin Microbiol Rev. 2010;23:740–80.
76. Rescorla FJ, West KW, Engum SA, Grosfeld JL. Laparoscopic splenic procedures in children: experience in 231 children. Ann Surg. 2007;246:683–7; discussion 687–688
77. Rice TW, Rubinson L, Uyeki TM, Vaughn FL, John BB, Miller RR 3rd, Higgs E, Randolph AG, Smoot BE, Thompson BT. Critical illness from 2009 pandemic influenza A virus and bacterial coinfection in the United States. Crit Care Med. 2012;40:1487–98.
78. Rogers ZR, Wang WC, Luo Z, Iyer RV, Shalaby-Rana E, Dertinger SD, Shulkin BL, Miller JH, Files B, Lane PA, Thompson BW, Miller ST, Ware RE. Biomarkers of splenic function in infants with sickle cell anemia: baseline data from the BABY HUG trial. Blood. 2011;117:2614–7.
79. Rosado MM, Gesualdo F, Marcellini V, Di Sabatino A, Corazza GR, Smacchia MP, Nobili B, Baronci C, Russo L, Rossi F, Vito RD, Nicolosi L, Inserra A, Locatelli F, Tozzi AE, Carsetti R. Preserved antibody levels and loss of memory B cells against pneumococcus and tetanus after splenectomy: tailoring better vaccination strategies. Eur J Immunol. 2013;43:2659–70.
80. Sadarangani M, Makani J, Komba AN, Ajala-Agbo T, Newton CR, Marsh K, Williams TN. An observational study of children with sickle cell disease in Kilifi, Kenya. Br J Haematol. 2009;146:675–82.
81. Safeukui I, Correas JM, Brousse V, Hirt D, Deplaine G, Mule S, Lesurtel M, Goasguen N, Sauvanet A, Couvelard A, Kerneis S, Khun H, Vigan-Womas I, Ottone C, Molina TJ, Treluyer JM, Mercereau-Puijalon O, Milon G, David PH, Buffet PA. Retention of Plasmodium falciparum ring-infected erythrocytes in the slow, open microcirculation of the human spleen. Blood. 2008;112:2520–8.
82. Serjeant GR. The spleen in sickle cell disease. In: Bowdler AJ, editor. The complete spleen: a handbook of structure, function, and clinical disorders. Totowa, NJ: Humana Press; 2001. p. 251–7.
83. Stein PD, Beemath A, Meyers FA, Skaf E, Olson RE. Deep venous thrombosis and pulmonary embolism in hospitalized patients with sickle cell disease. Am J Med. 2006;119(897):e7–11.
84. Svarch E, Nordet I, Gonzalez A. Overwhelming septicaemia in a patient with sickle cell/beta(0) thalassaemia and partial splenectomy. Br J Haematol. 1999;104:930.
85. Telfer P, Coen P, Chakravorty S, Wilkey O, Evans J, Newell H, Smalling B, Amos R, Stephens A, Rogers D, Kirkham F. Clinical outcomes in children with sickle cell disease living in England: a neonatal cohort in East London. Haematologica. 2007;92:905–12.
86. Topley JM, Rogers DW, Stevens MC, Serjeant GR. Acute splenic sequestration and hypersplenism in the first five years in homozygous sickle cell disease. Arch Dis Child. 1981;56:765–9.
87. Wandersee NJ, Punzalan RC, Rettig MP, Kennedy MD, Pajewski NM, Sabina RL, Paul Scott J, Low PS, Hillery CA. Erythrocyte adhesion is modified by alterations in cellular tonicity and volume. Br J Haematol. 2005;131:366–77.
88. Wang WC, Wynn LW, Rogers ZR, Scott JP, Lane PA, Ware RE. A two-year pilot trial of hydroxyurea in very young children with sickle-cell anemia. J Pediatr. 2001;139:790–6.
89. Wang WC, Ware RE, Miller ST, Iyer RV, Casella JF, Minniti CP, Rana S, Thornburg CD, Rogers ZR, Kalpatthi RV, Barredo JC, Brown RC, Sarnaik SA, Howard TH, Wynn LW, Kutlar A, Armstrong FD, Files BA, Goldsmith JC, Waclawiw MA, Huang X, Thompson BW. Hydroxycarbamide in very young children with sickle-cell anaemia: a multicentre, randomised, controlled trial (BABY HUG). Lancet. 2011;377:1663–72.
90. Weller S, Braun MC, Tan BK, Rosenwald A, Cordier C, Conley ME, Plebani A, Kumararatne DS, Bonnet D, Tournilhac O, Tchernia G, Steiniger B, Staudt LM, Casanova JL, Reynaud CA, Weill JC. Human blood IgM "memory" B cells are circulating splenic marginal zone B cells harboring a prediversified immunoglobulin repertoire. Blood. 2004;104:3647–54.
91. WHO. Guidelines for the treatment of malaria. 2nd ed. Geneva: World Health Organization; 2010.
92. William BM, Corazza GR. Hyposplenism: a comprehensive review. Part I: basic concepts and causes. Hematology. 2007;12:1–13.
93. Wright JG, Hambleton IR, Thomas PW, Duncan ND, Venugopal S, Serjeant GR. Postsplenectomy course in homozygous sickle cell disease. J Pediatr. 1999;134:304–9.

22 Primary Tumors of the Spleen

22.1 Introduction

- The spleen is located in the left upper quadrant hidden under the rib cage and this makes it not easily palpable unless it is enlarged.
- Tumors of the spleen are very rare when compared to the incidence of tumors in other intra-abdominal organs
- There are benign and malignant neoplasms of the spleen.
- The most common neoplasms of the spleen are:
  - Lymphomas
  - Vascular tumors
- The most common mesenchymal tumors of the spleen are:
  - Splenic hemangiomas
  - Splenic angiosarcomas
- Image-guided fine needle aspiration and/or biopsy of focal splenic lesions are being increasingly performed to obtain tissue for histopathological diagnosis of splenic tumors. This is important for planning optimal patient management.
- Tumors of the spleen can be divided into two main types:
  - Lymphoid tumors
  - Nonlymphoid tumors
- The most common nonlymphoid tumors of the spleen are the vascular tumors which include:
  - Benign and malignant hemangiomas
  - Littoral cell angiomas
  - Lymphangiomas
  - Hemangioendotheliomas
  - Fibrosarcoma
  - Neurinoma
  - Lipoma

A. H. Al-Salem, *The Spleen*, https://doi.org/10.1007/978-981-99-6191-7_22

- The lymphoid tumors include:
  - Hodgkin's and non-Hodgkin's lymphoma
  - Histiocytic lymphoma
  - Plasmacytoma
- Metastatic tumors to the spleen are rare and mainly originate from:
  - Melanoma
  - Breast carcinoma
  - Lung tumors

CLASSIFICATION OF SPLENIC TUMORS

- **LYMPHOID TUMORS**
  - **HODGKIN'S AND NON-HODGKIN'S LYMPHOMA**
  - **HISTOCYTIC LYMPHOMA**
  - **PLASMACYTOMA**
  - **CASTLEMAN'S TUMOR**
  - **LOCALIZED REACTIVE LYMPHOID HYPERPLASIA**
  - **INFLAMMATORY PSEUDOTUMOR**
- **NON-LYMPHOID TUMORS**
  - **BENIGN AND MALIGNANT HEMANGIOMAS**
  - **LITTORAL CELL ANGIOMA**
  - **LITTORAL CELL ANGIOSARCOMA**
  - **LYMPHANGIOMAS**
  - **HAEMANGIOENDOTHELIOMA**
  - **FIBROSARCOMA**
  - **NEURINOMA**
  - **LIPOMA AND ANGIOLIPOMA**
  - **MALIGNANT FIBROUS HISTOCYTOMA**
  - **LEIOMYOSCARCOMA**
  - **MALIGNANT TERATOMA**
  - **KAPOSI'S SARCOMA**
  - **HAEMANGIOPERICYTOMA**
  - **MALIGNANT HAEMANGIOSARCOMA**
- **METASTATIC TUMORS FROM:**
  - **MELANOMA**
  - **BREAST CARCINOMA**
  - **LUNG TUMORS**

22.2 Clinical Features

- The clinical features of splenic tumors are variable and depend on the site and size of the tumor.
- They may present with:
  - Upper abdominal quadrant discomfort, pain or tenderness.

- Laboratory investigations may reveal:
  - Anemia
  - Neutropenia
  - Thrombocytopenia
- Splenomegaly is the commonest clinical findings.
- Malignant tumors of the spleen may present with splenomegaly which is accompanied by:
  - Fever
  - Cachexia
  - Pleural effusion
- Massive splenomegaly (> 1000 gram) can be seen in some splenic tumors and this may cause displacement of adjacent abdominal viscera and pressure, leading to:
  - Dyspnea
  - Left shoulder pain
  - Constipation

22.3 Classification

- Primary splenic neoplasms can be divided into lymphoid neoplasms arising from the white pulp and vascular neoplasms, which arise from the red pulp.
- Primary vascular tumors:
  - Benign tumors:
    - Hemangioma
    - Lymphangioma
    - Hamartoma
  - Intermediate tumors:
    - Hemangioendothelioma
    - Hemangiopericytoma
    - Littoral cell angioma
  - Malignant hemangiosarcoma
- Splenic tumors are also classified as follows:
- Vascular tumors
  - Benign
    - Hemangioma
    - Lymphangioma
    - Littoral cell angioma
    - Hemangioendothelioma
- Malignant
  - Hemangiosarcoma
  - Littoral cell angiosarcoma
- Lymphoid tumors
  - Hodgkin's disease
  - Non-Hodgkin's lymphoma

  - Plasmacytoma
  - Castleman's tumor
  - Localized reactive lymphoid hyperplasia
  - Inflammatory pseudotumor
- Non-lymphoid tumors
  - Lipoma, angiolipoma
  - Malignant fibrous histiocytoma
  - Fibrosarcoma
  - Leiomyosarcoma
  - Malignant teratoma
  - Kaposi's sarcoma

22.4 Splenic Hemangioma

- Splenic hemangiomas remain the most common benign neoplasms of the spleen.
- Splenic hemangiomas are very rare.
- The reported incidence is 0.02% to 0.16%.
- In the majority of cases splenic hemangiomas are asymptomatic and found incidentally during laparotomy for other conditions or during radiological evaluation of other non-related complaints.
- Splenic hemangiomas are commonly seen in middle-aged adults and are seen equally in males and females.
- Splenic hemangiomas are commonly seen as an isolated lesion but they can be multiple or diffuse as seen in hemangiomatosis or be associated with generalized angiomatosis syndromes.
- Radiological features:
  - Splenic hemangiomas appear on CT scans are usually homogenous, hypodense or multicyclic.
  - They can be single or multiple lesions.
  - They may contain calcifications and on CT-scan this will be seen as punctate peripheral calcifications.
  - They generally demonstrate peripheral enhancement after intravenous contrast injection.
  - Abdominal ultrasound:

    Splenic hemangiomas on ultrasound appear as round, echogenic masses, with or without cystic areas.

    Smaller hemangiomas can be seen as discrete echogenic lesions but larger hemangiomas may have a more complex appearance.

    Splenic hemangiomas are considered as dysplasia with excessive vascular production.

    They represent an intermediate form, between true tumors and dysplasia and they resemble pseudotumor.

- Abdominal MRI:
  - Splenic hemangiomas are typically iso- to hypointense on T1-weighted images and hyperintense on T2-weighted images.
- Following contrast administration, splenic hemangiomas show varying patterns of enhancement:
  - Most lesions show peripheral enhancement in the arterial phase with progressive centripetal fill-in and persistent retention of contrast on the delayed phase.
  - The discontinuous peripheral puddles or globules of contrast enhancement are uncommonly seen in splenic hemangiomas.
  - A more continuous solid rim of enhancement is seen more often in the arterial phase.
  - Immediate homogeneous enhancement with persistent delayed enhancement can be seen
  - Peripheral enhancement with persistent lack of central enhancement on delayed images can sometimes be seen.
  - Cavernous hemangiomas typically show heterogeneous enhancement due to the presence of cystic non-enhancing areas within the lesion.
- Technetium-99 m labeled red blood cell (RBC) scans are also helpful in characterizing atypical lesions by demonstrating increased activity on delayed images.

- Pathological features
  - Macroscopically, hemangiomas appear as blood-filled cysts, occurring either singly or in groups.
  - The size of these lesions is variable.
  - Histologically, hemangiomas are divided into two types:
    - Capillary-type
    - Cavernous-type
  - Histologically, they are made up of vascular spaces lined by a single layer of endothelial cells.
  - Hemangiomas can be complicated by thrombosis and infraction as a result of abnormalities in their vascular supply.
  - Infracted hemangiomas when completely organized may resemble a fibroma or a healed, simple infract of the spleen.
  - Splenic hemangioma may arise from splenic venous structures, as well as splenic sinusoidal cells.
  - Hamartomas can be distinguished from hemangiomas by their usually more poorly defined margins.
- Although benign, large splenic hemangiomas may rupture causing intra-abdominal hemorrhage.
- The potential for malignant transformation of benign splenic hemangioma to angiosarcoma is unknown.
- Splenectomy is the treatment of choice for all cases of splenic hemangioma.

- Diffuse hemangiomatous of the spleen:
  - This is a very rare type of splenic hemangioma.
  - The spleen is diffusely replaced by blood vessels of varying sizes.
  - Typically, this is associated with generalized angiomatosis of other hematopoietic organs such as the bone marrow and the liver.
  - Diffuse splenic hemangiomatous is frequently associated with other hematologic abnormalities including:
    - Anemia
    - Thrombocytopenia
    - Coagulopathies such as microangiopathic hemolytic anemia

22.5 Lymphangioma

- Lymphangioma of the spleen is very rare.
- Splenic lymphangiomas are rare, benign, slow-growing tumors usually seen in childhood.
- Splenic lymphangiomas may present:
  - As an isolated lesion
  - As part of rare lymphangiomatosis syndrome involving multiple organs.
- It is commonly seen as part of disseminated lymphangiomatosis. This is characterized by multiple lymphangiomas involving other sites or organs, such as the bone, soft tissue, or viscera.
- Splenic lymphangiomas may be single or multiple.
- Splenic lymphangiomas are usually small multicystic and subcapsular.
- They are difficult to distinguish from either hemangioma or mesothelial cysts.
- Large splenic lymphangiomas may present as an abdominal mass in children.
- In adults they are often asymptomatic discovered as an incidental finding.
- Splenic lymphangiomas are usually seen as multiloculated cysts of varying size.
- Splenic lymphangiomas are usually subcapsular in location.
- Radiological features:
  - On ultrasound, splenic lymphangiomas are hypoechoic.
  - On CT scan, they are hypodense with few enhancing septa and occasional have a peripheral rim of calcification.
  - On MRI, splenic lymphangiomas show the signal intensity of simple fluid, but few cysts may demonstrate a high T1 signal due to proteinaceous or hemorrhagic content.
- Histology:
  - Splenic lymphangiomas are made up of thin-walled cystic structures of varying sizes.
  - They are lined by flat endothelium, filled with a pink eosinophilic proteinaceous material, and devoid of red blood cells.
  - In the past, these were mistaken with mesothelial cysts but immunophenotypic analysis showed evidence of mesothelial cysts lining cells rather than lymphatic lining cells.

- Splenic lymphangiomas tend to form foci of plump lining cells, often forming striking papillary projections.
- It is important to distinguish these papillary projection areas from angiosarcoma.

22.6 Angiosarcoma

- Primary angiosarcoma of the spleen is a very rare.
- It is an aggressive highly malignant tumor.
- It metastasizes early and has a poor prognosis.
- Angiosarcoma is the most common primary non-hematopoietic malignant tumor of the spleen.
- One of the common signs of angiosarcoma is splenomegaly.
- Spontaneous splenic rupture is common in patients with splenic angiosarcoma and this has been reported in approximately 25% of patients.
- Splenic angiosarcoma metastasizes early and most of the patients have metastasis at the time of presentation with the liver being the most common site of metastasis.
- It is commonly seen in patients who are 50–60 years of age.
- It is seen equally among males and females.
- Females on the other hand present at an older age than men.
- There is an association between splenic angiosarcoma and exposure to environmental factors such as:
  - Thorium dioxide
  - Ionizing radiation
  - Vinyl chloride
- Clinically, patients with splenic angiosarcoma present with:
  - Abdominal pain, typically localized to the left upper quadrant
  - Weakness and fatigue
  - Weight loss
  - Fever
  - Splenomegaly
  - Spontaneous splenic rupture with hemoperitoneum
- Investigations:
  - A normochromic, normocytic anemia is common in patients with splenic angiosarcoma
  - Thrombocytopenia
  - Low white blood cell counts
  - Some patients present with pancytopenia
  - Rarely, some patients have elevated white cells counts or thrombocytosis
  - Coagulopathy with elevated PT/PTT
  - Microangiopathic hemolytic anemia

- Radiological features:
  - On ultrasound, splenic angiosarcoma appear as multiple complex heterogeneous masses.
  - On contrast-enhanced CT, splenic angiosarcoma appear as multiple hyper vascular masses.
  - They have a heterogeneous appearance due to areas of hemorrhage and necrosis within the tumor.
  - Calcifications are rarely seen.
  - On MRI, splenic angiosarcoma appear heterogenous with areas of mixed high and low signal on T1- and T2-weighted images and heterogeneous enhancement is seen in the solid components of the tumor on contrast-enhanced images.
  - Intra- or peri splenic hemorrhage or frank hemoperitoneum can be seen on CT or MRI.
  - There is a high risk of massive hemorrhage in these tumors and percutaneous biopsy should be avoided.
- Histologically
  - Splenic angiosarcoma has an extremely variable histological appearance.
    - They usually have a spongiform or honeycomb-like proliferation.
    - This is made up of an irregularly anastomosing network of slit or capillary-like spaces.
    - This appearance is similar to that seen in other benign splenic vascular neoplasms but the presence of cytological atypia of the lining cells distinguishes these cases as angiosarcoma.
  - Histologically, these tumors have:
    - Mitotic figures
    - Multinucleated, bizarre, tumor giant cells
  - Immunohistochemical studies are important to confirm the vascular origin of these tumors.
- Prognosis
  - The prognosis of splenic angiosarcoma is poor.
  - The majority of patients rapidly succumb and dye of disseminated disease.
  - Most studies reveal that the majority of patients have median survivals of around 5–6 months and nearly all patients die within 3 years of diagnosis.
- Treatment
  - Splenectomy followed by radiation and/or chemotherapy is the treatment of choice
  - Metastasis is early and frequent in angiosarcoma, with rates of metastatic spread reported from 70% to 85%.
  - Common sites of metastatic disease are:
    - The liver
    - The lung
    - The lymph nodes
    - The Bones

  - Less frequent sites of metastatic disease are:
    - The brain
    - Soft tissue
    - Adrenal gland
  - The associated hematological abnormalities may also play a significant role in the patients' ultimate poor prognosis.
  - Patients with tumor size less than 5 cm at the time of diagnosis do better than those with larger size tumors.

22.7 Littoral Cell Angioma

- Littoral cell angioma is a rare vascular tumor arising from the littoral cell.
- Littoral cells make up the lining of the sinuses of splenic red pulp.
- Littoral cell angioma usually involves the spleen diffusely with multiple nodular masses of red pulp.
- Littoral cell angioma (LCA) of the spleen is a vascular tumor that is unique to the spleen.
- Littoral cell angioma arise from the lining cells of the splenic red pulp sinuses.
- Littoral cell angioma are commonly seen in middle age and affect both males and females equally.
- Presentation:
  - These tumors are commonly discovered incidentally during splenectomy or during radiological investigations of other unrelated abdominal conditions.
  - Sometimes these tumors cause splenomegaly leading to thrombocytopenia or anemia.
  - Littoral cell angioma is usually seen as multiple masses of varying sizes involving the entire spleen.
  - There are reports of patients with Littoral cell angioma presenting with fever that resolve after splenectomy.
  - Sometimes Littoral cell angioma is seen associated with other cancers and in these patients, it is discovered as an incidental finding during the radiological evaluation of the primary cancer.
- Radiological features
  - On CT-scan, Littoral cell angioma appear as:
    - Isodense on non-contrast CT
    - Hypodense on the early portal phase following contrast administration
    - Isodense on the delayed phase
  - On MRI, littoral cell angiomas may show low signal intensity on all sequences similar to siderotic nodules due to hemosiderin accumulation within neoplastic littoral cells.
- The diagnosis of littoral cell angioma should be suspected when multiple hypodense nodules are seen in a patient with splenomegaly and hypersplenism.
- Rarely malignant splenic vascular tumors are encountered with the morphological features of littoral cell angioma.

- These display malignant nuclear cytology and an infiltrative or solid growth pattern.
- Macroscopically
  - Littoral Cell Angiosarcoma appears as an ill-defined, non-lobulated pattern with alternating dark red-spongy nodules and interspersed solid white areas.
  - Littoral cell angioma appear as multiple, spongy, cystic blood filled, circumscribed nodules, ranging in size from 0.2 to 9 cm.
  - Less commonly, these lesions can be solitary or completely replace the splenic parenchyma.
- Microscopically
  - These lesions appear as a complex anastomosing network of vascular spaces of varying size, some with narrow compressed lumens; others are composed predominantly of large dilated blood-filled spaces.
  - A very characteristic finding of the LCAs is focal aggregates of eosinophilic globules 0.5–2 μ in size, which often entirely fill the tumor cell cytoplasm.
  - These globules originate from the phagocytized red blood cells, lymphocytes and plasma cells.
  - Several morphologic features of LCA, including the papillary projections and occasional solid areas, can make it difficult to differentiate Littoral cell angioma from angiosarcoma.
  - Their morphologic features as well as their immunohistochemical analysis are helpful in distinguishing Littoral cell angioma from angiosarcoma.

22.8 Hemangioendothelioma

- Hemangioendothelioma of the spleen is a vascular tumor with morphologic and clinical properties that are intermediate between those of the hemangioma and angiosarcoma.
- Their malignant potential is best described as borderline or intermediate.
- Their microscopic features include:
  - Ill-defined vascular spaces
  - epithelioid morphology with mild cellular atypia
  - A low mitotic rate
  - Absence of necrosis
- Clinical features:
  - The clinical features of hemangioendothelioma are variable and can present with:
    - Splenomegaly or a palpable abdominal mass
    - Chronic anemia
    - Sometimes they present with splenic rupture
  - The behavior of these tumors is difficult to predict from their morphological features.
  - They can present as several well-circumscribed hemangioendotheliomas with spindle and/or epithelioid morphology and no evidence of metastatic spread.

- Hemangioendothelioma can be more aggressive with extra-splenic spread.
- Careful attention to morphologic features and correlation with clinical findings in these tumors is important.
- The presence of atypical cytological features on histology calls for close follow-up and monitoring of these patients.

22.9 Lymphomas

- Primary involvement of the spleen by lymphoma is much less common than secondary involvement, accounting for less than 1% of all lymphomas.
- Primary splenic lymphoma usually represents non-Hodgkin lymphoma of B cell origin.
- Secondary involvement of the spleen is much more common. Splenic involvement is commonly seen in association with lymphadenopathy in the rest of abdomen.
- Radiologically four different patterns of involvement of splenic lymphoma have been described:
  - Diffuse infiltration, which manifests as splenomegaly
  - Small, focal or miliary nodules
  - Multiple large nodular lesions
  - Bulky solid masses
- On contrast-enhanced computed tomography (CT), Lymphoma nodules appear as low-density lesions.
- On magnetic resonance imaging (MRI) these lesions are isointense on precontrast T1- and T2-weighted sequences, but hypointense on post-contrast sequences.
- Diffuse infiltration or tiny (<1 cm) nodules are seen in 45–70% of splenic lymphoma and sometimes these nodules can be difficult to visualize on CT and MRI.
- Fluorodeoxyglucose (FDG)-positron emission tomography (PET) imaging can be helpful in identifying lymphomatous involvement in these cases.
- Treated lymphoma nodules or deposits may appear necrotic and eventually can show areas of calcification within them.

22.10 Hodgkin's Lymphoma

- Lymphoid malignancies are uncommon as primary splenic tumors.
- The spleen is often the site of secondary involvement by Hodgkin's disease.
- Histological examination following splenectomy is the most accurate method to confirm splenic involvement.
- When lymphoma does involve the spleen, either as a primary or secondary lymphoma, the white pulp of the spleen is involved first.
- Splenic involvement may be:
  - Diffuse involvement
  - Or as large, irregular tumor masses

- Splenomegaly in these patients does not always indicate a malignant involvement.
- On the other hand, a normal spleen size does not exclude malignancy.
- Splenic involvement is often associated with increased incidence of liver and bone marrow involvement.

22.11 Non-Hodgkin's Lymphoma

- Secondary splenic involvement in non-Hodgkin's lymphoma is reported in 50–80% in patients.
- Primary splenic involvement with non-Hodgkin's lymphoma is much less common.
- The presence of splenomegaly with other small lesions, possibly microscopic, in other organs, raises the possibility of primary splenic non-Hodgkin's lymphoma.
- The size and weight of the spleen in these patients is variable and does not correlate with the prognosis.
- Macroscopically these tumors appear as:
  - Diffuse involvement of the spleen
  - Small nodular tumors
  - Large nodular tumors
  - A single large tumor mass
  - Many infracts or blue subcapsular dots can be seen.
- Benign lymphoid lesions may also cause splenomegaly and may mimic lymphoma.
- Reactive lymphoid hyperplasia and inflammatory pseudotumor have been also described.
- Castleman's tumor:
  - Angio follicular lymphoid hyperplasia, also known as Castleman's tumor may also affect the spleen.
  - This may appear either as a solitary lesion or as a part of a diffuse lymphoid hyperplasia syndrome.
  - Castleman's tumor is thought to be a true lymphoid hamartoma, but does not resemble the follicular pattern of splenic hamartoma described by Berge.
- Primary plasmacytoma of the spleen is exceedingly rare and is not macroscopically recognizable, but is easily identified histologically as a plasma cell neoplasm.
- The spleen may also be involved as a primary, isolated site or as a part of generalized myelomatosis.

22.12 Non-lymphoid Tumors

- Primary mesenchymal tumors of the spleen are very rare. These include the followings:
  - Lipoma
  - Angiomyolipoma

  - Fibroma
  - Fibrosarcoma
  - Leiomyosarcoma
  - Malignant fibrous histiocytoma
- Lipoma of the spleen is very rare.
- Usually, it appears as round mass of lipoid tissue.
- Angiolipomas have rarely been reported to affect the spleen.
- Primary splenic malignant fibrous histiocytoma has also been described to affect the spleen.
  - Malignant fibrous histiocytoma is a distinct, polymorphic sarcoma.
  - It usually occurs in the soft tissues, but has also been found in a number of organs, including the spleen.
  - Primary splenic malignant fibrous histiocytoma usually causes massive splenomegaly.
  - This is a very aggressive tumor.
  - Malignant fibrous histiocytoma may be misdiagnosed as fibrosarcoma or leiomyosarcoma, both of which also occur in the spleen.
- Neurinoma or schwannoma of the spleen has also been described but it is rare.
- There is a report of a malignant teratoma of the spleen, with papillary carcinoma, spindle cell carcinoma and cartilaginous tissue.
- Kaposi's sarcoma has not been described as a primary splenic tumor.
  - This is seen in patients in whom the spleen is involved as part of a general sarcomatosis.
  - The incidence of Kaposi's sarcoma has greatly increased, mainly in association with human immunodeficiency virus infection (HIV) and the acquired immunodeficiency syndrome (AIDS).
  - It is a spindle cell malignant neoplasm.
  - It is characterized by vascular spaces not generally lined by endothelial cells and may contain blood.
- Splenic inflammatory pseudotumor
  - Splenic inflammatory pseudotumor is a rare benign tumor of unknown etiology.
  - These tumors may develop secondary to an unusual inflammatory response to injury such as infection.
  - It is usually asymptomatic discovered incidentally.
  - It may also present with symptoms.
  - On CT- scan it appears as a well-circumscribed mass, which may show areas of calcification.
  - Enhancement can be homogeneous or heterogeneous and delayed enhancement patterns have been described.
  - A definitive diagnosis is established only after surgery.
- Hamartomas or Splenomas of the spleen
  - These are rare, benign splenic tumors.
  - They are dysplastic nodules that may be misdiagnosed as neoplasms.
  - It is composed of malformed splenic red pulp elements without organized lymphoid follicles.

- The exact cause of hamartoma is unknown.
- Their incidence has been estimated at 3 in 200.000 splenectomies.
- They occur equally in males and females, at all ages, but particularly in older people.
- Hamartomas are usually discovered incidentally during evaluation of other unrelated medical conditions.
- Rarely, they present as a large mass or cause splenomegaly.
- Most of the reported cases have been found at autopsy.
- Radiological evaluation:

  On ultrasound, hamartomas are usually seen as well-circumscribed homogeneous solid masses.

  On CT-scan and MRI, hamartomas are isodense on non-contrast CT and are usually isointense on T1-weighted images and heterogeneously hyperintense on T2-weighted images.

  Following contrast administration, hamartomas show diffuse early enhancement and uniform delayed enhancement appearing isodense/isointense to the remainder of the spleen and hence on the delayed phase a contour abnormality may be the only visible finding.

  In larger lesions, heterogeneous enhancement may be observed with uniform and prolonged retention of contrast on delayed images.

  The persistent delayed enhancement is thought be due to stagnant contrast material within the sinusoids of the red pulp in the hamartoma.
- Typically, they appear as well-circumscribed lesion, which appears darker than the surrounding spleen.
- Their size varies from a few mms to many CMS.
- They are easily diagnosed microscopically
- The have a distinct border and slit like and tortuous endothelial-lined spaces.
- Sometimes they can cause splenomegaly, but hypersplenism is rare.
- Splenic hamartoma can also rupture causing life-threatening hemoperitoneum.
- A definitive diagnosis of hamartoma is based on percutaneous biopsy or following splenectomy.

22.13 Metastatic Tumors to the Spleen

- Metastatic tumors of the spleen are relatively rare.
- The overall incidence of splenic metastases was reported as 7.1%.
- The incidence of splenic metastasis from colon and rectal carcinomas was reported as 4.4% and 1.6% respectively.
- Common primary carcinomas that metastasize to the spleen include:
  - Carcinoma of breast
  - Carcinoma lung
  - Melanoma

- Retroperitoneal tumors and pancreatic carcinoma can also spread and invade the spleen via direct extension.
- Almost all primary tumors have been shown to be able to metastasize to the spleen but this rare.
- Most cases of metastatic tumors to the spleen are asymptomatic discovered by imaging investigations such as abdominal ultrasound (US) or computerized tomography (CT).
- Other presentations include:
  - Splenomegaly
  - Left pleural effusion
  - Spontaneous rupture causing hemoperitoneum
- Hematogenous spread is the most probable route of spread resulting in splenic metastases.
- Splenic metastases are seen in only 2–9% of untreated cancer patients and are seen as isolated splenic lesions in 5.2% of patients.
- Tumors that most commonly metastasize to the spleen include:
  - Melanoma
  - Tumors of the breast
  - Tumors of the lung
  - Tumors of the ovary
  - Tumors of the colon
  - Tumors of the stomach
  - Tumors of the pancreas
- Most metastases to the spleen are solitary but multiple masses have been reported.
- Diffuse infiltration of the spleen by secondaries is very rare.
- Radiological evaluation
  - On ultrasound, splenic metastasis is variable and most lesions appear hypoechoic.
  - On contrast-enhanced CT, splenic metastasis appears as hypodense lesions that are best appreciated on the portal venous phase.
  - On MRI, splenic metastasis may be difficult to identify on the precontrast T1- and T2 weighted sequences and are seen as hypointense lesions on T1-weighted images following contrast administration.
  - Cystic or necrotic degeneration can occur in splenic metastasis. This is usually seen in metastasis from melanoma and the lesions appear as unilocular or multilocular cystic lesions with septations.
  - Calcification in splenic metastasis is rare unless the primary tumor is a mucinous adenocarcinoma.
- Implants on the serosal surface of the spleen are seen in patients with peritoneal carcinomatosis, commonly from ovarian or gastrointestinal primary neoplasms.
- These implants are seen on CT and they cause indentation and scalloping of the surface of the spleen.
- Pseudomyxoma peritonei also often involves the spleen and is seen as scalloping of the splenic margin.

- Direct tumor invasion of the spleen is uncommon, but can be seen in tumors originating from the:
  - Pancreas
  - Stomach
  - Colon
  - Left kidney
  - Retroperitoneum

22.14 Non-neoplastic Splenic Lesions Mimicking Splenic Neoplasms

- Granulomatous diseases and sarcoidosis
  - Sarcoidosis is a multisystem disease and in 5–15% of patients there is abdominal involvement.
  - Sarcoidosis affecting the spleen usually manifest as splenomegaly and less commonly as multiple nodules of the spleen.
  - Splenic sarcoidosis can be associated with hepatic involvement and lymphadenopathy in some cases.
  - These nodules on CT-scan appear as hypodense lesions.
  - On MRI, they are typically hypointense and show minimal and delayed enhancement following contrast administration.
  - The nodular pattern of involvement can mimic lymphoma or metastatic disease to the spleen.
  - The presence of larger, confluent lymph nodes suggests the possibility of lymphoma.
  - The presence of a primary malignancy suggests metastatic tumors to the spleen.
- Peliosis of spleen
  - Peliosis of the spleen is a rare condition.
  - It is secondary to splenic sinusoidal dilation leading to formation of multiple cyst-like blood-filled cavities within the splenic parenchyma.
  - Splenic peliosis is usually discovered as an incidental finding in an asymptomatic patient.
  - Spontaneous splenic rupture has been reported in patients with peliosis.
  - On CT-scan, multiple small well-defined, hypodense cyst-like lesions are seen.
  - The cystic spaces contain blood components within them that might lead to the presence of fluid-fluid levels and variable signal intensity on MRI.
  - The lesions may or may not enhance after contrast administration.
  - The cystic nature of the lesions should be helpful in distinguishing peliosis from lymphoma or metastasis, which are typically solid in nature.

Further Reading

1. Kutoc JL, Fletcher C. Splenic vascular tumors. Semin Diagn Pathol. 2003;20(2):128–39.
2. Willcox TM, Speer RW, Schlinkert RT, Sarr MG. Haemangioma of the spleen: Presentation, diagnosis and management. J Gastrointest Surg. 2000;4:611–3.
3. Arber DA, Strickler JG, Chen YY, et al. Splenic vascular tumors: A histologic, immunophenotypic and virologic study. Am J Surg Pathol. 1997;21:827–35.
4. Goerg C, Schwerk WB, Goerg K. Splenic lesions: sonographic patterns, follow-up, differential diagnosis. Eur J Radiol. 1991;13:59–66.
5. Neuhauser TS, Derringer GA, Thompson LD, et al. Splenic angiosarcoma: A clinicopathologic and immunophenotypic study of 28 cases. Mod Pathol. 2000;13:978–87.
6. Ruck P, Horny HP, Xiao JC, et al. Diffuse sinusoidal haemangiomatosis of the spleen. A case report with enzyme-histochemical, immunohistochemical and electron-microscopic findings. Pathol Res Pract. 1994;190:708–14.
7. Morgenstern L, Bello JM, Bl F, et al. The clinical spectrum of lymphangiomas and lymphangiomatosis of the spleen. Am Surg. 1992;58:599–604.
8. Komatsuda T, Ishida H, Konno K, et al. Splenic lymphangioma: US and CT diagnosis and clinical manifestations. Abdom Imaging. 1999;24:414–7.
9. Iwabuchi A, Otaka M, Okuyama A, et al. Disseminated intraabdominal cystic lymphangiomatosis with severe intestinal bleeding. A case report. J Clin Gastroenterol. 1997;25:383–6.
10. Morphis LG, Arcinue EL, Krausee JR. Generalized lymphangioma in infancy with chylothorax. Pediatrics. 1970;46:566–75.
11. Arber DA, Strickler JG, Weiss LM. Splenic mesothelial cysts mimicking lymphangiomas. Am J Surg Pathol. 1997;21:334–8.
12. Ziske C, Meybehm M, Sauerbruch T, et al. Littoral cell angioma as a rare case of splenomegaly. Ann Hematol. 2001;80:45–8.
13. Schneider G, Uder M, Altmeyer K, et al. Littoral cell angioma of the spleen: CT and MR imaging appearance. Eur Radiol. 2000;10:1395–400.
14. Bisceglia M, Sickel JZ, Giangaspero F, et al. Littoral cell angioma of the spleen: An additional report of four cases with emphasis on the association with visceral organ cancers. Tumori. 1998;84:595–9.
15. Michal M, Skalova A, Fakan F, et al. Littoral cell angioma of the spleen. A case report with ulrtastructural and immunohistochemical observations. Zentralbl Pathol. 1993;139:361–5.
16. Falk S, Krishnan J, Meis JM. Primary angiosarcoma of the spleen. A clinicopathologic study of 40 cases. Am J Surg Pathol. 1993;17(10):959–70.
17. Sordillo EM, Sordillo PP, Hajdu SI. Primary angiosarcoma of the spleen: report of 4 cases. Med Pediatr Oncol. 1981;9:319–24.
18. Thompson WM. Angiosarcoma of the spleen: Imaging characteristics in 12 patients. Radiology. 2005;235:106–15.
19. Rosso R, Paulli M, Gianelli U, et al. Littoral cell angiosarcoma of the spleen. Case report with immunohistochemical and ultrastructural analysis. Am J Surg Pathol. 1995;19:1203–8.
20. Kaw YT, Duwaji MS, Knisley RE, et al. Hemangioendothelioma of the spleen. Arch Pathol Lab Med. 1992;116:1079–82.
21. Ferrozzi F, Bova D, De Chiara F. Hemangioendothelioma of the spleen: Imaging findings at color Doppler, US and CT. Clin Imaging. 1999;23:111–4.
22. Farrer-Brown G, Bennet MG, Harisson CV, Millet Y, et al. The diagnosis of Hodgkin's disease in surgically excised spleens. J Clin Pathol. 1981;25:294.
23. Abbott RM, Levy AD, Aguilera NS, Gorospe L, Thompson WM. From the archives of the AFIP: primary vascular neoplasms of the spleen: radiologic-pathologic correlation. Radiographics. 2004;24:1137–63.
24. Abdel-Wahab M, Abou-Elenin A, Sultan A, El-Ghawalpy N, Ezzat F. Lymphangiomatous cysts of the spleen: report of 3 cases and review of the literature. Hepatogastroenterology. 1998;45:2101–4.

25. Arber DA, Kamel OW, van de Rijn M, et al. Frequent presence of the Epstein-Barr virus in inflammatory pseudotumor. Hum Pathol. 1995;26:1093–8.
26. Asayama Y, Fukuya T, Honda H, et al. Chronic expanding hematoma of the spleen caused by angiomyolipoma in a patient with tuberous sclerosis. Abdom Imaging. 1998;23:527–30.
27. Autry JR, Weitzner S. Hemangiosarcoma of spleen with spontaneous rupture. Cancer. 1975;35:534–9.
28. Awamleh AA, Perez-Ordonez B. Sclerosing angiomatoid nodular transformation of the spleen. Arch Pathol Lab Med. 2007;131:974–8.
29. Bean MJ, Horton KM, Fishman EK. Concurrent focal hepatic and splenic lesions: a pictorial guide to differential diagnosis. J Comput Assist Tomogr. 2004;28:605–12.
30. Ben-Izhak O, Bejar J, Ben-Eliezer S, Vlodavsky E. Splenic littoral cell haemangioendothelioma: a new low-grade variant of malignant littoral cell tumor. Histopathology. 2001;39:469–75.
31. Brinkley AA, Lee JK. Cystic hamartoma of the spleen: CT and sonographic findings. J Clin Ultrasound. 1981;9:136–8.
32. Burke JS. Surgical pathology of the spleen: an approach to the differential diagnosis of splenic lymphomas and leukemias. Part I. Diseases of the white pulp. Am J Surg Pathol. 1981;5:551–63.
33. Chen KT, Bolles JC, Gilbert EF. Angiosarcoma of the spleen: a report of two cases and review of the literature. Arch Pathol Lab Med. 1979;103:122–4.
34. Coffin CM, Watterson J, Priest JR, Dehner LP. Extrapulmonary inflammatory myofibroblastic tumor (inflammatory pseudotumor). A clinicopathologic and immunohistochemical study of 84 cases. Am J Surg Pathol. 1995;19:859–72.
35. Dachman AH, Buck JL, Krishnan J, Aguilera NS, Buetow PC. Primary non-Hodgkin's splenic lymphoma. Clin Radiol. 1998;53:137–42.
36. de Jong PA, van Ufford HM, Baarslag HJ, et al. CT and 18 F-FDG PET for noninvasive detection of splenic involvement in patients with malignant lymphoma. AJR Am J Roentgenol. 2009;192:745–53.
37. Elsayes KM, Narra VR, Mukundan G, Lewis JS Jr, Menias CO, Heiken JP. MR imaging of the spleen: spectrum of abnormalities. Radiographics. 2005;25:967–82.
38. Falk S, Stutte HJ. Hamartomas of the spleen: a study of 20 biopsy cases. Histopathology. 1989;14:603–12.
39. Falk S, Stutte HJ, Frizzera G. Littoral cell angioma. A novel splenic vascular lesion demonstrating histiocytic differentiation. Am J Surg Pathol. 1991;15:1023–33.
40. Falk S, Krishnan J, Meis JM. Primary angiosarcoma of the spleen. A clinicopathologic study of 40 cases. Am J Surg Pathol. 1993;17:959–70.
41. Feigenberg Z, Wysenbeek A, Avidor E, Dintsman M. Malignant lymphangioma of the spleen. Isr J Med Sci. 1983;19:202–4.
42. Ford CD, Gabor F, Morgan R, Dabbas B. False-positive restaging PET scans involving the spleen in two patients with aggressive non-Hodgkin lymphoma. Clin Nucl Med. 2006;31:391–3.
43. Garvin DF, King FM. Cysts and nonlymphomatous tumors of the spleen. Pathol Annu. 1981;16(Pt 1):61–80.
44. Giovagnoni A, Giorgi C, Goteri G. Tumors of the spleen. Cancer Imaging. 2005;5:73–7.
45. Gupta MK, Levin M, Aguilera NS, Pastores GM. Littoral cell angioma of the spleen in a patient with Gaucher disease. Am J Hematol. 2001;68:61–2.
46. Karaosmanoglu DA, Karcaaltincaba M, Akata D. CT and MRI findings of sclerosing angiomatoid nodular transformation of the spleen: spoke wheel pattern. Korean J Radiol. 2008;9(Suppl):S52–5.
47. Kehoe J, Straus DJ. Primary lymphoma of the spleen. Clinical features and outcome after splenectomy. Cancer. 1988;62:1433–8.
48. Komaki S, Gombas OF. Angiographic demonstration of a calcified splenic hamartoma. Radiology. 1976;121:77–8.
49. Kraus MD, Fleming MD, Vonderheide RH. The spleen as a diagnostic specimen: a review of 10 years' experience at two tertiary care institutions. Cancer. 2001;91:2001–9.

50. Lam KY, Tang V. Metastatic tumors to the spleen: a 25-year clinicopathologic study. Arch Pathol Lab Med. 2000;124:526–30.
51. Lam KY, Yip KH, Peh WC. Splenic vascular lesions: unusual features and a review of the literature. Aust N Z J Surg. 1999;69:422–5.
52. Lee JC, Lien HC, Hsiao CH. Coexisting sclerosing angiomatoid nodular transformation of the spleen with multiple calcifying fibrous pseudotumors in a patient. J Formos Med Assoc. 2007;106:234–9.
53. Lee D, Wood B, Formby M, Cho T. F-18 FDG-avid sclerosing angiomatoid nodular transformation (SANT) of the spleen: case study and literature review. Pathology. 2007;39:181–3.
54. Li L, Fisher DA, Stanek AE. Sclerosing angiomatoid nodular transformation (SANT) of the spleen: addition of a case with focal CD68 staining and distinctive CT features. Am J Surg Pathol. 2005;29:839–41.
55. Martel M, Cheuk W, Lombardi L, Lifschitz-Mercer B, Chan JK, Rosai J. Sclerosing angiomatoid nodular transformation (SANT): report of 25 cases of a distinctive benign splenic lesion. Am J Surg Pathol. 2004;28:1268–79.
56. Metser U, Miller E, Kessler A, et al. Solid splenic masses: evaluation with 18 F-FDG PET/CT. J Nucl Med. 2005;46:52–9.
57. Meybehm M, Fischer HP. Littoral cell angiosarcoma of the spleen. Morphologic, immunohistochemical findings and consideration of histogenesis of a rare splenic tumor. Pathologe. 1997;18:401–5.
58. Meyer JE, Harris NL, Elman A, Stomper PC. Large-cell lymphoma of the spleen: CT appearance. Radiology. 1983;148:199–201.
59. Musgrave NJ, Williamson RM, O'Rourke NA, Searle JW. Test and teach. Incidentally discovered splenic vascular lesion. Littoral cell angioma of the spleen. Pathology. 2002;34:579–81.
60. Neuhauser TS, Derringer GA, Thompson LD, et al. Splenic angiosarcoma: a clinicopathologic and immunophenotypic study of 28 cases. Mod Pathol. 2000;13:978–87.
61. Okada M, Sato N, Ishii K, Matsumura K, Hosono M, Murakami T. FDG PET/CT versus CT, MR imaging, and 67 Ga scintigraphy in the posttherapy evaluation of malignant lymphoma. Radiographics. 2010;30:939–57.
62. Oliver-Goldaracena JM, Blanco A, Miralles M, Martin-Gonzalez MA. Littoral cell angioma of the spleen: US and MR imaging findings. Abdom Imaging. 1998;23:636–9.
63. Rabushka LS, Kawashima A, Fishman EK. Imaging of the spleen: CT with supplemental MR examination. Radiographics. 1994;14:307–32.
64. Dachman A. Radiology of the spleen. St. Louis, MO: Mosby; 1993.
65. Kamaya A, Weinstein S, Desser TS. Multiple lesions of the spleen: differential diagnosis of cystic and solid lesions. Semin Ultrasound CT MR. 2006;27:389–403.
66. Spier CM, Kjeldsberg CR, Eyre HJ, Behm FG. Malignant lymphoma with primary presentation in the spleen. A study of 20 patients. Arch Pathol Lab Med. 1985;109:1076–80.
67. Bhatia K, Sahdev A, Reznek RH. Lymphoma of the spleen. Semin Ultrasound CT MR. 2007;28:12–20.
68. Ahmann DL, Kiely JM, Harrison EG Jr, Payne WS. Malignant lymphoma of the spleen. A review of 49 cases in which the diagnosis was made at splenectomy. Cancer. 1966;19:461–9.
69. Thompson WM, Levy AD, Aguilera NS, Gorospe L, Abbott RM. Angiosarcoma of the spleen: imaging characteristics in 12 patients. Radiology. 2005;235:106–15.
70. Lucey BC, Boland GW, Maher MM, Hahn PF, Gervais DA, Mueller PR. Percutaneous nonvascular splenic intervention: a 10-year review. AJR Am J Roentgenol. 2002;179:1591–6.
71. Goerg C, Schwerk WB, Goerg K. Sonography of focal lesions of the spleen. AJR Am J Roentgenol. 1991;156:949–53.
72. Ramani M, Reinhold C, Semelka RC, et al. Splenic hemangiomas and hamartomas: MR imaging characteristics of 28 lesions. Radiology. 1997;202:166–72.
73. Wijaya J, Kapoor R, Roach P. Tc-99 m-labeled RBC scintigraphy and splenic hemangioma. Clin Nucl Med. 2001;26:1022–3.
74. Wadsworth DT, Newman B, Abramson SJ, Carpenter BL, Lorenzo RL. Splenic lymphangiomatosis in children. Radiology. 1997;202:173–6.

75. Urrutia M, Mergo PJ, Ros LH, Torres GM, Ros PR. Cystic masses of the spleen: radiologic-pathologic correlation. Radiographics. 1996;16:107–29.
76. Morgenstern L, McCafferty L, Rosenberg J, Michel SL. Hamartomas of the spleen. Arch Surg. 1984;119:1291–3.
77. Chou YH, Chiou HJ, Tiu CM, Chiou SY, Hsia CY, Tsay SH. Splenic hamartoma: presentation on contrast-enhanced sonography. J Clin Ultrasound. 2004;32:425–8.
78. Yu RS, Zhang SZ, Hua JM. Imaging findings of splenic hamartoma. World J Gastroenterol. 2004;10:2613–5.
79. Ohtomo K, Fukuda H, Mori K, Minami M, Itai Y, Inoue Y. CT and MR appearances of splenic hamartoma. J Comput Assist Tomogr. 1992;16:425–8.
80. Lee SH. Fine-needle aspiration cytology of splenic hamartoma. Diagn Cytopathol. 2003;28:82–5.
81. Levy AD, Abbott RM, Abbondanzo SL. Littoral cell angioma of the spleen: CT features with clinicopathologic comparison. Radiology. 2004;230:485–90.
82. Inada T, Yano T, Shima S, et al. Inflammatory pseudotumor of the spleen. Intern Med. 1992;31:941–5.
83. Irie H, Honda H, Kaneko K, et al. Inflammatory pseudotumors of the spleen: CT and MRI findings. J Comput Assist Tomogr. 1996;20:244–8.
84. Warshauer DM, Hall HL. Solitary splenic lesions. Semin Ultrasound CT MR. 2006;27:370–88.
85. Schon CA, Gorg C, Ramaswamy A, Barth PJ. Splenic metastases in a large unselected autopsy series. Pathol Res Pract. 2006;202:351–6.
86. Hahn PF, Weissleder R, Stark DD, Saini S, Elizondo G, Ferrucci JT. MR imaging of focal splenic tumors. AJR Am J Roentgenol. 1988;150:823–7.
87. Williams L, Kumar A, Aggarwal S. Calcified splenic metastasis from gastric carcinoma. Abdom Imaging. 1995;20:312–4.
88. Taylor AJ, Dodds WJ, Erickson SJ, Stewart ET. CT of acquired abnormalities of the spleen. AJR Am J Roentgenol. 1991;157:1213–9.
89. Warshauer DM, Lee JK. Imaging manifestations of abdominal sarcoidosis. AJR Am J Roentgenol. 2004;182:15–28.
90. Warshauer DM. Splenic sarcoidosis. Semin Ultrasound CT MR. 2007;28:21–7.
91. Tsokos M, Erbersdobler A. Pathology of peliosis. Forensic Sci Int. 2005;149:25–33.
92. Shimono T, Yamaoka T, Nishimura K, et al. Peliosis of the spleen: splenic rupture with intraperitoneal hemorrhage. Abdom Imaging. 1998;23:201–2.
93. Maves CK, Caron KH, Bisset GS 3rd, Agarwal R. Splenic and hepatic peliosis: MR findings. AJR Am J Roentgenol. 1992;158:75–6.
94. Ng KK, Lee TY, Wan YL, et al. Splenic abscess: diagnosis and management. Hepatogastroenterology. 2002;49:567–71.

Splenic Artery Aneurysm 23

23.1 Introduction

- Aneurysms of the visceral arteries are relatively rare.
- Approximately 25% of visceral arteries aneurysms present with rupture resulting in a high mortality of 8.5%.
- Early diagnosis and treatment are important to avoid the associated morbidity and mortality.
- Splenic artery aneurysms are rare and account for 60% of all visceral artery aneurysms.
- The estimated prevalence of splenic artery aneurysms is about 0.8% in the population.
- Splenic artery aneurysm was first reported by Beaussier in 1770.
- In 1844, Parker reported another case of splenic artery aneurysm.
- The first successful treatment of a ruptured splenic artery aneurysm was reported by McLeod in 1940.
- Splenic artery aneurysm is much more common in females than males and the incidence is four times higher in women than in men (a female: male ratio of 4:1).
- Splenic artery aneurysms usually present in the fifth and sixth decade of life with a mean age of presentation at 52 years. The majority of splenic artery aneurysms occur in patients above the age of 50 (80%).
- Splenic artery aneurysm can rupture during pregnancy. This is a rare but serious obstetrical complication.
- Splenic artery aneurysm rupture during pregnancy was first reported in the literature by the Danish obstetrician Sylver Saxtorph in 1803.
- Pregnancy is considered as one of the most significant risk factors for the rupture of a splenic artery aneurysm.
- It was reported that 50% of splenic artery aneurysm ruptures occur in pregnant women.
- About 95% of all pregnancy-associated splenic artery aneurysm are attributed to rupture of the aneurysm.

A. H. Al-Salem, *The Spleen*, https://doi.org/10.1007/978-981-99-6191-7_23

- The ruptures occur predominantly in multiparous women, 40% of whom are grand multiparous.
- The highest risk of rupture for a splenic artery aneurysm is during the third trimester.
  - This accounting for 69% of ruptures and typically occurs in the last two weeks of pregnancy.
  - 12% of ruptures occur during the first and second trimesters of pregnancy
  - 13% of ruptures occur during labor.
  - 6% of ruptures occur postpartum.
- Patients with splenic artery aneurysm are more often asymptomatic and the majority of these aneurysms are small in size, less than 2 cm in diameter.
- Splenic artery aneurysms are saccular and located in the middle of the splenic artery or in its distal segment.
- About 20% of splenic artery aneurysm presents with symptoms including:
  - Vague epigastric or left upper quadrant abdominal pain.
  - The pain can radiate to the left shoulder.
  - The classic presentation of a ruptured splenic artery aneurysm occurs in 25% of cases.
  - The presentation of a ruptured splenic artery aneurysm which is also known as the double rupture phenomenon occurs in two stages:

    At first, the initial hemorrhage is limited when blood clots block the foramen of Winslow, temporarily containing the hemorrhage to the lesser sac. This is followed 6–96 hours later by free rupture into the peritoneal cavity with hemodynamic instability that can lead to sudden maternal and fetal death.
- The risk of splenic artery aneurysm rupture in pregnancy is 2–3%.
- The incidence of rupture is 3.6 occurrences per 100,000 pregnancies.
- The mortality outside of pregnancy ranges from 10–25%.
- The mortality during pregnancy increases markedly.
  - The maternal mortality increases up to 75%
  - The fetal mortality increases to as high as 95%.
- The highest risk of rupture is during pregnancy.
- The risk of rupture appears to be strongly associated with the diameter of the aneurysm.
- The size of the splenic artery aneurysm is usually more than 2.5 cm in diameter in most patients at the time of rupture.
- The diameter of the aneurysm however does not reflect the probability of rupture.
- 70% of splenic artery aneurysm ruptures during pregnancy are misdiagnosed as:
  - Uterine rupture
  - Placental abruption
- Splenic artery aneurysm should be considered in the differential diagnosis of any pregnant woman with abdominal pain, especially in those who are hemodynamically unstable.

23.2 Etiology

- The exact cause of splenic artery aneurysms is not known.
- The most common pathologic finding is a defect of the tunica media, with loss of elastic fibers and smooth muscles.
- Although atherosclerosis is commonly observed in patients with splenic artery aneurysms but this may not be a primary cause.
- The increased blood flow in the splenic artery during pregnancy is a contributing factor.
- The pathogenesis of splenic artery aneurysm is not well known.
  - An increase in circulating estrogen, progesterone, and relaxin levels during pregnancy is thought to increase blood flow and blood pressure within the splenic artery.
  - This will weaken the arterial wall specially if there is pre-existing atherosclerosis leading to dilatation of the artery and formation of the aneurysm.
- Splenic artery aneurysms are more frequently seen in patients with:
  - Arteria fibromuscular dysplasia
  - Portal hypertension
  - Infections
  - Congenital anomalies
  - Transplanted liver
  - Pancreatic carcinoma
- Splenic artery aneurysms are much more common in women than men, particularly multiparous.
  - This is a result of hormonal and hemodynamic changes during pregnancy that promotes intimal hyperplasia of the vessel and subsequent fragmentation, facilitating the development of aneurysm.
- Splenic artery aneurysms are also seen in patients with inflammatory changes of the arterial wall including:
  - Polyarteritis nodosa
  - Bacterial endocarditis
  - Enzymatic digestion of the wall by a previous attack of pancreatitis.

23.3 Clinical Features

- It is estimated that splenic artery aneurysm occurs in approximately 0.1% of all adults.
- Splenic artery aneurysm constitutes 60% of all visceral arterial aneurysms
- Splenic artery aneurysm is the third most common abdominal aneurysm after aortic and iliac artery aneurysms.
- It is estimated that 6% to 10% of splenic artery aneurysms will rupture.
- It is estimated that 25% to 40% of splenic artery aneurysms ruptures will occur during pregnancy, especially during the third trimester. This is one of the causes of maternal and fetal mortality.

- Maternal mortality for ruptured splenic artery aneurysm is around 75%, and fetal mortality is up to 95%.
- In nonpregnant patients the mortality rate for splenic artery aneurysm rupture is about 25%.
- The main risk factors for splenic artery rupture include:
  - Portal hypertension
  - Pregnancy
- Patients with a splenic artery aneurysm are usually asymptomatic.
- Splenic artery aneurysm is typically seen in a multiparous woman.
- The average number of pregnancies per patient was reported as 4.5.
- Other associated conditions include:
  - Portal hypertension
  - Congenital vascular diseases
  - Connective tissue diseases
  - Trauma
- Splenic artery aneurysms:
  - Splenic artery aneurysms are usually asymptomatic at diagnosis.
  - They may present acutely as an emergency with rupture inside the peritoneal cavity (Hemoperitoneum). This will lead to hemodynamic instability and shock.
  - Sometimes they rupture into the pancreatic duct, a condition known as hemossucus pancreatic.
  - Rupture occurs most commonly in the third trimester of pregnancy and this may signal a warning by a sentinel hemorrhage in 20% to 30% of patients.
  - During pregnancy, splenic artery aneurysms are prone to rupture in the third trimester or in the postpartum period because of the high circulating levels of estrogen, progesterone, and relaxin, in addition to maternal physiological changes, which increase arterial stress and weakens the arterial wall.
  - Splenic artery aneurysms may present as an asymptomatic abdominal mass.
  - Or discovered as an incidental finding during evaluation of other conditions or complaints.
  - Splenic artery aneurysms can also present with:
    - Vague epigastric pain
    - Left upper quadrant abdominal pain
    - Left shoulder pain
- Rupture of splenic artery aneurysms is associated with a high mortality rate.

23.4 Investigations

- Splenic artery aneurysms are usually asymptomatic discovered during evaluation of other abdominal unrelated conditions.
- Abdominal ultrasound will reveal an intraabdominal cystic lesion.
- This should be further investigated with other imaging techniques.

- An abdominal CT- scan or a magnetic resonance angiography is the investigation of choice to diagnose splenic artery aneurysm.
- HYPERLINK "https://www.sciencedirect.com/topics/medicine-and-dentistry/radiologic-studies" \o "Learn more about Radiologic studies from ScienceDirect's AI-generated Topic Pages" Radiologic investigations:
  - The majority of splenic artery aneurysms are diagnosed at the time of laparotomy or discovered during evaluation of other unrelated conditions.
  - Plain abdominal radiographs may raise suspicion of a splenic artery aneurysm which appears as a classic calcified ring in the left upper abdominal quadrant.
  - Abdominal ultrasound is the preferred initial investigation to diagnose splenic artery aneurysm during pregnancy. It is noninvasive, cheap and devoid of radiation.
  - Digital subtraction angiography was considered the "gold standard" for the diagnosis of splenic artery aneurysm.
  - Multi slice CT angiography is also valuable in diagnosing splenic artery aneurysm.
  - Plain abdominal X-rays

    On plain abdominal film, an oval calcification with a central lucent area is a very specific sign for splenic artery aneurysm.
  - Abdominal ultrasound

    This is a simple non-invasive investigation

    In pregnancy, ultrasound with Doppler study is preferred to minimize radiation exposure to the fetus.
  - Angiography

    This is the gold standard investigation when the patient is stable.

    It is associated with an increased risk of radiation exposure to the fetus during pregnancy.

23.5 Treatment

- The indications to treat splenic artery aneurysms include:
  - Specific symptoms such as epigastric abdominal pain, the left upper quadrant pain and backpain.
  - Splenic artery aneurysms in women of childbearing age.
  - The presence of portal hypertension and liver transplantation.
  - Pseudoaneurysms of splenic artery of any size.
  - Splenic artery aneurysms with diameter greater than 2.5 cm.
- In patients older than 60 years, follow-up CT scan every six months is recommended.
- Patients at the greatest risk include:
  - Patients with splenic artery aneurysms greater than 2 cm
  - Symptomatic patients
  - Transplant patients
  - Splenic artery aneurysms associated with inflammatory processes

  - Splenic artery aneurysms identified in women of childbearing age and pregnant women.
  - Splenic artery aneurysms in pregnant women represent a great risk for both the mother and the fetus.
- Therapeutic options:
  - Simple vascular ligation which can by done by open or laparoscopic technique
  - Splenectomy when the aneurysm is close to the spleen
  - Artery embolization
  - Stent placement
- All symptomatic splenic artery aneurysms require surgical treatment.
- For incidental and asymptomatic aneurysms, surgical treatment is indicated:
  - If the splenic artery aneurysm is larger than 2.5 cm in diameter
  - If the patient is pregnant or of childbearing age
- Because of the high morbidity and mortality associated with ruptured aneurysm, treatment for pregnant women and women of childbearing age consists of elective splenic artery ligation and resection of the aneurysm with or without splenectomy.
- Treatment of suspected ruptured aneurysm is immediate laparotomy.
- In the treatment of splenic artery aneurysms, every attempt should be made to preserve the spleen. Splenectomy is reserved for those aneurysms:
  - Found near the hilum of the spleen.
  - Found during emergency surgery because of ruptured splenic artery aneurysm.
- The operative treatment differs by location of the splenic artery aneurysm.
  - Proximal splenic artery aneurysms are excised after proximal and distal ligation.
  - Mid-splenic artery aneurysms are excluded by proximal and distal ligation of the splenic artery and all collateral vessels. In these cases, the spleen blood flow will depend on the short gastric arteries.
  - Distal or hilar splenic artery aneurysms are the most common and are treated with excision of the aneurysm and splenectomy.
  - Laparoscopic ligation and splenectomy for splenic artery aneurysms have been reported with success.
  - Patients who are not fit or cannot tolerate an operation are treated with transcatheter embolization. This results in splenic infarction but known to be associated with subsequent complications including:

    Distant embolization

    Arterial disruption

    Arterial recanalization with potential future rupture
- The elective operative mortality of splenic artery aneurysm ranges between 0.5 and 1.3%.
- The general approach to splenic artery aneurysms is an early elective treatment when recognized, rather than watchful waiting.
- This is to minimize the risk of rupture and the associated high mortality.
- Elective repair of splenic artery aneurysm should be considered for:
  - Symptomatic splenic artery aneurysm regardless of size
  - Asymptomatic splenic artery aneurysm larger than 2 cm

- Asymptomatic splenic artery aneurysm during pregnancy
- Asymptomatic splenic artery aneurysm in women of childbearing age
- The surgical options for the repair of splenic artery aneurysms include:
  - Open surgical approach
  - Laparoscopic approach
  - Embolization
- The surgical technique depends on the location of the splenic artery aneurysms, with 80% located in the distal portion of the splenic artery.
- Regardless of the approach, every attempt should be made to preserve the spleen.
- In the pregnant patient with a symptomatic splenic artery aneurysm, the aim is immediate resuscitation and laparotomy with splenectomy or splenopancreatectomy and ligation of the splenic artery.
- In some cases where there are aneurysmal adhesions to the pancreas, splenectomy, pancreatectomy or both may be necessary.
- The open surgical approach:
  - This remains the preferred approach for splenic artery aneurysms repair.
  - This approach is used mainly when the aneurysm is located in the distal part of the splenic artery.
  - It is also the preferred approach in cases of giant aneurysms where other approaches, such as endovascular treatment, would not be suitable.
- The endovascular approach:
  - This approach has recently gained popularity.
  - It is a less invasive approach especially in patients who are a high surgical risk.
  - It is however associated with complications:
    - Distant embolization
    - Splenic infarction
    - Arterial disruption
    - Arterial recanalization with potential future rupture
- The laparoscopic approach:
  - This is a safe alternative for cases of elective splenic artery aneurysm repair.
  - It has several advantages including:
    - Rapid recovery
    - Shorter hospital stay
    - Less postoperative pain when compared with the open approach.
  - The laparoscopic approach is also suitable and safe to treat pregnant women with splenic artery aneurysms.
  - Compared with the open approach, laparoscopic splenic artery aneurysm repair has a lower risk of preterm labor in these patients due to minimal manipulation of intra-abdominal contents.
  - For stable pregnant women with a splenic artery aneurysm ≥2 cm in diameter, laparoscopic splenic aneurysm repair is the recommended method of treatment in the first or second trimester.
- Pneumococcal vaccine, H influenzas vaccine and meningococcal vaccine should be given 2 weeks before elective splenectomy and immediately following emergency splenectomy.

Further Reading

1. Al-Habbal Y, Christophi C, Muralidharan V. Aneurysms of the splenic artery—a review. Surgeon. 2010;8:223–31.
2. Lynch MJ, Woodford NWF. Rupture of a splenic artery aneurysm during pregnancy with maternal and foetal death: a case report. Med Sci Law. 2008;48:342–5.
3. Lang W, Strobel D, Beinder E, Raab M. Surgery of a splenic artery aneurysm during pregnancy. Eur J Obstet Gynecol Reprod Biol. 2002;102:215–6.
4. Sadat U, Dar O, Walsh S, Varty K. Splenic artery aneurysms in pregnancy—a systematic review. Int J Surg. 2008;6:261–5.
5. Pavlis T, Seretis C, Gourgiotis S, Aravosita P, Mystakelli C, Aloizos S. Spontaneous rupture of splenic artery aneurysm during the first trimester of pregnancy: report of an extremely rare case and review of the literature. Case Rep Obstet Gynecol. 2012;2012:528051.
6. Gourgiotis S, Alfaras P, Salemis NS. Spontaneous rupture of splenic artery aneurysm in pregnancy: a case report. Adv Med Sci. 2008;53:341–3.
7. Moon E, Tam M, Nikano R, Karuppasamy K. Prophylactic antibiotic guidelines in modern interventional radiology practice. Semin Intervent Radiol. 2010;27:327–37.
8. McMahon DP, Ward WH, Harwood JL, Moore EM. An institutional review of splenic artery aneurysm in childbearing-aged females and splenic artery aneurysm rupture during pregnancy. Is screening justified? Mil Med. 2012;177:96–8.
9. Samamé J, Kaul A, Garza U, Echeverria A, Galvani C. Laparoscopic aneurysm resection and splenectomy for splenic artery aneurysm in the third trimester of pregnancy. Surg Endosc. 2013;27:2988–91.
10. Spittel JA Jr, Fairbairn JF II., Kincaid OW, ReMine WH. Aneurysm of the splenic artery. JAMA. 1961;175:452.
11. Beaussier M. Sur un anévrisme de l'artère splénique: Dont les parois se sont ossifiées. J Med Clin Pharmacol (Paris). 1770;32:157.
12. Sheps SG, Spittel JA Jr, Fairbairn JF II, Edwards JE. Aneurysms of the splenic artery with special reference to bland aneurysms. Proc Staff Meet Mayo Clin. 1958;33:381.
13. Stanley JC, Fry WJ. Pathogenesis and clinical significance of splenic artery aneurysms. Surgery. 1974;76:898.
14. Bedford PD, Lodge B. Aneurysm of the splenic artery. Gut. 1960;1:312.
15. Trastek VF, Pairolero PC, Joyce JW, Hollier LH, Bernatz PE. Splenic artery aneurysms. Surgery. 1982;91:694.
16. Owens JC, Coffey RJ. Aneurysm of the splenic artery, including a report of 6 additional cases. Int Abstr Surg. 1953;97:313.
17. Busuttil RW, Brin BJ. The diagnosis and management of visceral artery aneurysms. Surgery. 1980;88:619.
18. De Vries JE, Schattenkerk ME, Malt RA. Complications of splenic artery aneurysm other than intraperitoneal rupture. Surgery. 1982;91:200.
19. Boijsen E, Efsing H-O. Aneurysm of the splenic artery. Acta Radiol. [Diagn.] (Stockh.). 1969;8:29.
20. Barrett JM, Van Hooydonk JE, Boehn FH. Pregnancy-related rupture of arterial aneurysms. Obstet Gynecol Surv. 1982;37:557.
21. Manalo-Estrella P, Barker AE. Histopathologic findings in human aortic media associated with pregnancy: A study of 16 cases. Arch Pathol. 1967;83:336.
22. Wexler BC. Vascular degenerative changes in the uterine arteries and veins of multiparous rats. Am J Obstet Gynecol. 1970;107:6.
23. Barrett JM, Caldwell BH. Association of portal hypertension and ruptured splenic artery aneurysm in pregnancy. Obstet Gynecol. 1981;57:255.
24. Ho WC. Rupture of splenic artery aneurysm in pregnancy. Can Med Assoc J. 1977;117:123–6.
25. Macfarlane JR, Thorbjarnarson B. Rupture of splenic artery aneurysm during pregnancy. Am J Obstet Gynecol. 1966;95:1025.

26. O'Grady JP, Day EJ, Toole AL, Paust JC. Splenic artery aneurysm rupture in pregnancy: A review and case report. Obstet Gynecol. 1977;50:627.
27. Mehrotra D, Di Benedetto R, Theriot E, Moreland J, Mehta P. Spontaneous rupture of splenic artery aneurysm: Sixth instance of both maternal and fetal survival. Obstet Gynecol. 1983;62:665.
28. Derchi LE, Biggi E, Cicio GR, Bertoglio C, Neumaier CE. Aneurysms of the splenic artery: Noninvasive diagnosis by pulsed Doppler sonography. J Ultrasound Med. 1984;3:41.
29. Probst P, Castañeda-Zuñiga WR, Gomes AS, Yonehiro EG, Delaney JP, Amplatz K. Nonsurgical treatment of splenic-artery aneurysms. Radiology. 1978;728:619.
30. Schwartz PE, Sterioff S, Mucha P, Melton LJ III, Offord KP. Postsplenectomy sepsis and mortality in adults. JAMA. 1982;248:2279.
31. Madoff DC, Denys A, Wallace MJ, Murthy R, Gupta S, Pillsbury EP, et al. Splenic arterial interventions: anatomy, indications, technical considerations, and potential complications. Radiographics. 2005;25(Suppl 1):S191–211.
32. Larraín D, Fava M, Espinoza R. Splenic artery aneurysm: case report. Rev Med Chil. 2005;133(8):943–6.
33. Kenningham R, Hershman MJ, McWilliams RG, Campbell F. Incidental splenic artery aneurysm. J R Soc Med. 2002;95(9):460–1.
34. Dorman BA, Carney WI. Aneurysm of the splenic artery. Dis Chest. 1965;48:78–82.
35. Greene DR, Gorey TF, Tanner WA, Lane BE, Collins PG. The diagnosis and management of splenic artery aneurysms. J R Soc Med. 1988;81(7):387–8.

MIX
Papier aus verantwortungsvollen Quellen
Paper from responsible sources
FSC® C105338

If you have any concerns about our products,
you can contact us on
ProductSafety@springernature.com

In case Publisher is established outside the EU,
the EU authorized representative is:
Springer Nature Customer Service Center GmbH
Europaplatz 3, 69115 Heidelberg, Germany

Printed by Libri Plureos GmbH
in Hamburg, Germany